POST-COVID TOURISM AND HOSPITALITY DYNAMICS

Recovery, Revival, and Re-Start

Advances in Hospitality and Tourism

POST-COVID TOURISM AND HOSPITALITY DYNAMICS

Recovery, Revival, and Re-Start

Edited by

Umendra Narayan Shukla, PhD

Sharad Kumar Kulshreshtha, PhD

AAP | APPLE ACADEMIC PRESS

First edition published 2024

Apple Academic Press Inc.
1265 Goldenrod Circle, NE,
Palm Bay, FL 32905 USA

760 Laurentian Drive, Unit 19,
Burlington, ON L7N 0A4, CANADA

CRC Press
2385 NW Executive Center Drive,
Suite 320, Boca Raton FL 33431

4 Park Square, Milton Park,
Abingdon, Oxon, OX14 4RN UK

Library and Archives Canada Cataloguing in Publication

Title: Post-COVID tourism and hospitality dynamics : recovery, revival, and re-start / edited by Umendra Narayan Shukla, PhD, Sharad Kumar Kulshreshtha, PhD.
Names: Shukla, Umendra Narayan, 1968- editor. | Kulshreshtha, Sharad Kumar, 1978- editor.
Series: Advances in hospitality and tourism book series.
Description: First edition. | Series statement: Advances in hospitality and tourism book series | Includes bibliographical references and index.
Identifiers: Canadiana (print) 2023048610X | Canadiana (ebook) 20230486126 | ISBN 9781774915127 (hardcover) | ISBN 9781774915134 (softcover) | ISBN 9781003429722 (PDF)
Subjects: LCSH: Tourism. | LCSH: COVID-19 Pandemic, 2020-
Classification: LCC G155.A1 P65 2024 | DDC 338.4/791—dc23

Library of Congress Cataloging-in-Publication Data

Names: Shukla, Umendra Narayan, 1968- editor. | Kulshreshtha, Sharad Kumar, 1978- editor.

Title: Post-covid tourism and hospitality dynamics : recovery, revival, and re-start / Edited by Umendra Narayan Shukla, PhD, Sharad Kumar Kulshreshtha, PhD.

Description: First edition. | Palm Bay, FL : Apple Academic Press, 2024. | Series: Advances in hospitality and tourism | Includes bibliographical references and index. | Summary: "The chapter authors offer post-COVID tourism sector recovery strategies, helping to provide a roadmap for stakeholders of sustainable tourism practices after COVID-19. The book also presents detailed analyses of post-COVID trends in the travel, tourism, and hospitality sector along with case studies from several countries of the world. Topics also cover post-COVID pandemic travel preferences, post-COVID safety protocols, sustainable tourism practices, post-COVID-19 public policies for tourism, and more. Specific tourism and hospitality sectors are considered, including wine tourism, MICE (meetings, incentives, conferences, and exhibitions) tourism, regional tourism markets, food delivery services, and others. The book also explores innovations and digital solutions for tourism and hospitality in the COVID-19 pandemic. The book defines the realities of post-COVID-19 and global strategies to recover, revive, and re-start the tourism and hospitality industry, adapting to new normal era with digital innovations, re-imaging, re-designing, and re-skilling tourism and hospitality practices. Post-COVID Tourism and Hospitality Dynamics: Recovery, Revival, and Re-Start provides valuable information for professionals and researchers in hospitality and tourism as well as for faculty and students in the field"-- Provided by publisher.

Identifiers: LCCN 2023030389 (print) | LCCN 2023030390 (ebook) | ISBN 9781774915127 (hardcover) | ISBN 9781774915134 (paperback) | ISBN 9781003429722 (ebook)

Subjects: LCSH: Tourism--Management. | Hospitality industry--Management. | COVID-19 Pandemic, 2020---Influence. | Sustainable tourism--Management.

Classification: LCC G155.A1 P675 2024 (print) | LCC G155.A1 (ebook) | DDC 910.68--dc23/eng/20230721

LC record available at https://lccn.loc.gov/2023030389

LC ebook record available at https://lccn.loc.gov/2023006326

ISBN: 978-1-77491-512-7 (hbk)
ISBN: 978-1-77491-513-4 (pbk)
ISBN: 978-1-00342-972-2 (ebk)

Dedication

This book is dedicated to all those stakeholders of the tourism and hospitality industry who have faced hardship in saving their existence during this crisis and really strived for recovery, revival, and restart during and after this COVID-19 unprecedented effect around the world.

ABOUT THE ADVANCES IN HOSPITALITY AND TOURISM BOOK SERIES

Editor-in-Chief:
Mahmood A. Khan, PhD
Professor, Department of Hospitality and Tourism Management,
Pamplin College of Business,
Virginia Polytechnic Institute and State University,
Falls Church, Virginia, USA
Email: mahmood@vt.edu

This series reports on research developments and advances in the rapidly growing area of hospitality and tourism. Each volume in this series presents state-of-the-art information on a specialized topic of current interest. These one-of-a-kind publications are valuable resources for academia as well as for professionals in the industrial sector.

BOOKS IN THE SERIES:

Food Safety: Researching the Hazard in Hazardous Foods
Editors: Barbara Almanza, PhD, RD, and Richard Ghiselli, PhD

Strategic Winery Tourism and Management: Building Competitive Winery Tourism and Winery Management Strategy
Editor: Kyuho Lee, PhD

Sustainability, Social Responsibility and Innovations in the Hospitality Industry
Editor: H. G. Parsa, PhD
Consulting Editor: Vivaja "Vi" Narapareddy, PhD
Associate Editors: SooCheong (Shawn) Jang, PhD,
Marival Segarra-Oña, PhD, and Rachel J. C. Chen, PhD, CHE

Managing Sustainability in the Hospitality and Tourism Industry: Paradigms and Directions for the Future
Editor: Vinnie Jauhari, PhD

Management Science in Hospitality and Tourism: Theory, Practice, and Applications
Editors: Muzaffer Uysal, PhD, Zvi Schwartz, PhD, and
Ercan Sirakaya-Turk, PhD

Tourism in Central Asia: Issues and Challenges
Editors: Kemal Kantarci, PhD, Muzaffer Uysal, PhD, and Vincent Magnini, PhD

Poverty Alleviation through Tourism Development: A Comprehensive and Integrated Approach
Robertico Croes, PhD, and Manuel Rivera, PhD

Chinese Outbound Tourism 2.0
Editor: Xiang (Robert) Li, PhD

Hospitality Marketing and Consumer Behavior: Creating Memorable Experiences
Editor: Vinnie Jauhari, PhD

Women and Travel: Historical and Contemporary Perspectives
Editors: Catheryn Khoo-Lattimore, PhD, and Erica Wilson, PhD

Wilderness of Wildlife Tourism
Editor: Johra Kayeser Fatima, PhD

Medical Tourism and Wellness: Hospitality Bridging Healthcare (H_2H)©
Editor: Frederick J. DeMicco, PhD, RD

Sustainable Viticulture: The Vines and Wines of Burgundy
Claude Chapuis

The Indian Hospitality Industry: Dynamics and Future Trends
Editors: Sandeep Munjal and Sudhanshu Bhushan

Tourism Development and Destination Branding through Content Marketing Strategies and Social Media
Editor: Anukrati Sharma, PhD

Evolving Paradigms in Tourism and Hospitality in Developing Countries: A Case Study of India
Editors: Bindi Varghese, PhD

The Hospitality and Tourism Industry in China: New Growth, Trends, and Developments
Editors: Jinlin Zhao, PhD

Labor in Tourism and Hospitality Industry: Skills, Ethics, Issues, and Rights
Abdallah M. Elshaer, PhD, and Asmaa M. Marzouk, PhD

Sustainable Tourism Development: Futuristic Approaches
Editor: Anukrati Sharma, PhD

Tourism in Turkey: A Comprehensive Overview and Analysis for Sustainable Alternative Tourism
Editor: Ahmet Salih İKİZ

Capacity Building Through Heritage Tourism: An International Perspective
Editor: Surabhi Srivastava, PhD

Medical Travel Brand Management: Success Strategies for Hospitality Bridging Healthcare (H2H)
Editors: Frederick J. DeMicco, PhD, RD, and Ali A. Poorani, PhD

Strategic Management for the Hospitality and Tourism Industry: Developing a Competitive Advantage
Editor: Vincent Sabourin, PhD

Tourist Behavior: Past, Present, and Future
Editors: Narendra Kumar, PhD, Bruno Barbosa Sousa, PhD, and Swati Sharma, PhD

The Hospitality and Tourism Industry in ASEAN and East Asian Destinations: New Growth, Trends, and Developments
Editors: Jinlin Zhao, PhD, Lianping Ren, D.HTM, and Xiangping Li, PhD

Event Tourism in Asian Countries: Challenges and Prospects
Editor: Shruti Arora, PhD

Small Island and Small Destination Tourism: Overcoming the Smallness Barrier for Economic Growth and Tourism Competitiveness
Robertico Croes, PhD

Ecotourism and Sustainable Tourism: From Principle to Practice in the New Millennium
Editors: Kelly S. Bricker, PhD

Post-COVID Tourism and Hospitality Dynamics: Recovery, Revival, and Re-start
Editors: Umendra Narayan Shukla, PhD, and Sharad Kumar Kulshreshtha, PhD

Brand Co-Creation Tourism Research: Contemporary Issues and Challenges
Editor: Raouf Ahmad Rather, PhD

Dynamics of the Tourism Industry: Post-Pandemic and Post-Disaster Perspectives and Strategies
Editors: Gül Erkol Bayram, PhD, and Anukrati Sharma, PhD

ABOUT THE SERIES EDITOR

Mahmood A. Khan, PhD, is a Professor in the Department of Hospitality and Tourism Management, Pamplin College of Business at Virginia Tech's National Capital Region campus. He has served in teaching, research, and administrative positions for the past 35 years, working at major U.S. universities. Dr. Khan is the author of several books and has traveled extensively for teaching and consulting on management issues and franchising. He has been invited by national and international corporations to serve as a speaker, keynote speaker, and seminar presenter on different topics related to franchising and services management. Dr. Khan has received the Steven Fletcher Award for his outstanding contribution to hospitality education and research. He is also a recipient of the John Wiley & Sons Award for lifetime contribution to outstanding research and scholarship; the Donald K. Tressler Award for scholarship; and the Cesar Ritz Award for scholarly contribution. He also received the Outstanding Doctoral Faculty Award from Pamplin College of Business. He has served on the board of governors of the Educational Foundation of the International Franchise Association, on the board of directors of the Virginia Hospitality and Tourism Association, as a trustee of the International College of Hospitality Management, and as a trustee on the Foundation of the Hospitality Sales and Marketing Association's International Association. He is also a member of several professional associations.

ABOUT THE EDITORS

Umendra Narayan Shukla, PhD

Umendra Narayan Shukla, PhD, is a Professor at the Institute of Tourism and Hotel Management Dr. Bhimrao Ambedkar University, Agra, Uttar Pradesh, India, where he is also Director of the Institute of Tourism and Hotel Management and Dean of Faculty of Management. He was formerly Head in the Department of Travel and Tourism Management and Department of Hotel Management. Dr. Shukla has 28 years of teaching experience and has contributed to research as well. He is a member of the board of studies and academic committees of tourism at various universities in India. He has visited many countries, such as Sri Lanka, Indonesia, Malaysia, and Italy, for the presentation of his research papers at many national and international conferences. He has been invited as an expert as well as resource person by many forums of academicians to deliver talks on sustainable and eco-tourism, adventure tourism, personality development, spiritual management, and so on. He has contributed research chapters and articles in more than 32 books, magazines, and journals of national and international importance. He has guided many research scholars in the field of tourism and hospitality. He is a lifetime member of different tourism bodies and social organizations in India. He has edited the book *Emerging Dynamics of Indian Tourism and Hospitality,* while another book, *Incredible Tourism of India: Potentials, Policies and Practices* by Copal Publishing Group, India, is under publication. Dr. Shukla has completed his qualifications for his MSc, MTA, and PhD degrees in Tourism Management.

Sharad Kumar Kulshreshtha, PhD

Sharad Kumar Kulshreshtha, PhD, is an Assistant Professor at the Department of Tourism and Hotel Management, North-Eastern Hill University (A Central University) Shillong—Meghalaya (India). He has more than 15 years of teaching and research experience. He has acclaimed publications and research papers in various books and journals of national repute. He has participated in and presented many research papers in national and international seminars and conferences. Recently, he presented a paper at the International Conference on Wine Market and Cultures of Consumption at the School of Hotel and Tourism Management, the Hong Kong Polytechnic University, Hong Kong. He is supervising PhD research scholars in the areas of event tourism, wine tourism, and avitourism of northeast India. He has also been invited as a resource person at various universities and business associations of India to deliver talks as a subject expert. He is the guest editor of a special Issue on "Tourism in Times of COVID-19," published in the journal *Anais Brasileiros de Estudos Turísticos* (ABET), indexed in Emerging Sources Citation Index (ESCI) and ISI Web of Science. He has edited books titled *Emerging Dynamics of India Tourism and Hospitality: Transformation and Innovation*; *Global Developments in Healthcare and Medical Tourism*; *Managing Festivals for Destination Marketing and Branding; Virus Outbreaks and Tourism Mobility: Strategies to Counter Global Health Hazards; Incredible Tourism of India: Potentials, Policies and Practices;* and *Wellness Tourism in Asia: Emerging Destinations for Rejuvenate, Relax, and Retreat* (forthcoming); among others. Recently in April, 2023 edited a book on *Managing Festivals for Destination Marketing and Branding,* published by IGI Global. U.S.A.

CONTENTS

CONTRIBUTORS

Pooja Agrawal
Department of Commerce and Management, Vishwakarma University, Pune, Maharashtra, India

Albattat Ahmad
Graduate School of Management, Post Graduate Center, Management and Science University, Malaysia

Sunildro L. S. Akoijam
Assistant Professor (Senior Scale), Department of Management, North-Eastern Hill University (NEHU), A Central University, Tura Campus, Tura, Meghalaya, India

Nagihan Cakmakoglu Arici
Republic of Turkey, Turkish Consulate General in Stuttgart, Stuttgart, Germany

Noor Farahiyah Azman
School of Hospitality and Creative Arts, Management and Science University, Malaysia

T. R. Bandaru
Department of Management, North-Eastern Hill University (NEHU), Tura Campus, Tura, Meghalaya, India

Sachin Kumar Behera
Research Scholar, Department of Tourism and Hotel Management, North-Eastern Hill University, Shillong, Meghalaya, India

Bivek Datta
Nepal College of Travel and Tourism Management, Lazimpat, Kathmandu, Nepal

Claudia Corrêa De Almeida Moraes
Professor at the Faculty of Tourism and Hospitality and Coordinator of the MBA in Service Management at the Universidad Federal Fluminense (UFF), Vice Leader of the Research Group Tourism, Management and Territory, São Paulo, SP, Brazil

Isabela De Fátima Fogaça
Professor at the Tourism Course (Bachelor's degree) at the Federal Rural University of Rio de Janeiro (UFRRJ) and the Postgraduate Program in Heritage, Culture, and Society (PPGPACS) at UFRRJ, Coordinator of the Tourism and Leisure Observatory of the Baixada Verde Tourist Region, Rua Bambina, Rio de Janeiro, RJ, Brazil

Mehmet Ertas
Department of Tourism, Faculty of Tourism, Pamukkale University, Denizli, Turkey

Omvir Gautam
Department of Commerce and Management, Vishwakarma University, Pune, Maharashtra, India

Daniel Guttentag
Assistant Professor, Hospitality and Tourism Management, College of Charleston, United States of America

Andy Heyes
Stenden Hotel Management School, Leeuwarden, The Netherlands

Ut Lon (Billy) Im
School of Tourism Management and School of Hospitality Management, Macao Institute for Tourism Studies, Macao SAR, China

Julie Kellershohn
PhD, Assistant Professor Ted Rogers School of Hospitality and Tourism Management,
Ryerson University, Toronto, ON, Canada

Maximiliano E. Korstanje
Professor, Lecturer, Department of Economics, University of Palermo, Mario Bravo, Buenos Aires, Argentina

Mohit Kukreti
University of Technology and Applied Sciences (CAS), Ibri, Oman

Sharad Kumar Kulshreshtha
North-Eastern Hill University, Shillong, Meghalaya, India

Ashok Kumar
Research Scholar, Department of Tourism & Hotel Management, North-Eastern Hill University, Shillong, Meghalaya, India
Senior Lecturer, Institute of Hotel Management, Shillong, Meghalaya, India

Cindia Ching Chi Lam
School of Tourism Management and School of Hospitality Management,
Macao Institute for Tourism Studies, Macao SAR, China

Erdan Ma
School of Tourism and Hospitality Management, Yunnan University of Finance and Economics, China

Byron Marlowe
Don Smith Distinguished Professor of Hospitality Management, Director, Wine and Beverage Business, The School of Hospitality Business Management, Carson College of Business Washington State University Tri-Cities Campus, United States

Amitabh Mishra
University of Technology and Applied Sciences (CAS), Ibri, Oman

Patita Paban Mohanty
SOA University, Khandagiri, Bhubaneswar, Odisha, India

Arslan Munir
Kansas State University, Manhattan, KS, United States

Sanjay Nadkarni
Emirates Academy of Hospitality Management, Dubai, UAE

Justin Matthew Pang
RMIT University, Royal Melbourne University of Technology (RMIT) University, Hanoi, Vietnam

Karen Peesker
Assistant Professor Ted Rogers School of Hospitality and Tourism Management,
Ryerson University, Toronto, ON, Canada

Eva Rachmawati
Department of Forest Resources Conservation and Ecotourism, Faculty of Forestry and Environment, IPB University, Indonesia

Rajeshramasamy
Assistant Professor, Department of Tourism & Hospitality Management, Mizoram Central University, Aizawl, India

Fatimah Zahra Ramly
Ramly School of Hospitality and Creative Arts, Management and Science University, Shah Alam, Malaysia

Samik Ray
Ex-Faculty of Department of Folklore, University of Kalyani, Kalyani, West Bengal, India
Travel and Tourism Management in MPTI (Kolkata), Kolkata, West Bengal, India
WTCC School of Trade and Commerce (Kolkata), Kolkata, West Bengal, India

Andreia Maria Roque
Instituto de Desenvolvimento do Turismo Rural Brasil Rural, São Paulo, Brazil

Shwetasaibal Samanta Sahoo
Faculty of Management Studies, Sri Sri University, Cuttack, Odisha, India

Gino A. Sangma
Department of Agri-Business Management & Food Technology, North-Eastern Hill University (NEHU), Tura Campus, Tura, Meghalaya, India

Manisha Seal
Assist. Prof. Department of Tourism and Travel Management, Jyoti Nivas College Autonomous, Bangalore, Karnataka, India

Hugues Seraphin
Department of Marketing, Faculty of Business, Law & Digital Technologies, University of Winchester Business School, Winchester, UK

Naman Sharma
Department of Management, Indian Institute of Foreign Trade, Kolkata, West Bengal, India

U. N. Shukla
Dr. Bhim Rao Ambedkar University, Agra, Uttar Pradesh, India

Amarjit Singh
Extension Lecturer, Department of Tourism Management, Government College Kaithal Haryana, Affiliated to Kurukshetra University Kurukshetra, Haryana, India

Kuldeep Singh
Assistant Professor, School of Hospitality & Tourism, GD Goenka University, Gurugram, India

Wayne W. Smith
Professor, Ted Rogers School of Hospitality and Tourism Management, Ryerson University, Toronto, ON, Canada

Divya Talreja
Research Scholar, School of Studies, Tourism and Travel Management (SOSTTM), Jiwaji University, Gwalior, Madhya Pradesh, India

Chai Ching Tan
Rattanakosin International College of Creative Entrepreneurship, Bangkok, Thailand

Fábia Trentin
Professor at the Faculty of Tourism and Hospitality at Universidade Federal Fluminense and the Master's Program in Tourism, Coordinator of the Policy, Governance, and Tourism Laboratory (LabPGTUR), Rua Noronha Torrezão, Santa Rosa, Niterói, RJ, Brazil

Harsh Varma
UNWTO, Madrid, Spain

Adrian Wee
British University, Hưng Yên, Vietnam

Ahmad Sayuti Yahya
School of Hospitality and Creative Arts, Management and Science University, Shah Alam, Malaysia

ABBREVIATIONS

ACI	Airports Council International
ADR	average daily rate
AI	artificial intelligence
ASD	autism spectrum disorder
ASI	Archaeological Survey of India
ASPA	average spending per arrival
AVE	average variance extracted
BLE	Bluetooth low energy
BMC	BioMed Central
BSC	balanced scorecard
CBT	community-based tourism
CCTV	closed-circuit television
CLIA	Cruise Lines International Association
DMC	Destination Management Companies
DMO	Destination Management Organization
DXB	Dubai International Airport
EDM	electronic dance music
EUL	Emergency Use Listing
EVD	Ebola virus disease
FTA	Family Travel Association
FTA	foreign tourist arrivals
GRID	green, resilient, and inclusive development
HEPA	high-efficiency particle air
IATA	International Air Transport Association
IBCCES	International Board of Credentialing and Continuing Education Standards
ICAO	International Civil Aviation Organization
ICT	information and communication technology
ICU	intensive care units
ILO	International Labour Organization
IMO	International Maritime Organization
IoHT	Internet of Human Things
IoT	Internet of Things
MCO	movement control orders

MERS	Middle Eastern Respiratory Syndrome
MFOAs	mobile food ordering apps
MICE	Meetings, Incentives, Conventions, Exhibitions
NCSI	National Centre for Statistics and Information
NFC	near-field communication
NPI	nonpharmaceutical interventions
PHEIC	Public Health Emergency of International Concern
POS	point-of-sale
PPE	personal protective equipment
PPP	public-private partnership
RBV	resource-based view
RevPAR	revenue per available room
RFID	radio-frequency identification
RLG	regional level guide
RVs	recreational vehicles
SARS	severe acute respiratory syndrome
SDGs	sustainable development goals
SEM	structural equation modeling
SMEs	small, medium enterprises
SOP	standard operating procedures
SSS	screening, sanitization and social distancing
STR	Smith Travel Research
TCDM	tourism crisis and disaster management
TGFI	Tourist Guides' Federation of India
VFR	visit their friends and relatives
VLC	visual light communication
VOX	voice-operated exchange
VR	virtual reality
WB	World Bank
WBG	World Bank Group
WFH	work from home
WHO	World Health Organization
WTTC	World Travel and Tourism Council

FOREWORD 1

The edited volume titled *Post-COVID Tourism and Hospitality Dynamics: Recovery, Revival, and Restart* is an excellent compendium of timely articles discussing the nature of tourism after COVID-19. The editors, *Professor U. N. Shukla* and *Professor S. K. Kulshreshtha*, have carefully chosen the manuscripts that would create impact and synergy in this regard.

The characteristic features of tourism post-COVID are already evident in certain areas. For instance, an increased thrust upon safety and security, cleanliness, responsible conduct, and sustainability are visible on various fronts. The meaning of tourism itself has gone through a fundamental redefinition. Despite the clamor for sustainability, what it meant to travel before COVID was all about competitive consumption. The predominant belief that guided mass tourism was that you could not consume unless you exclude someone else from the opportunity. During COVID, the virtualization of the experiential process of tourism attractions happened when natural and cultural attractions worldwide began to offer virtual tours. Even physical site visits have become of shorter durations and with more limiting impacts. The guest–host exchange is an integral aspect of tourism. However, this exchange has to be positive and mutually beneficial.

If businesses, tourism policymakers, and tourists plan for a reboot to the "way tourism always used to be" before COVID-19, that is, detrimental to progress in tourism as a social phenomenon; also, as a society, we would be collectively blamed for our neglect to make use of the best opportunity to reinvent tourism for better. Fortunately, the case studies from different national contexts presented in this book indicate the willingness and courage to reimagine a new kind of tourism that is focused on sustainability goals. Among other things, I could notice a bold embrace of diversity, equity, and inclusiveness in people, processes, and products. These case studies truly enrich this book and provide insights, from Brazil to Indonesia, on how actual businesses craft their policies and practices to bravely face the new post-COVID world.

The COVID crisis reiterated that tourism is not a single industry, and approaches to dealing with the crisis cannot be one toolkit for all. The chains of industries and sectors constituting tourism are so complex that a diverse array of crisis management approaches must function in a very coordinated fashion. Until COVID hit, tourism management was considered to be a trivial profession, something anyone with a decent experience (e.g., travel booking, hotel front office, tour guiding, etc.) could handle effectively. The realization of the nuanced nature of management in the tourism industry must lead to the creation of institutions of higher learning and professional development that prepare future leaders for crises like this.

Every industry has gone through the COVID transition in some way. In some of those industries, these changes will become an integral part of their continued operations in the future. Say, in addition to remote work, automation and AI came to the forefront of some industries and business leadership found in these critical bases of competitive advantage. These and other technological augmentations would be the hallmarks of leading firms in these industries. Tourism, however, is very different. There is some way the "high touch" needs to be brought back, at least in areas related to authentic experiences of attractions. Beyond this, tourism is a force of peace, and it must bring people together. Tourism cannot happen in environmental bubbles without losing some of its core experiential quality. Social distancing expectations during COVID have somewhat dented this premise, and I hope social distancing would slowly give way for safe interactions that also respect safety considerations. With more and more people getting vaccinated, this is not a farfetched possibility anymore.

—Babu George, PhD, DBA, EdS
Professor and Associate Dean
School of Business, Christian Brothers University
Memphis, TN, USA
Email: bgeorge@cbu.edu

FOREWORD 2

COVID-19 pandemic has devastated the global tourism and hospitality industries. The outbreak of the COVID-19 pandemic has severely impacted organizations and entails a change in business dynamics. The impact of the pandemic on these industries is so severe that most businesses are now striving quite hard to resurrect their functioning in the so-called 'new normal'. The relationship between businesses and consumers is also altering due to restrictions from COVID-19 lockdowns and to ensure the social distancing protocols. The world is moving beyond instantaneous survival employing new ways; some of these changes may persist and become part of everyday life, while others may revert with the recovery. During revival and renaissance, businesses explore different innovative means to complete their business transactions. The whole world is working in close collaboration to restart and resume tourism activities through concentrated efforts.

In its Global Tourism Recovery Report (May 28, 2020), the United Nation's World Tourism Organization highlights seven priorities essential for tourism recovery. These are, to provide liquidity and protect jobs, recover confidence through safety and security, public–private collaboration for an efficient reopening, open borders with responsibility, harmonize and coordinate protocols and procedures, added value jobs through new technologies, innovation, and sustainability as the new normal. To build a better future for the industry, first, we need to improve traveler confidence; second, understand the emerging market trends to understand the drivers of demand; third, to build more resilient and inclusive tourism sectors, leveraging interest in sustainability. To ensure this, the tourism stakeholders and consumers should come closer to strengthen sustainability and resilience and also share benefits more equitably.

The present title is a welcome addition to the body of hospitality and tourism knowledge, reflecting upon the varied facets related to recovery, revival, and restart of the tourism and hospitality sectors globally. With its impactful divergent nature of contents authored by multidisciplinary and truly, global experts, this title is an extremely appropriate resource for students, research scholars, academicians, industry practitioners, and trade professionals to understand their preparedness to survive, sustain, and to ensure smooth industry practices during the so-called new normal.

—Professor Saurabh Kumar Dixit
Department of Tourism and Hotel Management,
North-Eastern Hill University, Shillong, India.

ACKNOWLEDGMENTS

Om Sarve Bhavantu Sukhinah Sarve Santu Niraamayaah |
Sarve Bhadraanni Pashyantu Maa Kashcid-Duhkha-Bhaag-Bhavet |
Om Shaantih Shaantih Shaantih ||

Meaning:
1: Om, May All be Happy,
2: May All be Free from Illness.
3: May All See what is Auspicious,
4: May no one Suffer.
5: Om Peace, Peace, Peace.

First and foremost, our heartfelt thanks to the almighty God Shri Krishna whose blessings and path of karma motivated us for the successful completion of this book, especially focused on Post-COVID Tourism and Hospitality Dynamic: Recovery, Revival, and Restart.

We deeply acknowledge our sincere gratitude and regards to our respected teacher Prof. S. C. Bagri, from H. N. B. University, Garhwal (Uttarakhand) for his blessing and motivation for completing this book.

Our most special thanks to Prof. Mahmood A. Khan, book series editor: Advances in Hospitality and Tourism, Apple Academic Press, who has considered and given consent for the title of this book project.

Our special thanks to Prof. Maximiliano E. Korstanje, book series editor "Tourism Security-Safety and Post Conflict Destinations" Emerald Publication, for extending his helping hand for achieving the depth and meaning in this book.

Our sincere thank goes to all authors and co-authors whose interest, initiative, integrity, commitment, and hard work along with their valuable contributions added the steps in the ladder of this successful edited work. We highly appreciate their quality chapters, timely contribution, and expertise to this book.

We are also grateful to our faculty colleagues who were always inspiring and encouraging us for this indelible work.

Our deepest gratitude to Apple Academic Press and all team members for their constant support to edit this book successfully. Without their support and guidance the completion of this task would have been merely a dream.

No words are sufficient to express our debt of gratitude for our parents who are always blessing us to do something meaningful in this materialistic world. We express our gratitude toward our beloved family members, especially Seema and Sakshi.

Last but not least, we are also thankful to our friends and relatives, staying in India and abroad, for their continuous heart touch in the final shaping of this book.

Thanking you

Editors

—Professor Umendra Narayan Shukla

Dr. Sharad Kumar Kulshreshtha

PREFACE

The COVID-19 pandemic was one of the most painful and difficult experiences for all humans and a problematic global scenario on the social and economic front. It disrupted the economic system badly and also affected millions of lives, livelihoods, liberty, employment, and caused losses in business due to global lockdown. Despite the advancement of health, medical science, rapid growth of medical technological development, multispecialty hospitality with experienced doctors and nursing team staff, and advanced health-infrastructure system in developed and developing countries, nobody was able to control this invisible infection which resulted in the faster spreading of the virus infection, severely attacking billions of people around the world.

The symptoms of this disease seemed like some normal pneumonia that very much resembles the SARS disease. For the medical science fraternity, it was a new variant of virus infection. The doctors were trying to find out the origin and actual root cause, source of this severe disease as well as keep monitoring and controlling the disease and its symptoms. The whole world panicked and was cautious to protect themselves against this critical pathogenic disease. Although COVID-19 is still affecting many places throughout the world, the vaccination drives and immunization programs are going on phase-wise worldwide and have shown good results by providing health shields to protect against this virus and have been able to control the virus infection. After the first, second, and third waves of this coronavirus, there is still a possibility of mutation of coronavirus, which can infect in a new form and with changing symptoms.

The COVID-19 pandemic has harshly impacted tourists' arrival and the tourism economy globally. The strict COVID-19 protocols, issue of global travel advisories for international tourism by the government and various health agencies, and the emergence of new-norm health safety and security, all are responsible for the global economic crisis and the impact on those dependent on the tourism industry. During this time, all the countries imposed global lockdowns, quarantines, masking, social distancing, and major restrictions on national and international arrivals. These reasons caused the complete closure of operations of the tourism industry globally. According to UNWTO, international tourist arrivals declined by 74% in 2020 compared with 2019.

The ending year of 2019, the whole year of 2020, and the year of 2021 have been painful moments for humanity as a whole with billions of new experiences due to this severe, critical, and the worst pandemic of the 21st century, that is, COVID-19. Nobody could imagine this unknown and unseen virus attack that became a major cause of change in the global scenario. What was happening seemed like scenes from some Hollywood movies like Jurassic Park or Star Wars that we watch and think about the future through the lenses of advanced technology development. It was really surprising that for a very short period, the whole world was in the grip of this mysterious virus attack. People disconnected socially and economically from their routine.

The impact of the virus outbreak has been experienced in every corner of the world, through the introduction of a term that is most common in industrial relation, that is, "Lockdown," which was not imposed by a company or a factory but by the government of respective countries with very strict protocols. Under the health safety guidelines, the World Health Organization issued a health advisory related to the COVID-19 quarantine period, regular hand washing and sanitization, use of masks, and social distancing were some immediate actions to prevent the spread of this pandemic. In this regard, people were locked at home with fear, stress, unrest, and some with hope toward normal. But this normal was not easy to adapt for everyone. There was a complete lockdown situation globally which directly affected almost every sector of the economy.

There was immediate direct demand for medicines, medical equipment, health-care services, essential goods, and services, as well as IT-enabled services. Technology played a catalytic role during and after this pandemic and for contactless services. According to International Labour Organization, there are some sectors which have been badly affected due to this COVID-19, that is, transport services including port, civil aviation, railways, roads, media, food retails, and public services including health and education, entertainment, leisure, tourism, and hospitality services.

Throughout the world, tourism and hospitality professionals, associations, experts, and academicians had gone through online webinars, conferences, and conclaves on this issue with a very in-depth and insightful discussion about the impact and effects of COVID-19 on the global tourism, hospitality, and aviation industry. The global and local communities are very enthusiastic about becoming as normal as earlier. During and after this pandemic, many countries have unlocked and reopened their markets, shopping malls, tourist spots and destinations, amusement parks, hotels, resorts, bars and clubs, restaurants, airports, and railways with certain mandatory protocols. Simultaneously, continuous efforts and initiatives were being taken to develop the COVID-19 vaccine by the scientist community. After a number of trials, finally, vaccines were developed and immediately governments started to vaccinate their citizens. All these experiences has shown us a new way of thinking to survive, sustain, and grow in the face of difficulties which are directly connected with our future, that is, "new normal."

In this context, this present book, *"Post-COVID Tourism and Hospitality Dynamics: Recovery, Revival, and Re-start,* published under the book series Advances in Hospitality and Tourism, has been compiled with scholarly chapters from academicians, research and scholars, industry practitioners.

The COVID-19 pandemic has affected the travel and tourism market globally resulting in a loss of revenue due to the absence of tourist mobility and badly affecting the livelihood of billions of people who are directly or indirectly connected with this industry. Presently, the tourism and hospitality industry are passing through a transition phase, looking forward to the immediate recovery and growth in it after COVID-19 and new normal situations. The new normality will focus on innovative approaches of reviving business processes, strategy plans to better connect ongoing market needs with the inclusion of strict health-safety measures, and standard operating procedures as part of responsible tourism practices. This book will prove to be a torch bearer for uncertain global pandemic situations and enlighten students, research scholars, academicians, industry practitioners, and trade professionals to acquire new knowledge and cutting-edge information related to readiness and preparation to survive, sustain, and smooth industry practices in the era of new normalcy.

OBJECTIVES OF THE BOOK

The objective of this book is to collect information and gain in-depth knowledge of this global pandemic issue that has created havoc for mankind and tourism and hospitality industry and to offer it to academia and industry practitioners from the tourism and hospitality disciplines. In brief, the book will focus on the following objectives:

- To highlight the issues, challenges, and innovations in the tourism and hospitality industry globally during and after COVID-19.
- To project all changes in the tourism and hospitality industry through case studies of various regions of the world after new normalcy.

- To focus on the application of smart technology solutions and digital practices for efficient travel and hospitality business operations after COVID-19.
- To emphasize the coverage of the priorities, preparations, future strategies in new normalcy.
- To explore new business models, new entrants, and innovative practices in the tourism and hospitality sector.
- To provide realistic information as a future road map for stakeholders of sustainable tourism practices after the COVID-19 period.

ORGANIZATION OF THE BOOK

The book has both theoretical and practical implications for researchers, academicians, policymakers who wish to have insights into such virus outbreaks and their impact and revival of tourism mobility. The book contains 17 chapters relating to various virus outbreaks and their impact, sharing knowledge about tourism industry experience, learning, outcomes, and practices during the pandemic from country to country. A chapter-by-chapter brief description follows:

Chapter 1 focuses on global tourism industry scenario, with hard realities, impact, and effects of COVID-19 pandemic with a focus on recovery, revival, and restart of the tourism and hospitality industry locally and globally.

Chapter 2 focuses on understanding the post-COVID-19 pandemic travel preferences and perception of Omani travelers' which includes a mix of citizens and expatriate residents. Primary and secondary data methodology was employed for collecting and analyzing data. Data from both qualitative and quantitative sources were gathered and analyzed.

Chapter 3 highlights the experience design guide and wine tourism exposure experience audit for reopening post-COVID-19. There is a review of the 4E Framework for Wine Tourism that includes education, entertainment, escapism, and esthetics.

Chapter 4 aims to provide details on how various protocol modifications have fostered a hotel design that is more ASD-friendly in terms of both social and physical safety. It is suggested that hotels take advantage of such circumstances by finding ways to maintain these protocols postpandemic to facilitate tourism opportunities for individuals with ASD and their families, giving the hotels a potentially lucrative competitive advantage.

In Chapter 5, the authors have extensively discussed the current state of the luxury hotel and tourism industries of Dubai and examined the impacts of COVID-19 on the Emirate of Dubai, and an analysis of the crisis management measures and possible ways to deal with the impact. It also suggests future planning possibilities and scenarios heading into the future.

Chapter 6 explains the real strategies and initiatives that have been taken during and after COVID-19 to revive the tourism and hospitality sector globally. The chapter also suggests strategic measures to restart the sector by adopting flexible policies and strategies, restoring tourists' confidence, introducing new health regimens, undertaking product diversification, working out new business models, reducing or deferring taxation, providing fiscal and monetary incentives, improving the physical condition of tourism resources, and motivating upskilling and reskilling of personnel.

Chapter 7 portrays strategies for the recovery of the tourism industry in the new normal era after the COVID-19 pandemic. The study also provides advice to accommodation facilities, transportation facilities, tour operators and travel agents, and governments and provides predictions about how tourists' behavior will change.

Chapter 8 deals with the challenges and future of tourism research as well as the opportunities of scholarship to be finely integrated to shape more resilient destinations. It also focuses on some voices claimed for the rebirth of the new tourism industry, a more ecofriendly version of tourism that helps to mitigate the negative impacts of climate change. It emphasizes the urgency to re-elaborate on new methods to study tourism in a world characterized by low mobilities.

Chapter 9 extends the understanding of the main impacts, coping strategies, and potential outcomes carried out by destination practitioners during and after the COVID-19 outbreak. Using a hermeneutic approach, by reviewing the extended literature of crisis management, pandemic, tourism destination governance, transformational leadership, this study provides recommendations about long-term strategies to tackle COVID-19 recently and similar pandemic crises in the future.

Chapter 10 discusses the changes in Indonesian community preferences in doing their tourism or leisure activities due to the COVID-19 pandemic based on the writer's quick research. The study results indicate that COVID-19 has changed community preferences in tourism behavior.

Chapter 11 reveals three important things regarding resort kids' clubs or mini-clubs: first, families write a significant number of reviews regarding resort mini-clubs on TripAdvisor; second, the experience of children at resort mini-clubs contributes to the overall experience of families; and finally, reviews on resort mini-clubs contribute to the overall performance (customer satisfaction) of resorts.

Chapter 12 is exploratory in nature, and based upon a qualitative research approach. This study identifies the impact of the pandemic on the Indian tourism and travel industry and formulates a strategic framework to reinstate and rejuvenate the tourism industry postpandemic by emphasizing global travel regulations and health prerequisites for domestic and international travelers.

Chapter 13 highlights the continued development and evolution of Vietnam's war against the COVID-19 global pandemic virus by both its government and industries. The study also depicts how the Vietnam governments and industries are constantly adapting to the current state of affairs in the country and in a COVID-19-ridden world. The chapter also projects their innovative ways to sustain their businesses and to adhere to new safety and hygienic practices that have been imposed by the state government and international regulatory bodies.

Chapter 14 aims to clarify the competence-related terms and organize them in a strategy-mapping structure that enables the practitioners (the communities in this study) and the local and national policymakers (the government), as implications, to strategically and operatively understand the roles and practical usages of competences; to test, using community-based cases, and by the use of structural equation modeling (SEM) statistics; and to provide the theoretical-empirical model-matching support for the proposed model.

Chapter 15 adopts bibliographic and documentary research of a qualitative nature to analyze how governments, and consequently the tourist trade, acted to reduce the spread of COVID-19, the intermediate conflicts generated by divergent or converging interests, and to incorporate the issues arising from this situation (health and economic crises in tourism) to maintain the well-being of its citizens and the tourism sector.

Chapter 16 attempts to identify and analyze the challenges, prospects, and strategies for reviving and recovering domestic tourism in India post-COVID-19 pandemic. In this regard, all stakeholders of the tourism industry particularly the state government and central government have to discharge their roles effectively in framing protourist policies that will rebuild the confidence of safety from COVID-19 among domestic tourists.

Chapter 17 deals with the necessary cultural shifts in the tourist guide-service paradigm from old to new normal to make travelers confident enough and comfortable to get travel experience amid any kind of global health calamity like COVID-19. This chapter involves a descriptive study based on both naturalistic and participant overt observations drawn primarily from interactions with professional tourist guides of India and the self-experience of the author.

Chapter 18 intends to showcase the impact of distancing on various service providers. There are both positive and negative impacts which are described in an elaborate manner. Critical standard operating procedures and practices are also discussed. Examples from practices being followed in movie theatres, shopping malls, events, and so on, and case studies are included to make the article interesting and understandable.

Chapter 19 aims to determine differences in perceptions of wine tourism between the wine producers of Meghalaya during COVID-19. A survey of 24 wineries is conducted to investigate wine tourism perception between wineries according to their production levels and cellar door sales. The results indicate that many wine producers have positive attitudes toward wine tourism.

Chapter 20 investigates whether the government's efficacy will increase if the government performs well in dealing with the pandemic. The results of the study are analyzed using SPSS 22.0. Percentage and frequency are calculated to analyze the responses restricted to India with a specific focus on Indian respondents. The study concludes with some strategies and recommendations and future implications for the government.

Chapter 21 outlines how travel demand is influenced by technology and investigates how that pandemic stimulated the stakeholders to go for electronic tourism acceptance. COVID-19 outbreak recommendations show that a positive stimulus for the adoption of automation technologies by travel and tourism companies.

Chapter 22 identifies and investigates the impact of technological innovations facilitating contactless services in the hospitality industry as well as identifies various areas where the technological innovations can be incorporated post-COVID-19. The study also provides future directions for hotels' operations post-COVID-19 and also highlights various emerging research areas that can facilitate contactless services in the hospitality industry.

Chapter 23 focuses on post-COVID-19 sustainable measures along with present challenges for the tourism industry to be appended by futuristic researchers, policymakers, and disaster management organizations to be prepared for such futuristic catastrophic endemics.

Chapter 24 focuses on Foods Delivery Apps and examines the factors affecting customer satisfaction with meal delivery applications among MSU students using the 7Ps of the marketing mix. The theoretical framework focuses on the seven Ps of the marketing mix, which includes the product, physical proof, people, promotion, pricing, place, and process, which resulted in a positive and statistically significant relationship between customer satisfaction with meal delivery applications.

Chapter 25 aims to reflect the role that tourism can assume toward meeting the challenges proposed by Agenda 2030. This study adopted as its research universe the participating members of the Conscious Rural Tourism Network, a network in Brazil that brings together entrepreneurs, service providers, farmers, and related technicians, advocating to adopt corporate social responsibility policies for the safety of all health and sustainability in favor of modern Brazilian rural tourism, responsible and professional, without losing rurality.

Chapter 26 focuses on the post-COVID-19 recovery road map with respect to all the regional tourism markets. The global approach is suggested by various experts from academia, industry, destination management companies, travel agents, and tour operators so that a complete transformation of the tourism industry may become a reality in the future.

Chapter 27 of the current study, corporate social responsibility (CSR) that is normally perceived to generate non-operational value at strengthening corporate image is also found to serve as an effective tool that mediates the effect of work stress on the undesirable work behaviors of employees. This finding broadens the knowledge on the value of CSR, when management can further utilize CSR in their strategic plans to cushion the struggles of the employees, especially during a crisis.

Finally, this book elucidates the realities of post-COVID-19 and global strategies to recover, revival, and restart the tourism and hospitality industry toward shifting to a new normal era with digital innovations, reimaging, redesigning, and reskilling tourism and hospitality practices to restart and reopening global tourism destination with new positive hope and prospects. We are thankful to the chapter's contributors and reviewers who have worked with dedication and commitment. Our sincere thanks to *Sandra Jones Sickels*, VP, Editorial and Marketing, Apple Academic Press, and *Professor Mahmood A. Khan*, Series Editor (Advances in Hospitality and Tourism), and the entire team for this opportunity to edit this book at this crucial time.

REFERENCES

https://www.who.int/emergencies/diseases/novel-coronavirus-2019?gclid=EAIaIQobChMIiNaD9-08wIVQYNLBR0H8QPOEAAYASAAEgLFQvD_BwE

https://www.ilo.org/global/topics/coronavirus/sectoral/lang--en/index.htm

BOOK ENDORSEMENTS

I would like to begin by congratulating the entire team, who have put in their soul and heart into writing this useful piece of work, *Post-COVID Tourism and Hospitality Dynamics: Recovery, Revival, and Restart* and have come up with this fantastic idea of accumulating and bringing ongoing thoughts on paper and in life.

COVID, as we all know has been harsh, not just to lives, but to livelihood as well. Industries like tourism have been hit majorly by the pandemic. Having said that, we are delighted that things are coming back to normal and so is the tourism industry. This book talks about a fresh start to revive and restart tourism and hospitality dynamics to once again ignite the soul of the travel industry and make it flourish as always. As an avid reader, I would like to mention that this book not only talks about the restructuring of the system but also connects with you on a level that helps you understand the challenges and queries of what a traveler may have gone through.

My heartiest wishes to Apple Academic Press USA to have initiated such a great step toward addressing the restarting and revival of tourism amid this pandemic. I wish them all the very best for this beautiful work and for their future projects that will benefit society as a whole.

—Jyoti Mayal
President
Travel Agents Association of India (TAAI)

Dr. Sharad Kumar Kulshreshtha and Professor Umendra Narayan Shukla have expertly edited a much-needed and long-awaited compendium of extraordinary scholarly work designed to review, reflect, reimagine, and reinvent the global hospitality industry in the post-COVID-19 world. The broad range of topics demonstrates upon each page the thoughtfulness and intense research of the contributors. I commend you to purchase this book and adapt it for your students.

—Professor Joe Goldblatt, Emeritus Professor
Queen Margaret University, Edinburgh, Scotland

COVID-19 has undoubtedly been one of the largest shocks to the global tourism and hospitality system that we have ever seen. This book provides a range of insights on recovery from such shocks that may have important long-term lessons and insights to responses to future crises.

—Prof. C. Michael Hall
University of Canterbury, New Zealand

COVID-19 related travel restrictions have impacted every aspect of tourism in every country around the globe. This book offers a multi-dimensional perspective on tourism recovery efforts that address critical issues, from tourism employee management to destination governance. I congratulate the authors and the editors on putting together such a diverse collection of contributions that address practical issues but also engage with and expand the tourism literature in important ways.

—Prof. Ulrike Gretzel, PhD
University of Southern California, USA

This book is published at the right time – a time when the world is recovering from the COVID-19 pandemic. It highlights the approaches and measures that have been taken by the countries, states, corporate companies, and more in combating the COVID-19 virus. As we enter a new post-COVID chapter, this book bears testimony of how the world came together to overcome one of the most challenging periods of the tourism scene in the 21st century.

—Dr. Justin Matthew Pang
RMIT University, Vietnam

Doubtless, the COVID-19 pandemic put the tourism industry into a paralleled gridlock, affecting not only present but future business in all sections of the industry. In this must-read book, the editors offer a new, innovative, and thoughtful initiative to re-think tourism just after this disaster. Chapters gathered for this volume are chiefly written by skilled professional researchers concerned with the future of tourism in a post-COVID-19 context.

—Prof. Maximiliano Korstanje
University of Palermo, Argentina

The global tourism and hospitality industry faced very tough times during COVID-19. This book will definitely serve the various stakeholders of this industry with better enlightenment and insightful features focused on post-COVID recovery, revival, and re-start of tourism and hospitality industry worldwide. Both the editors and authors have done incredible work to better shape this book. I congratulate the editors and my best wishes for the great success of this upcoming book.

—Prof. Amit Singh

Department of management studies, Babasaheb Bhimrao Ambedkar University (A central University), Lucknow (U.P.) India

CHAPTER 1

GLOBAL SCENARIO OF POST-COVID-19 TOURISM AND HOSPITALITY DYNAMICS: EXPLORATORY STUDY

U. N. SHUKLA[1] and SHARAD KUMAR KULSHRESHTHA[2]

[1]*Dr. Bhimrao Ambedkar University, Agra, Uttar Pradesh, India*

[2]*North-Eastern Hill University, Shillong, Meghalaya, India*

ABSTRACT

The COVID-19 has emerged as the most severe virus infection in Wuhan, China which badly impacted the tourism industry and global economy too (Soliku et al., 2021). The tourism and travel industry is going through a very severe crisis due to this immediate virus outbreak which spread all over the world. According to the UNWTO, a complete lockdown has been imposed all over the world, which caused 100% travel restrictions on tourism destinations and affected travel mobility (2020). Travel, tourism, hospitality, and aviation have become the fastest growing and emerging sectors globally before this COVID-19 pandemic, As per the IMF report on tourism, it has contributed approximately 10% of GDP and 320 million jobs globally. The tourism industry had to reduce their staff strength with staff salary reduction and as a result, millions of tourism human resources with various professional skills at different levels have lost their jobs (Sandhya, 2021). This COVID-19 crisis has created a very adverse situation for the tourism sector globally. This vacuum has badly affected the mobility of leisure,

Post-COVID Tourism and Hospitality Dynamics: Recovery, Revival, and Re-Start.
Umendra Narayan Shukla & Sharad Kumar Kulshreshtha (Eds.)

business, and special interest tourists with essential travels, such as health and medical, education, religious, events and festivals, and visiting their friends and relatives. The present chapter highlights a very insightful view of the global tourism and hospitality industry with focuses on the adverse effect ascended due to COVID-19 crisis, with the hardship faced by tourism stakeholders worldwide with a positive hope that a new normal will mobilize faster recovery, smooth revival, and easy restart of the tourism industry and will bring back happiness, prosperities among gusts and hosts community locally and globally.

1.1 OVERVIEW

COVID-19 pandemic has badly affected everyone's life socially and economically. Actually, in a true sense, this pandemic has changed our life in manifolds. There is a great shift from the old normal to a new normal. In spite of technological advancement and new developments, the rate of adoption dependence on this transformation was average and very limited. But after the lockdown, this crisis has given an instant action toward serious thinking to adopt these technological changes. As a result, many e-businesses have come into existence with new business models. There have been many significant accountable aspects for tourism industry resilience, which include (i) governance, (ii) smart and intelligent system, (iii) local association, and (iv) consumer and employee confidence (Sharma et al., 2021). The emerging area for tourism resilience in crisis management is the holistic approach. Sustainable tourism development strategies can be a feasible solution that stimulates faster and mass tourist movements and may help the faster momentum of tourism business cycles for the growth and development of the tourism industry (Palacios-Florencio et al., 2021).

1.2 CATALYST OF TOURISM AND HOSPITALITY INDUSTRY

- World's fastest growing service industry as per UNWTO
- Major employment generators
- Income generation
- Livelihoods
- Multiplier effect on the economy.
- Regional development

1.2.1 INTERNATIONAL COOPERATION OF GLOBAL AGENCIES

In this severe global crisis, the World Health Organization (WHO) had to declare "*Public Health Emergency of International Concern*"*(PHEIC)* on January 30, 2020. In this pandemic situation, both international agencies UNWTO and WHO were coordinating and cooperating meticulously, standing in unity and harmony, and very actively engaged in prompt decision-making for the well-being of the global community and have immediately taken these following urgent considerations:

i) Travel restrictions
ii) Measures to support tourism
iii) UNWTO global tourism dashboard
iv) Latest impacts assessment
v) Courses, Webinars, and Resources
vi) Tourism recovery technical assistance package
vii) Recommendations for action
viii) Tourism recovery
ix) Sustainability as the new normal
(x) Restarting tourism

The whole world has gone through the immediate unknown infectivity. The severe virus outbreak has caused millions of infections and casualties. This virus infection got transmitted faster and people become positive for COVID-19 due to its symptoms and some people showed negative results due to good immunity, health and hygiene factors. The international tourism arrival has decreased by 73% in 2020 and has badly affected the global tourism industry (UNWTO, 2021).

The COVID-19 pandemic has not only affected the health and happiness of society but also severely affected the economic condition of the individual and corporates worldwide due to lockdowns and heavy restrictions. It directly impacted the socialization activities which affected every human being. They immediately felt being lonely, emotionally weak, stressed, and sad. All these conditions directly disturbed the mental and physical well-being of the society. All these have led to great economic and financial losses, price hikes, low and no demand and supply of many products and services, loss of most near and dear loveable friends and family members, loss of livelihoods, leisure, luxury, lifestyle, etc. Afterward, a number of immediate questions arose which are directly linked to our future of survivability, sustainability, safety, security, and socioeconomic feasibility.

Tourism has contributed 10.3% of Gross Domestic Product and generated 330 million job opportunities worldwide in the year 2019. It has been observed that the global tourism industry has achieved a positive growth rate in the last decade due to factors, such as corporate travel incentive program, higher disposable income, low-cost carriers, online payments and digital meeting platforms, and the ease of tourist e-visa regimes. It has been observed that the South Asian economy had a positive impact due to faster tourism growth by generating $234 billion which represented 6.6% of the region's GDP in 2019 (WTTC 2020).

Globally, there have been a number of initiatives taken to revive and restart in sustainable and systematic manner to reopen the tourism industry. In this connection, a very significant and sincere initiative taken by WTTC and UNWTO, that is, "*Safe Travels*" stamp ensuring standardized health and hygiene protocols which were also acceptable across the travel and tourism industry, including resorts, hotels, restaurants, airlines, cruises, travel agents and tour operators, shopping malls, railways, coaches, and taxi operators. In this regard, many countries and cities, such as Cancun. Egypt, Portugal, Seville, and Turkey also adopted these protocols established by the WTTC.

TABLE 1.1 5Ps That Affected Tourism and Hospitality Industry During COVID-19 the Most.

People:	Global citizen, society, service providers, consumer, customer, tourist, traveler, guest, tourism entrepreneurs
Pleasure:	Leisure, recreation, entertainment, adventure, wellness and mindfulness
Place:	Natural, cultural destination, rural and urban destination, fairs and festivals destinations, religious destinations
Process:	Service delivery, manufacturing, supply chain, logistic, marketing and distribution
Productivity:	Resource mobilization, factors of production, that is, capital (interest) accumulation (men, material, money, method, machines), land (rent), labor (wages and salary of human resource), firm, industry (profit), tourism arrivals, tourism revenue, tourism employment

Source: Authors.

The COVID-19 is the most unprecedented global crisis that has affected billions of people, resulting in unproductive business and industry growth, loss of livelihoods of millions of people, migration, depression and unemployment, inability to pay for standing liability, and difficulty in managing working capital and adequate revenue margin. All such circumstances have led to very critical and challenging situations for earning livelihoods for those who were working in large, small, medium, and microtourism and hospitality enterprises. Travel is the most essential activity for tourism, therefore, any factor that hampers traveling and touring may have a deep impact on the tourism industry which also includes COVID-19 (Yeh, 2020). This uncertainty and related "travelers" worries will be the symbol of travel and tourism during the time of COVID-19. There are mixed recovery paths for countries across the world, which focus on reopening and revival strategies of tourism destination and tourism operating procedure with set COVID-19 protocols, and all will have to be tailor-made to the specific situation (WEF, 2020).

TABLE 1.2 Guidelines Issues by Various International Agencies for Post-COVID Tourism Revival.

Global organizations	Key actions for post-COVID revival
UNCTAD	The report focuses on the economic losses in the tourism sector that occurred due to COVID-19
UNWTO	Unites the tourism industry to immediate response to the challenges raised after COVID-19
WTTC	Safe Travels Stamp for standardized protocols at tourism destinations around the world
PATA	Developed five dimensions of response, health, economic resilience, support industry, policy revival, and sustainability to counter COVID-19
IMF	New normal and sustainable tourism models and new technologies for faster recovery
ICAO	Managing and monitoring health and safety response and recovery for the aviation sector
IATA	Developed new experience travel technologies for the future of travel. And better solutions for aviation management
WHO	Focused on health emergencies, urgent medical flights, and medical evacuations); travel of essential personnel (such as emergency responders. border closure or suspension of travel, etc.
ILO	Decent work and socially responsible tourism policy framework to guide action to promote decent jobs in the tourism sector
IMO	Supporting seafarers on the frontline of COVID-19
WBG	World Bank Group (WBG), we are taking a green, resilient, and inclusive development (GRID) approach
OECD	Tourism Policy Responses to the coronavirus (COVID-19)
ACI	ACI World provides resources, initiatives, and guidance to help airports during COVID-19 closely connected with WHO and ICAO, and global partners
CLIA	Cruise Lines International Association

Source: Authors.

Figure 1.1: 5S approach involves safety in travel, a seamless system, sustainable practices, and smart approach for Post-COVID tourism revival.

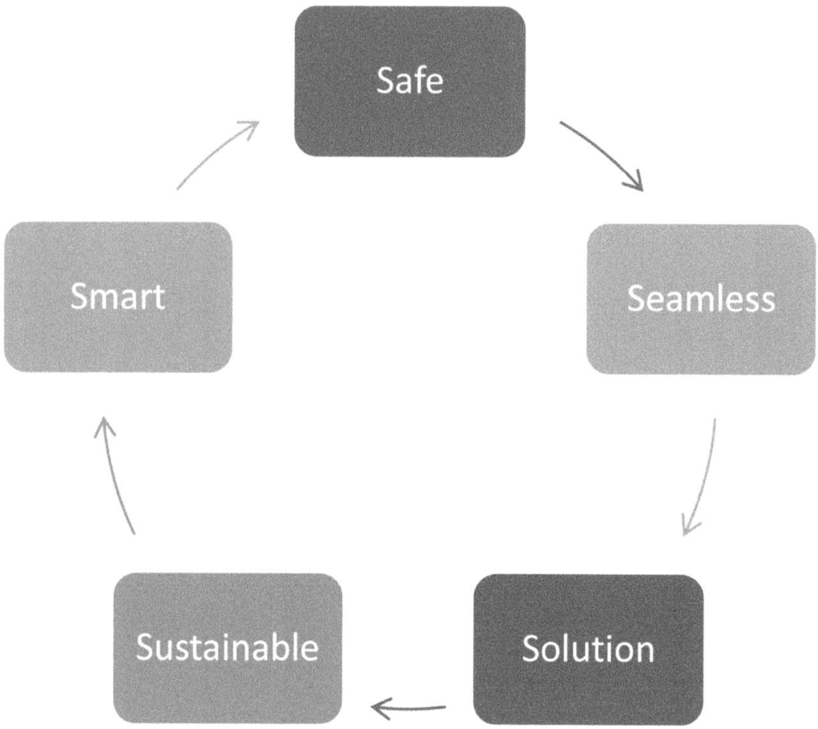

FIGURE 1.1 5S approach post-COVID tourism revival.

1.3 OLD NORMAL VS. NEW NORMAL

Experiencing travel in the new normal is really different as the old normal. The World Health Organization (WHO) has also a major emphasis on "physical distancing" rather than "social distancing" because its purpose is to maintain physical distance which controls or stops the virus transmission and to stay socially connected by maintaining safe distance while meeting together.

"Namste," the Indian greeting style has been very much appreciated by the global community. Throughout the world, there have been many practices developed and adopted during the COVID period to discourage handshaking practices for maintaining safety and hygiene protocols. The sign of greeting

and meeting "Namaskar" or "Namaste," has become a non-touch gesture and an alternative for greeting a person..

Hence, immediate action has been taken to minimize the severe impact of COVID pandemic, and many countries have strictly followed the government directives as well as encouraged the restart with new practices, that is, "new normal" for the global tourism industry. In this regard, there have been lots of focuses on differentiating, inclined toward redefining, redesigning sustainable tourism practicing models, and with the intent of investing in new cutting-edge technologies which could support the faster recovery of the tourism sector. Therefore, there is some expectation that the "New normal" will start with unique practices and an emphasis on more cautious, concerns, careful, and contactless future tourism practices globally.

TABLE 1.3 Old Normal vs. New Normal.

Aspects	Old Normal	New Normal
Focus of travel	Long-haul destinations	Short-haul destinations
Priorities	Leisure, business, and special interest tourism with longer duration	Health and hygiene, less-crowded destinations
Business model	Offline and online business models	Major businesses adopted online business practices and contactless services
Marketing mix	Inclusion of maximum mixes to attract tourists. There were hardly any SOPs in practices	Safety health and hygiene and SOP are some major marketing mixes
Digitalization	There were offline and online business processes	Digital transformations and innovations
Tourism services	Physical and online	Contactless tourism services
Types of tours	Culture, adventure, leisure, and business.	Rural, Eco, and Wellness and most essential Tours with SOPs
Tourism services value chain	Close and mostly focused on mass tourists' perspectives	Developed new and safe protocols to build-up trust for safety and hygiene, vaccine passport, quarantine, isolation, safe tours itinerary value-oriented tourism services

Source: Authors

Figure 1.2 illustrates the four essential Cs and creates a practical flow of decision-making with a cautious, concerned manner, careful, and through contactless travel delivery services which are directly inter-linked in New Normal.

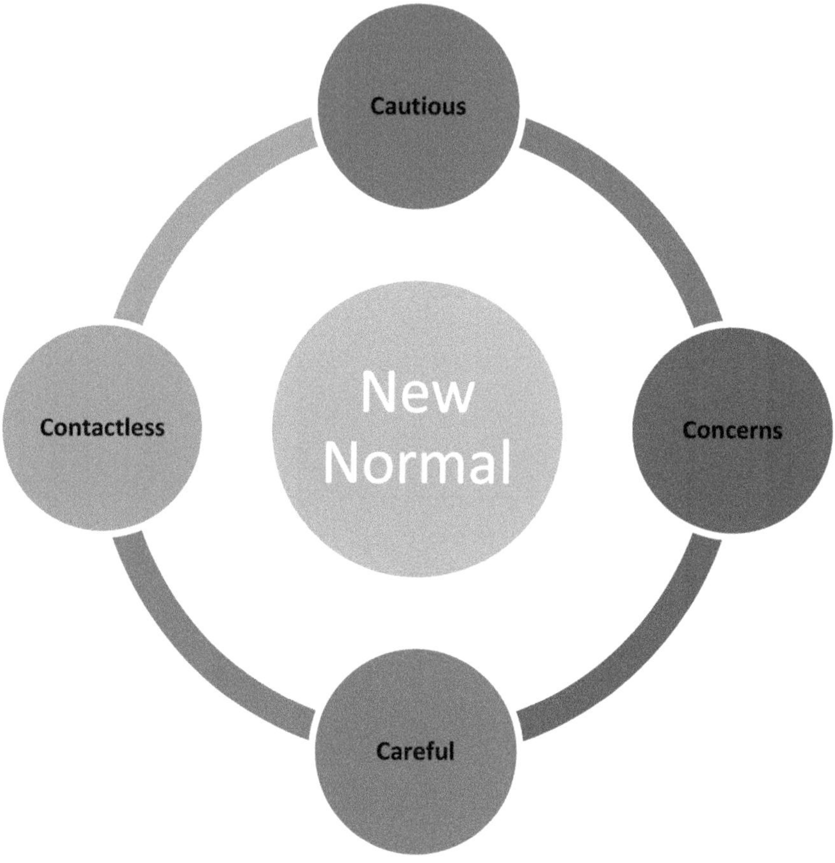

FIGURE 1.2 Essential 4Cs for the new normal.

1.4 RECOVERY

The World Tourism Organization (UNWTO) has launched a new *Tourism Recovery Tracker* to support the global tourism industry to suggest various initiatives with the following SOPs and contactless services with the intervention of innovative technologies. This represents the latest concrete action undertaken by the United Nations' specialized agency as it leads the response of global tourism and guides recovery.

The recovery of global tourism industry will be sluggish and will be depending on the respective countries' travel reopening norms and protocols. As a result, many countries have already started to develop their own recovery plans to restart their tourism industry with stakeholders.

TABLE 1.4 Global to Local (GLOCAL) Economic Problem.

Industry point of view	
What to do?	The industry has no business due to strict COVID protocols, lockdown. Industry can plan to redesign services. Focus on short-haul domestic tourism services. Discussion forums, brainstorming sessions, regular communication with tourism and government agencies
When to do?	This is the right time for regular and mutual consultation on economic issues with industry stakeholders and the international tourism association.
Where to do?	All international, national, and regional tourism destinations and hotels resorts need to rethink and redevelop effective services
How to do?	Deliver high-quality service experience and safety measures in cost-effective and affordable manner. Need to focus on reskilling, redesigning, and innovations
Whom to do?	Focus on potential customers who are looking forward and willing to essential travels and holidays during this high time
Why to do?	Sustaining and operating the business in the new normal with revival and restart strategies
Tourist and Guest's Point of View	
What to plan for travel leisure recreation?	Nowadays tourists are looking for safe and less-crowded destination
When to plan travel for leisure trips?	When the situation is normal, less restriction on travel and a focus on vaccination drive before traveling
Where to plan to travel, stay?	Safe and less-crowded tourist destination, short-haul domestic tourist destinations, where tourism and hotel staff are already vaccinated
Which will be safe or nearby destinations?	Local rural destination, spas, and resorts, hill stations, wellness center
How to travel for holidays?	After vaccination, with all essential safety measures planned
Whom to contact for safe travel and tourism?	Vaccinated and certified tour operators and hoteliers

Source: Authors.

According to *Mckinsy & Company*, five key drivers are likely to impact the tourism recovery trajectory, which are (i) attractive local destinations, (ii) air travel, (iii) health hygiene and safety measures, (iv) business tourism, and (v) sustainable practices.

A number of courses have been launched to provide the essential travel recovery plans covering in-depth information, videos, and case studies to assist tourism and hospitality entrepreneurs to tackle this unprecedented crisis. These courses also focus on the effective recovery plan with strong advice for reopening borders and restrictions for travel mobility. This course will help prepare you to be ready for recovery (Griffith University, Australia)

In this context, the post-pandemic scenario has been constantly moved toward rural and eco-tourism-related micro and small tourism enterprises and also promotes short-haul tourism destinations with encouragement on sustainable tourism practices as an alternative source of local livelihoods of environment conservation.

Although the recovery of the tourism industry is pacing growth gradually worldwide, still it is waiting for the overall recovery after this pandemic, and it will take some time to recover with mass mobility to restart traveling and staying for particular destination. There is also a very good sign that after two years, people have now restarted planning to explore to visit safe and less-crowded destinations and hotels. Many countries have started their airlines and allowed tourists to travel with "vaccine passport." The international tourism is still seeing very little growth in inbound and outbound tourism and is only purposive and essential tourism, and its major focuses are on regional and domestic tourism destinations.

1.5 POST-COVID-19 TOURISM RECOVERY GLOBAL ACTION PLAN

i. Encourage tourists to visit and revisit short-haul destinations.
ii. Assurance of health safety and hygiene.
iii. Develop contactless service procedures.
iv. Opening borders with proper protocols
v. Schemes for MSMEs, entrepreneurs and self-employed
vi. Financial support for reopening the business.
vii. Recalling employees with honor and incentives.
viii. Partnership with local regional vendors.
ix. Creativity and innovation.
x. Follow-up procedure and protocols.
xi. Technology-enabled services.
xii. Value-addition services.
xiii. Future sustainability.

Apart from these, there are a number of initiatives have been taken to better encounter this post-COVID situation for effective and efficient tourism recovery and restart which are discussed below:

TABLE 1.5 15 Cs for Tourism and Hospitality Industry Post-COVID Revival.

1. Confidence	There is an urgent need to develop confidence among tourists to travel to safe destinations, assurances by government and tourism agencies
2. Criteria	There is a need to follow strict criteria for health, safety, and hygiene (mask, hand sensitization, social distancing including vaccination) for tourist entry in respective countries, states, and regions
3. Communication	There is a need to develop effective and efferent communication system for all industry stakeholders
4. Campaign	The campaigns for safe, short hauls, rural, eco less-crowded destinations with new experiences
5. Changes	All essential changes for a shift from old normal to new normal
6. Control	There is a need to control mass tourist arrival and over tourism situation for any destination
7. Concerns	All essential concerns and safety measures of tourists and guest should better take care of for their best satisfaction
8. Community	Proactive community participation and involvement for new normal tourism preparation
9. Customize	Tourism services and tour packages for greater scope customization as per the need of tourists
10. Care	Provide good care and concerns for comfort and safe feelings among the tourists /guests for enriching experiences, revisit, and word of mouths
11. Crisis	There is a need to learn lessons about this COVID, and form and perform effective strategies for handling such crisis management in the future
12. Casualties	Casualties can be avoided through proper health system
13. Capacity	The capacity building needs to develop among tourism and hospitality services providers
14. Contactless	Major focus on transformation for contactless technologies in hotels and airports and destinations
15. Cashless	Digital payments through various online modes have increased and a huge shift toward cashless transactions for tourism hospitality services

Source: Authors.

a. **Six Common Post-Disaster Recovery Measures:** (1) getting support from the higher level government, (2) launching new marketing campaigns, (3) developing new tourism products, (4) focusing on key generating markets, (5) communicating with the major media, and (6) forming alliances/cooperating with industry partners (Zhang et al., 2009).

b. **UNWTO Tourism Recovery Tracker:** In this crucial time, the recovery tracker encourages and supports the collaborative effort by a group of partners including the International Civil Aviation Organization (ICAO), Forward Keys, STR, Sojern, TCI Research, and AIRDNA (UNWTO), etc.
c. **UNWTO Inclusive Recovery Guide:** UNWTO has also approached the UNESCO for better recovery plans and to work together to address the *issue related to cultural tourism. A*nother issue relating to the sociocultural impacts of COVID-19 focuses on the situation changes.
d. **Participatory Governance for Reviving Cultural Tourism:** UNWTO has initiated an approach to bring together artists, creators, tourism, and culture professionals, the private sector, and local communities for the effective participation in cultural tourism. To achieve this objective, there have been continuous efforts for sharing and starting an open discussion, information transfer, and focus on instantaneous ground solutions. This participation will address the urban–rural tie-up for the sustainability of culture and tourism.
e. **Innovation, Digitalization, Sustainability, and Partnerships:** Rebuilding the tourism and hospitality industry with innovation, digitalization, sustainability, and collaborative partnerships is the most urgent responsibility of tourism stakeholders for better future opportunities during and post-COVID.
f. **Financial Support for Tourism Business Recovery:** In June 2021, the Union Finance Ministry, Government of India announced the relief package to revive tourism and other industries in the country to stimulate economic growth. This Scheme is for tourism stakeholders and registered tourist guides under the Ministry of Tourism. Under this scheme, the travel and tourism stakeholders will be eligible to get a loan up to Rs. 10 lakh each, while tourist guides can avail loan upto Rs 1 lakh each *(Ministry of Tourism, Govt of India 2021).*
g. **Resilience Local Communities and Tourism Businesses during Post-COVID:** This transformation has led on leveraging its impact on destinations visited and building more resilient community and tourism stakeholders of respective destinations.

Figure 1.3 "*Recovery focus for Tourism, Hospitality, and Events*" includes preparing plans to support the sustainable recovery of tourism, promoting the digital transition and move to a greener tourism system, and rethinking tourism for the future.

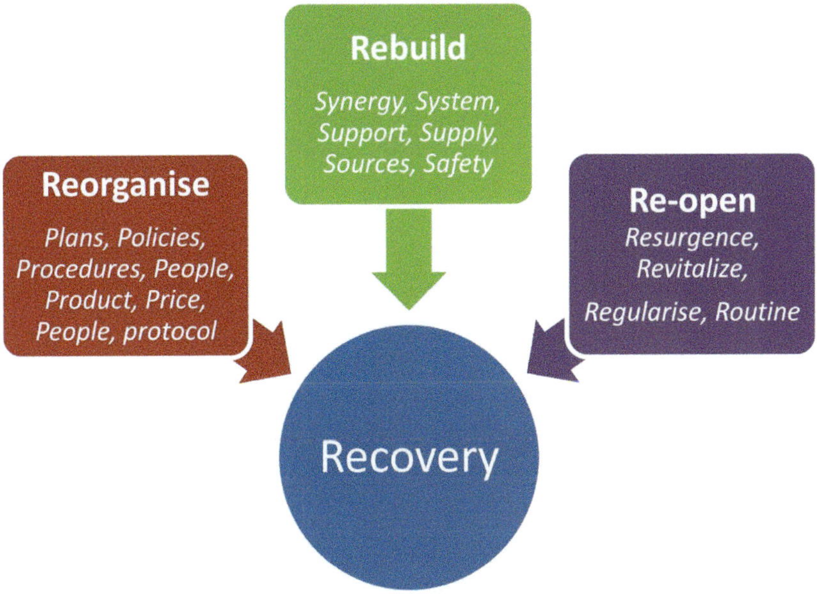

FIGURE 1.3 Recovery focus for tourism, hospitality, and events.

1.6 POST-COVID TOURISM AND HOSPITALITY INDUSTRY REVIVAL

Tourism and hospitality is the most affected sector due to this COVID-19 pandemic. There have been a number of initiatives taken to revive this industry, but the impact of COVID mutation still continues throughout the world. Some countries have opened their international border for travel trade and tourism activities. The main focus of this revival is basically on four aspects, that is, policies, protocols, partnership, and packages.

1. **Policies:** *As tourism is the most sensitive sector, there have been special emphasis on setting tourism policies during COVID-19. From time to time, international and national organization were very active to frame policies for tourism revival. The formulation of policies and their implementation vary widely from country to country. In these policies, travel advisories, entry and exit, international travel, hotel quarantine period, travel restrictions, etc. are included.*
2. **Protocols:** *International agencies UNWTO and WHO have issued essential guidelines for health safety measures for ttravelers. All guests/visitors/passengers will be required to complete a medical*

and travel declaration related to general health, chronic and other conditions, and medication, physical impairments, symptoms prior to 30 days. COVID-19 history and disclosure, temperature monitoring for all guests and passengers will be taken on arrival/ boarding/ pick-up/check-in. Mask wearing, hand sanitizing and correct hand washing, sanitizing hotel, airlines, and travel agency staff for health hygiene and sanitizing practices are the strategies implemented to fight against COVID.

Figure1.4 "Revival Approach after Post-COVID" describes Four Ps, including Policies, Protocols, Partnership, and Packages which reflect integrative collaboration among various stakeholders, that is, Government, Tourism Institutions/Agencies, Industry Associations, and Local Community, etc.

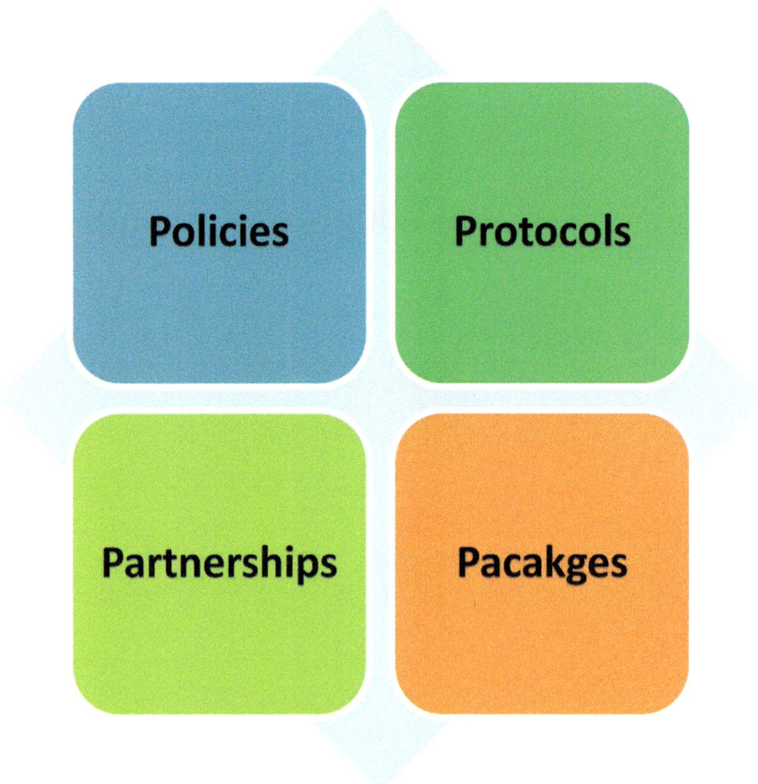

FIGURE 1.4 Four Ps revival approach after post-COVID.

3. **Partnership:** *The international, national, and regional organizations are very active to share opportunities with tourism stakeholders worldwide. This partnership is essential for sustaining the livelihoods of people who directly and indirectly depend on the tourism and hospitality industry. These initiatives are taken by rebuilding, redesign, and reskilling the tourism sector with innovations and transformation inclusion of resilient communities' partnerships.*
4. **Packages:** *There have been many efforts taken to boost the financial support to the registered Tourism Stakeholders which include New Loan Guarantee Scheme for Covid Affected Sectors, issuing free tourist visas, and vaccine passport scheme. Travel agencies and hotels have also customized special tour packages for travel and tourism.*

1.6.1 THE FIVE KEY FACTORS FOR TOURISM AND HOSPITALITY SECTOR

1. Safety: creating awareness and implementation
2. Health: issues and follow-up effective health guidelines
3. Hygiene: it must be the top priority
4. Brands: those who follow these SOPs will be able to sustain
5. Value: value for money will be the new focus area

Source: Brand Equity, The Economic Times e-paper national daily dated April 30, 2020.

1.7 POST-COVID TOURISM AND HOSPITALITY INDUSTRY RESTART

Restarting of the tourism industry is really challenging from country to country. According to the UNWTO's, the most important priorities are vaccines and reopening the border will drive more tourism.

1.7.1 HOW TO RESTART TOURISM

Figure 1.5 Restart Global Model for Tourism and Hospitality Sector, including three significant steps to restart tourism, that is, strategic preparation, redesigning system, and health and hygiene measures.

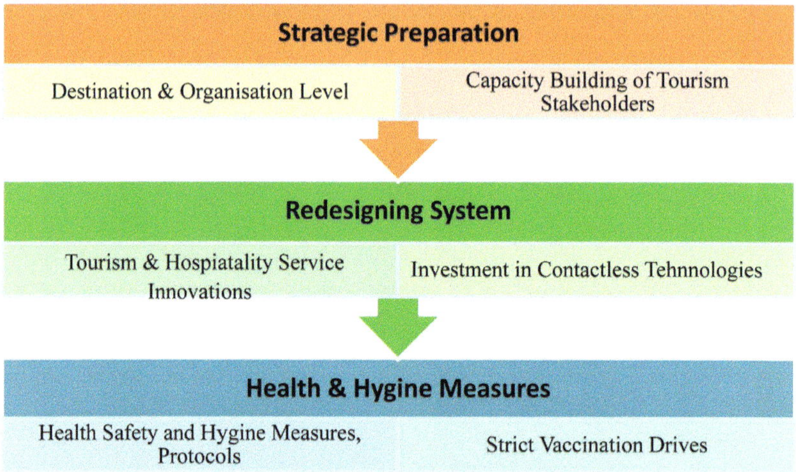

FIGURE 1.5 Restart global model for tourism and hospitality sector.

1.8 UNLOCK: 4.0: REOPENED HOTELS WITH MANDATORY COVID-19 PRECAUTIONS

- *Hotels and resorts maintain guests' information related to their travel history, health condition, and vaccination.*
- *Feel safe motivation and regain trust for travel.*
- *Thermal scanning and hand sanitization.*
- *Avoid large gatherings and maintain social distancing norm.*
- *Strictly follow-up standard operating procedures (SOPs) which include online transactions, digital payments, downloading service Mobile Apps, Scan QR Code, online information, generate e-bills and placing dispenser sanitizers in entry areas and spraying for luggage, common areas, and following hygiene standards for each department.*
- *Reinventing tourism value chains.*
- *Hygiene environment and contactless services.*
- *Revitalizing loyalty programs in new normal*
- *Customized tour packages for short hauls holiday vacations.*

1.9 GLOBAL POST-COVID TOURISM INITIATIVES

1. **World Bank Group's GRID Approach:** *The World Bank Group has envisioned green, resilient, and inclusive development (GRID) approach for post-COVID world (Pangestu et al., 2021).*

2. **Responsible Reopening Plan for Amusement Parks:** *California's amusement parks have initiated Responsible Reopening Plan during COVID for encouraging the visitors.*
3. **CLIA Initiative Cruise Industry:** *The cruise industry has started with properly following COVID-19 protocols including health and hygiene safety, cleanliness, regular screening, covid testing, vaccination, hand sanitizing, masking, physical and social distancing.*
4. **Portugal National Tourist Authority's initiative for Clean and Safe Stamp** *with focus on managing health safety guidelines, which were issued by the National Tourist Authority, according to National Health Authority advisories to avoid high-risk virus outbreaks.*
5. **IUCN and GIZ Local Support Program:** *The sincere efforts of the International Union for Conservation of Nature and the German Development Agency have launched a program to support the local communityinvolvedinruralandcommunitytourismpracticesinandaround protectedandconservedareasworldwide,speciallyforpost-covidrecovery.*
6. **UNWTO "Best Tourism Village":** *In connection with the restart, UNWTO has launched "Best Tourism Village" to recognize the villages for their innovative practices and to maintain sustainability to achieve sustainable development goals with objectives and better* opportunities to share their unique culture, delicious food, and *local beverages. The village acts as a catalyst for the rural economy supporting micro and small medium enterprises and innovative rural start-ups.*
7. **Global Vaccination Drive:** WHO has validated several COVID-19 vaccines for use (given Emergency Use Listing). These vaccines were properly developed to ensure standards of quality, safety, and efficacy by using clinical trial data, and maintain standard quality control processes for further manufacture of these vaccines. In this regard, the following vaccines have obtained Emergency Use Listing (EUL) on 26 November 2021 as mentioned below:

COVID-19 vaccines	Date of EUL
The Pfizer/BioNTech *Comirnaty*	*31 December 2020*
*The SII/**COVISHIELD** and **AstraZeneca**/AZD1222 vaccines*	*16 February 2021*
The Janssen/*Ad26.COV 2.S developed by Johnson & Johnson*	*12 March 2021*
The Moderna *COVID-19 vaccine (mRNA 1273)*	*30 April 2021*
The Sinopharm *COVID-19 vaccine*	*7 May 2021*
The Sinovac-*CoronaVac*	*1 June 2021*
*The Bharat Biotech BBV152 **COVAXIN** vaccine*	*3 November 2021*

Source: World Health Organization (WHO, 2021).

8. **The Success of COVID Vaccination in India:** According to the Ministry of Health and Family Welfare, the Government of India and my Gov.in India has achieved a great milestone under the dynamic leadership of Prime Minister Shri Narendra Modi to cover the journey of 200 crore vaccination on July 16, 2022.

PLATE 1.1

Source: https://www.mygov.in/covid-19

1.10 FUTURE STRATEGIES FOR POST-COVID TOURISM AND HOSPITALITY INDUSTRY

- *The customized pricing for budget tourists also may increase the demand for such tourist destinations*
- *High demand for budget and luxury destinations with special care of hygiene and cleanliness.*
- *To regain the competitive advantages of destinations, all tourism countries should attempt to focus more on high-value tourists.*
- *Develop and promote less-crowded destination.*
- *Health and safety measures.*
- *After the reopening of international borders, there is an urgent need to develop contactless tourism services delivery by promoting digital*

tourism services, that is, e-booking, e-payment, e-marketing with greater specialization of smart and innovative tourism platforms.

- *Determining factors of pricing which directly is responsible for competition, price, and income inequality.*
- *Elasticity in job creation in the tourism sector. There is a need to focus more on multi-skilling, training, and development, to absorb pro-people in the industry in an inclusive manner.*
- *Adapt to various skills to more value labor efficiency by upskilling and reskilling programs.*
- *Continuous monitoring of the tourism industry offerings and future of destination.*
- *Reorganization of the industry to better respond post-COVID tourism demand.*
- *More focus on networking to strengthen client and business relationship.*
- *Marketing research and policy implementation.*
- *Instant alleviation of economic and social impact that arise after COVID.*

Figure 1.6 Stakeholder's inclusive partnership focuses on an inclusive approach among stakeholders for effective coordination and partnership to find a better solution for post-COVID tourism restart.

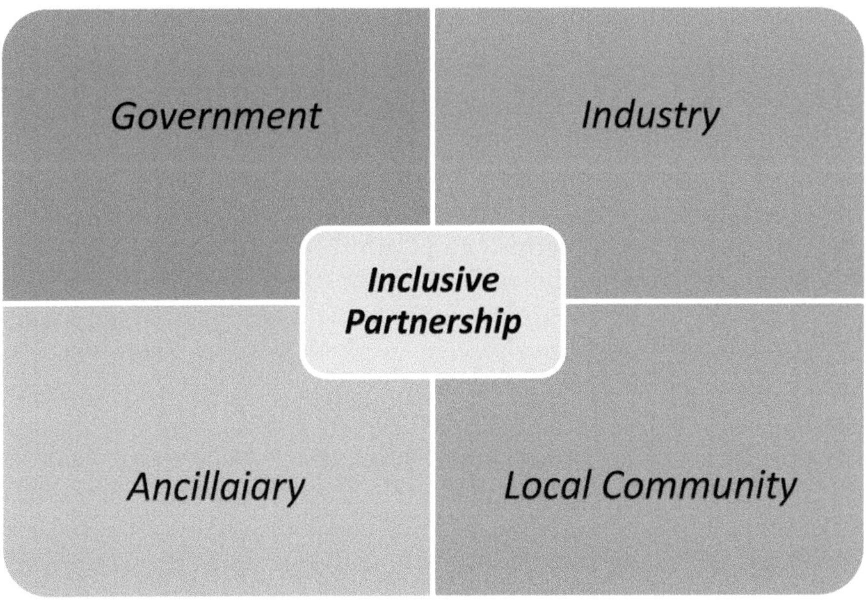

FIGURE 1.6 Stakeholder's inclusive partnership.

TABLE 1.6 Priorities for Future Travel.

Present focus	Continuous monitoring and focus on the present and real-time issues and challenges-related pandemic situation
Positive approach	There is an urgent need for positive and optimistic thinking with practical and possible solutions
Possible initiatives	Explore the possibility of new ideas for revival and restart
Practical approach	Be practical on safety standards and protocols by a clear realistic approach
Priority setting	Set the most important first
Prospective view	Look forward present and future opportunities
Potential markets	Assess market need
Policies orientation	Redefine, redesign existing policies
Practice persuasion	Innovative practices
Protection of people	Maintain social distancing and essential protocols to protect health and safety of host and guest community
Projection of future	Future plan, strategies, goals, safety measures, and quality tourist experience

Source: Authors.

1.11 CONCLUSIONS

The impact of COVID-19 pandemic has been observed and experienced globally in the economic front drastically with severe effects of tourism business, including small, medium, and large tourism enterprises, that is, travel agencies, tour operators, hospitality services, and transportation services including aviation, railways, cruising. United Nations' World Tourism Organization has already declared 100% travel restrictions globally. In this regard, most countries had to close their destinations and had implemented strict travel restrictions for tourist mobility. According to the *UNWTO Guidelines to Restart Tourism*, responsible tourism recovery for the global community and their health and wealth well-beings includes public health, social inclusion, and livelihoods. It is also related to eco-biodiversity, environmental conservation, climate change aspects, maintaining a circular economy through *reuse, recycle, and reduce* and effective governance and public finance management. COVID continues with new variants and mutations. After reopening the tourism sector, once again, a new health threat has emerged in the form of new COVID-19 variant, that is, "Omicron" to disturb the usual way. It was found in South Africa in November 2021, which was the fastest spreading virus earlier. The World Health Organization is very proactive in issuing guidelines and health advisory with health safety protocols to all countries, especially those countries that have recently opened their tourism destinations. There is further strengthening to rework and re-plan covid policies, standard operating procedures, and inclusion of digital and contactless technologies, as well as training the working force to serve in such pandemic situations. There is no doubt that the COVID-19 and its related new variant have affected the tourism sector badly, but resilient with safety procedures, re-skilling, redesigning services, and technological transformation are some immediate measures to revive and restart the tourism industry. There is an urgent need to focus on effective strategies as well as sound recovery plans for the resilience of tourism and hospitality sector including some corrective measures, with set standard protocols and sustainable practices to better encounter the COVID-19 situation. There is also a need to support the uplift of vulnerable stakeholders (micro and small entrepreneurs) whose livelihood very much depends on tourist footfalls in their respective destination.

KEYWORDS

- **travel**
- **tourism**
- **hospitality**
- **post-COVID**
- **recovery**
- **revival**
- **restart**

REFERENCES

Sandhya, H. The Survival of Travel and Tourism Industry amidst the Covid 19 Pandemic—Challenges and Opportunities of the Indian Tourism Sector. *Atna J. Tour. Studies* **2021,** *16* (1). https://doi.org/10.12727/ajts.25.3

Zhang, H. Q.; Li, M.; Xiao, H. Tourism Recovery Strategies and Their Implications: A Multiple Case Study Approach. *J. China Tour. Res.* **2009,** *5* (3), 273–286, DOI: 10.1080/19388160903152811

Orîndaru, A.; Popescu, M.-F.; Alexoaei, A. P.; Căescu, Ş-C.; Florescu, M. S.; Orzan, A.-O. Tourism in a Post-COVID-19 Era: Sustainable Strategies for Industry's Recovery. *Sustainability* **2021,** *13* (12), 6781. MDPI AG. Retrieved from http://dx.doi.org/10.3390/su13126781

Palacios-Florencio, B.; Santos-Roldán, L.; Berbel-Pineda, J. M.; Castillo-Canalejo, A. M. Sustainable Tourism as a Driving force of the Tourism Industry in a Post-Covid-19 Scenario. *Soc. Indicat. Res.* **2021,** 1–21. Advance Online Publication. https://doi.org/10.1007/s11205-021-02735-2

Reddy, M. V.; Boyd, S. W.; Nica, M. Towards a Post-Conflict Tourism Recovery Framework. *Ann. Tour. Res.* **2020,** *84*, 102940. https://doi.org/10.1016/j.annals.2020.102940

Sharma, G. D.; Thomas, A.; Paul, J. Reviving Tourism Industry Post-COVID-19: A Resilience-Based Framework. *Tour. Manage. Persp.* **2021,** *37*, 100786. https://doi.org/10.1016/j.tmp.2020.100786

Yeh, S-S. Tourism Recovery Strategy Against COVID-19 Pandemic. *Tour. Recreat. Res.* **2020**. DOI: 10.1080/02508281.2020.1805933

Soliku, O.; Kyiire, B.; Mahama, A.; Kubio, C. Tourism Amid COVID-19 Pandemic: Impacts and Implications for Building Resilience in the Eco-Tourism Sector in Ghana's Savannah Region. *Heliyon* **2021,** *7* (9), e07892. https://doi.org/10.1016/j.heliyon.2021.e07892

WEB REFERENCES

https://www.unwto.org/unwto-tourism-recovery-tracker (accessed on Sept 30, 2021).

https://www.mckinsey.com/industries/travel-logistics-and-transport-infrastructure/our-insights/covid-19-tourism-spend-recovery-in-numbers (accessed on Sept 30, 2021).

https://www.mckinsey.com/industries/travel-logistics-and-transport-infrastructure/our-insights/covid-19-tourism-spend-recovery-in-numbers (accessed on Sept 30, 2021).

https://wttc.org/Research/To-Recovery-Beyond (accessed on Sept 30, 2021).

https://www.griffith.edu.au/engage/professional-learning/tourism-recovery (accessed on Oct 4, 2021).

https://journals.sagepub.com/doi/full/10.1177/1354816620933712 (accessed on Oct 4, 2021).

https://www.hotelmanagement.net/operate/wttc-ey-examine-path-to-tourism-recovery (accessed on Oct 4, 2021).

https://webunwto.s3.eu-west-1.amazonaws.com/s3fs-public/2020-06/BFN_V4.pdf (accessed on Oct 7, 2021).

https://www.g20-insights.org/policy_briefs/the-covid-19-crisis-and-tourism-response-and-recovery-measures-to-support-the-tourism-sector-in-oecd-countries/ (accessed on Oct 7, 2021).

https://www.gov.ie/en/publication/be0cb-tourism-recovery-plan-2020-2023/

file:///C:/Users/Dell/Downloads/90006_80801fc3-a69b-4faf-843b-be9bff4d6a0f%20(1).pdf (accessed on Oct 12, 2021).

https://restoreyoureconomy.org/clientuploads/2013/09/Tourism-White-Paper.pdf (accessed on Oct 12, 2021).

https://tiac-aitc.ca/Tourism_Recovery.html (accessed on Oct 15, 2021).

https://www.caribbean-council.org/tourisms-recovery-holds-key-to-future-caribbean-growth/ (accessed on Nov 7, 2021).

https://www.weforum.org/agenda/2020/08/travel-and-tourism-recovery-south-asia-covid19-pandemic-economy-india-nepal-bhutan-sri-lanka/ (accessed on Nov 7, 2021).
https://www.unwto.org/unwto-tourism-recovery-tracker (accessed on Nov 7, 2021).
https://webunwto.s3.eu-west-1.amazonaws.com/s3fs-public/2021-02/210225-culture-and-tourism-en_0.pdf?3cUCtl9hZumn8yxYjkUdKGxlQ2Z.lGGz (accessed on Nov 11, 2021).
https://reopencaamusementparks.com/rrp/ (accessed on Nov 11, 2021).
https://www.cdc.gov/coronavirus/2019-ncov/community/large-events/considerations-for-events-gatherings.html (accessed on Nov 11, 2021).
https://webunwto.s3.eu-west-1.amazonaws.com/s3fs-public/2020-05/UNWTO-Priorities-for-Global-Tourism-Recovery.pdf (accessed on Nov 16, 2021).
https://blogs.worldbank.org/voices/tourism-post-covid-world-three-steps-build-better-forward (accessed on Nov 16, 2021).
https://www.ey.com/en_us/bbc/post-covid-19-travel--how-to-prepare-for-tourism-s-rebound (accessed on Nov 16, 2021).
https://www.ey.com/en_us/bbc/post-covid-19-travel--how-to-prepare-for-tourism-s-rebound (accessed on Nov 18, 2021).
https://www.oecd.org/coronavirus/policy-responses/rebuilding-tourism-for-the-future-covid-19-policy-responses-and-recovery-bced9859/ (accessed on Nov 18, 2021).
https://www.imf.org/external/pubs/ft/fandd/2020/12/impact-of-the-pandemic-on-tourism-behsudi.htm (accessed on Nov 18, 2021).
https://www.mckinsey.com/industries/travel-logistics-and-infrastructure/our-insights/covid-19-tourism-spend-recovery-in-numbers (accessed on Nov 19, 2021).
https://wttc.org/COVID-19/SafeTravels-Global-Protocols-Stamp (accessed on Dec 3, 2021).
https://www.un.org/sites/un2.un.org/files/sg_policy_brief_covid-19_tourism_august_2020.pdf (accessed on Dec 4, 2021).
https://www.iata.org/contentassets/c0f61fc821dc4f62bb6441d7abedb076/guidance-for-post-covid-restart-of-operations-cbta-training-solutions.pdf (accessed on Dec 3, 2021).
https://www.iata.org/en/programs/covid-19-resources-guidelines/safely-restarting/ (accessed on Dec 5, 2021).
https://www.oecd.org/coronavirus/policy-responses/tourism-policy-responses-to-the-coronavirus-covid-19-6466aa20/ (accessed on Dec 5, 2021).
https://cruising.org/en/cruise-industry-covid-19-facts-and-resources (accessed on Dec 5, 2021).
https://webunwto.s3-eu-west-1.amazonaws.com/s3fs-public/2020-06/PORTUGAL_.mp4 (accessed on Dec 6, 2021).
https://www.unwto.org/restarting-tourism (accessed on Dec 6, 2021).
https://unctad.org/webflyer/covid-19-and-tourism-update (accessed on Dec 6, 2021).
https://pib.gov.in/PressReleasePage.aspx?PRID=1731924 (accessed on Dec 7, 2021).
https://www.who.int/emergencies/diseases/novel-coronavirus-2019/question-and-answers-hub/q-a-detail/coronavirus-disease-(covid-19)-vaccines?adgroupsurvey={adgroupsurvey}&gclid=Cj0KCQiAt8WOBhDbARIsANQLp96fTcbA5ODiB2JyURv4PXbcs5Z3o0QTCUkkgt9x3B6lY6GFGC-vfg0aAon8EALw_wcB (accessed on Dec 7, 2021).
https://www.visitportugal.com/en/content/clean-and-safe (accessed on Dec 7, 2021).
https://www.iucn.org/news/protected-areas/202109/iucn-and-german-government-support-post-covid-recovery-through-sustainable-tourism-initiative (accessed on Dec 7, 2021).
https://www.oneplanetnetwork.org/programmes/sustainable-tourism/covid-19-responsible-recovery/how (accessed on Dec 15, 2021).

https://pib.gov.in/PressReleasePage.aspx?PRID=1731924 (accessed on Dec 15, 2021).

https://www.unwto.org/tourism-villages/en/news/unwto-opens-call-for-best-tourism-villages/

https://transformingindia.mygov.in/covid19/major-milestone-in-200-crore-vaccination-journey/ (accessed on Sept 14, 2022).

https://transformingindia.mygov.in/covid19/worlds-largest-and-fastest-vaccination-drive-2/ (accessed on Sept 14, 2022).

CHAPTER 2

A STUDY OF POST-COVID PANDEMIC TRAVEL PREFERENCES: OMAN FOCUS

MOHIT KUKRETI and AMITABH MISHRA

University of Technology and Applied Sciences, CAS, Ibri Oman

ABSTRACT

This chapter focuses on understanding the post-COVID-19 pandemic travel preferences and perception of Omani travelers, which is a mix of citizens and expatriate residents. Primary and secondary data methodology was employed for collecting and analyzing data. Data from both qualitative and quantitative sources were gathered and analyzed. The sample survey includes 425 respondents from the Sultanate of Oman, consisting of a mixed population of citizens and expatriate residents. The findings conclude that most of the respondents are willing to travel to reunite with their families and friends post-COVID-19. They are more inclined toward staycations and want to go for domestic and local trips. Though they will be very interested in traveling post-pandemic, the sales campaign and promotional deals may not be the primary motivator for them to travel. Overall, safety, hygiene, dependable health, and sanitation arrangement will play a crucial role in the residents' decisions about traveling post-COVID-19 pandemic.

2.1 INTRODUCTION

The beginning of the year 2020 was catastrophic for the promising travel and tourism industry worldwide due to the COVID-19 pandemic (Terziyska and Dogramadjieva, 2021; Uğur and Akbıyık, 2020; Jones et al., 2020). Before

Post-COVID Tourism and Hospitality Dynamics: Recovery, Revival, and Re-Start.
Umendra Narayan Shukla & Sharad Kumar Kulshreshtha (Eds.)

COVID-19, there were many similar epidemics, such as SARS, MERS, and Ebola in the previous decades, but the world had not experienced such a disastrous effect on the global economy (Naumov et al., 2020). There is much research that discovered this unusual occurrence and its impact on people and the travel and tourism industry (Brouder et al., 2020; Sigala, 2020; Li et al., 2020). Several tourism organizations suffered massive losses because of negligible travel consumption (Wen et al., 2020).

A similar phenomenon is noticed in Oman also. The COVID pandemic has caused direct losses of an estimated 500 million Omani Rial in September 2020 to Oman's tourism sector. The hotel occupancy in the country fell by 60.2%, which is to the tune of RO 70 million in October 2020 (Zainab Al Nasseri, 2020).

Oman is a beautiful tourism-rich country in the Middle East Region of Asia. It is located on the southeastern coast of the Arabian Peninsula (Claereboudt, 2018). Oman is officially known as the Sultanate of Oman. The Sultanate is the oldest independent country among the Arabian countries (experienceoman.om, 2021). The country has an excellent network of roads regionally, nationally, and internationally, with the maximum allowed speed limit of 120 km per hour. This makes the movement of people and goods faster and easy. The government has been keen to provide the highest level of cleanliness, hygiene, facilities/amenities, and overall infrastructure. In 2010, the United Nations Development Program ranked Oman as the most improved nation globally concerning development during the past 40 years (UNOCHA, 2018).

The Sultanate of Oman is considered one of the world's hottest geographical regions. A big part of the country is covered with hot desert. The Sultanate has a long coastal stretch of almost 3000 km (Ministry-of-Tourism, 2021). Most parts of the country have a dry tropical climate, but the Dhofar region of the Sultanate is known for Monsoon rains from mid-June to mid-September. This period is known as the Khareef season.

In the Sultanate, the tourism sector contributes significantly to its economy. Tourism is one of the important sectors that the Sultanate of Oman focuses on in the current 5-year plan. The Sultanate is categorized as a High-Income Economy, and its rank is 68 in the list of the most peaceful nations in the world as per the Global Peace Index (Global Peace Index, 2020). According to the National Centre for Statistics and Information (NCSI), inbound tourism spending increased to around OMR 685 million in 2019, from OMR 364.7 million in 2015. The statistical data provided by ROP, Visa section reveal that there has been a 41.6% growth in inbound tourism.

In 2019, 1.8 million inbound visitors arrived in the Sultanate, compared with 1.2 million visitors in 2015. The Sultanate has a well-developed infrastructure that supports the tourism industry. At the end of 2019, there were 491 hotel facilities, including 26 five-star hotels, compared with 318 hotel facilities in 2015, including 11 five-star hotels. In 2019, hotel profits increased to OMR 276.6 million, up from OMR 226.9 million in 2015 (Times-Of-Oman, 2020).

The Sultanate of Oman is known for its spectacular tourist attractions. Its deserts, mountains, beaches, heritage, arts and customs, wadis, and historic buildings make Oman different from the other Arab countries. Salalah, located in Dhofar region, is one of the most important tourist destinations in the Sultanate. During Khareef season, thousands of visitors from especially the Middle East and some other countries visit the destination. Monsoon makes Salalah's weather different from the rest of the country. Salalah is not unique in only in Oman but also in the Gulf region.

The variations in the topography of Oman make it an exciting tourist destination to visit. Apart from this, Omanis are known for hospitality. They are humble, kind, friendly, and always ready to assist. Because of the versatility in climate and tourism resources, many visitors visit the country every year. Tourism is growing well in Oman. According to the Tourism Statistical Bulletin from the National Centre for Statistics and Information (NCSI), in Oman in 2019, approximately, 3.5 million visitors visited the Sultanate worldwide. It indicates a substantial increase from 3.1 million in 2017.

In the first month of 2020, all the sectors/industries were moving smoothly. But in the middle of February 2020, the Ministry of Health confirmed the entry of the novel coronavirus in the country. On 24th February 2020, the government confirmed the first two cases in Oman when two Omani citizens were tested positive after coming back from Iran (Oman-Observer, 2020). In the beginning, the maximum number of positive cases and deaths occurred in the expatriate community (Times of Oman, 2020). The majority of cases were initially registered in Muscat, the Sultanate's capital. Because of the magnitude of the novel coronavirus, the government announced a complete lockdown in Muscat on April 10, 2020 (Oman-Observer, 2020). Lockdown was eventually begun in other governorates of the Sultanate as well (TAS-News-Service, 2020).

As a result of the lockdown and continued spread of the novel coronavirus, the whole tourism industry came to a halt. Parks, beaches, hotels, restaurants, museums, forts, palaces, etc., had been closed. Reduced business incomes, cuts in disposable income, insufficient purchasing power, and unemployment were significant due to pandemic outbreaks (Al-Hasni, 2021). The government, industries, and people were hoping that shortly the pandemic will end and everything will again become normal. Overall, there has been a considerable loss the tourism industry has faced in Oman due to the current pandemic. However, there is a limited number of empirical studies on how the COVID-19 pandemic has impacted Oman's tourism sector (Al-Hasni, 2021). The contributions of this study are manifold. It emphasizes that people in Oman will travel with their families and friends post-COVID-19 and are more inclined toward staycations and domestic and local trips. The study emphasizes that the sales campaign and promotional deals may not be the primary motivator for people to travel. Overall, safety, hygiene, dependable health, and sanitation arrangement will play a crucial role in the residents' decisions about traveling post-COVID-19 pandemic in Oman.

2.2 LITERATURE REVIEW

Regardless of the crisis, the tourism industry is known for its ability to recover and adapt (Ritchie et al., 2014). The new coronavirus (COVID-19) that emerged in 2019 caused an unprecedented scenario that brought the entire planet to a halt. The ramifications of the pandemic are still unknown, but it is expected to have a long-term influence (Glossings et al., 2020).

The mechanism by which a person chooses, arranges, and interprets motivations into a meaningful and coherent image is known as perception (Haarhoff, 2018). Perception thus refers to how tourists perceive or interpret their senses for future decisions. Tourist's perception is always considered an influencing factor in deciding their tourism plan. The tourists' perception depends on various factors, such as demographic, personal, social, psychological, economic. The study of residents' perception of travel behavior has always been considered a great interest in tourism-related research. This perception of travel behavior has gone through major and significant changes after the onset of the COVID-19 pandemic. Most businesses worldwide have gone through ups and downs during the pandemic (Nguyen et al., 2020). With it, the perception of residents for tourism, their criteria for choosing destinations, attitudes, and priority to travel have also changed (Ivanova et al., 2020). This catastrophe has increased fear, tension, and confusion like other crises which happened previously (Chebli and Ben Said, 2020).

Molloy et al. (2020) examined the changes in travel behavior in Switzerland due to the COVID-19 pandemic. Many previous studies opined that during the pandemic, human mobility and interaction patterns had been the main contributing factors in the spread of infection (Abdullah et al., 2020). Different countries' governments provide various control and preventive measures and guidelines to avoid infection and safeguard ourselves, which becomes our behavior with time. This type of behavior chiefly depends on demographic characteristics, such as age, gender, marital status, race, income, education, and the person's perceived risk of infection (Li et al., 2020). The research results done by J. Lee et al. also highlight that people with a higher level of educational qualification and high income tend to have a more positive perception of travel.

Numerous tourism studies have also shown that during the outbreak of H1N1, older travelers were ready to delay their travel plans compared with the young travelers (18–35) (Ivanova et al., 2020). The researchers like A. Bhati have highlighted that hygiene, disinfection, and a reliable health system in any destination widely influence tourist's decisions. People with high health considerations, hygiene, and mental health problems have a high perception of their travel behavior post-COVID-19 (Bhati et al., 2020).

According to Kotler (2000), customer preferences show consumer choices among a large range of current service goods. An affection (inclination of the heart) for something is called preference. Preference may also be described as a person's decision to favor or hate a product. Meanwhile, customer preferences are defined as subjective (individual) tastes as evaluated by the utility of various items, according to Indarto (2011). Consumer preferences may be influenced by a variety of factors, such as product quality, pricing, advertising, and packaging.

Maslow (1943) considered safety a vital decisive factor in forming travel perception (Chebli and Ben Said, 2020). Tourists want to travel to safer countries. Previous research after EBOLA, SARS, and MERS has also shown that travel is constrained by risk factors and negatively affects tourists' perception (Chalk and Seitz, 2005). The outbreaks of diseases and crises bring uncertainty and insecurity which affect tourism. Several academic publications are devoted to this theme. This chapter takes up a qualitative approach to reach the following objectives:

- To understand the post-COVID-19 pandemic travel preferences and perception of travelers in Omani.
- To find out accommodation preferences, traveling style, preferred tourism activities, selected places to visit, sanitation, and safety preferences of the travelers in Oman.

2.3 METHODOLOGY

To focus on understanding the travel preferences and perceptions of the travelers in Oman, mainly the Omani citizens and expatriate residents, the authors designed a questionnaire survey. Data from both qualitative and quantitative sources were gathered and analyzed. The primary data were collected through a "Google Forms" questionnaire survey administered during the first 3 weeks of January 2021 by administering it online in the main cities of Oman, including Muscat, Sohar, Nizwa, Salalah, Sur, and Ibri (Figure 2.1). The main reason for choosing these six cities is that these cities are the main metropolitan cities of Oman where the majority of the expatriate residents and Omani citizens live. The population of Oman is a mix of expatriate residents and Omani citizens, however, the study does not differentiate between the respondents and it treats all the respondents alike. The questionnaires were distributed among respondents in these cities through the google platform and social media. The survey was distributed among 500 respondents. However, the authors received complete responses from a total of 425 respondents. Data were analyzed using Qualitative Descriptive Analysis. Bearing in mind the large size of respondents, the experimental distribution of responses was considered more or less normal (Kim, 2013). The questionnaire consisted of demographic characteristics, such as age, gender, education, marital status, travel preferences, intentions, and the type of tourism-intended activities. Additionally, questions regarding the tourist preferences on the kind of stay, level of hygiene, the purpose of visit, and motivations were also included.

2.4 RESULTS AND DISCUSSIONS

This research study examined the traveler's preferences and perceptions post-COVID-19 pandemic in Oman. Due to the majority of the male working population in Oman, male citizens and residents are the more common respondents (66.1% from the total sample), followed by females (33.9%). About 87% of the respondents are employed for wages, followed by self-employed, homemakers, and students, and 95.2% of the respondents are married; 69.4% of the respondents are from 41 to 55 years of age and they are more experienced and mature; 27.4% of the respondents are from 26 to 40 years of age. A total of almost 75% of residents and 25% of Omani citizens responded to this survey. The majority of the respondents, which is 20%, are from Ibri, followed by 18.82% from Muscat, 16.47% of respondents from

Nizwa and Sohar, 14.11% are from Sur and Salalah. The following Table 2.1 and Figure 2.1 shows the distribution of the respondents from the six major cities where the survey was administered.

TABLE 2.1 City-Wise Distribution of the Respondents.

Name of the City	No. of Respondents	% of Respondents
Muscat	80	18.82
Nizwa	70	16.47
Sur	60	14.11
Sohar	70	16.47
Salalah	60	14.11
Ibri	85	20.0
Total	**425**	**100**

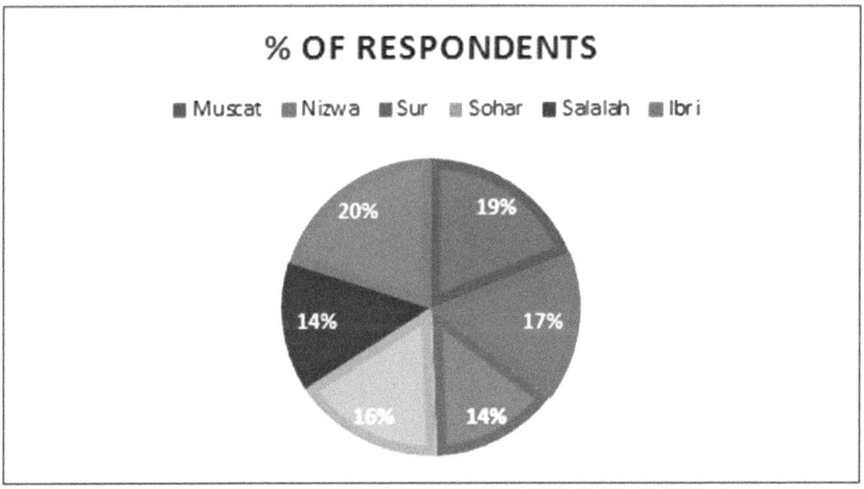

FIGURE 2.1 The % of respondents from the six major cities.

The most substantial question regarding the demand for the tourism industry (Intention to travel post-covid-19 as a holiday traveler) in Figure 2.2 shows that 55% of the responding citizens and residents specify that they will be going to travel and 27% of citizens and residents have a suppressed demand, and 18% of citizens and residents constitute no demand.

Further investigation to understand that the demand for travel is important, especially related to tourist attractions and hospitality products of

Oman, such as Resorts, City Hotel, and Business Hotel. The overall demand will also impact Muscat international airport traffic and hotel occupancy.

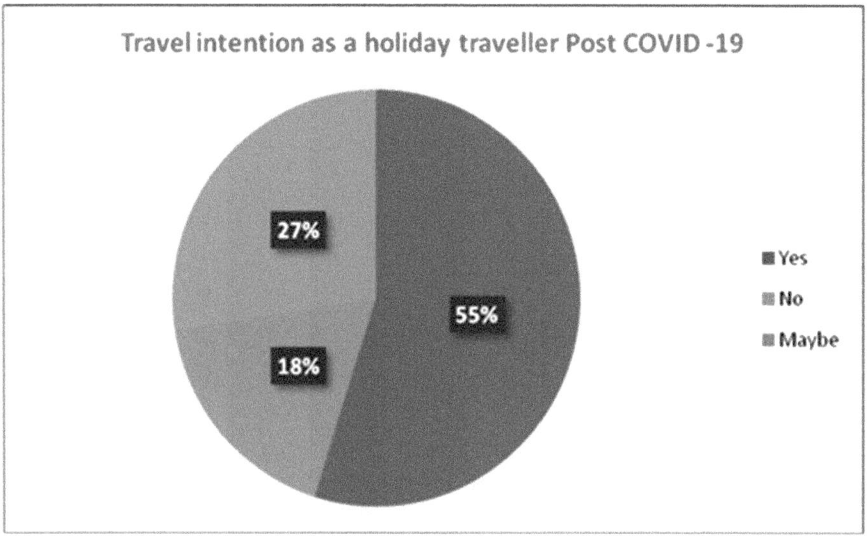

FIGURE 2.2 Travel intention as a holiday traveler post-COVID-19.

With regard to the primary motivation or purpose for post-COVID-19 travel, the authors observe that the majority of citizens and residents (56.5%) wish to travel to visit their Friends and Relatives (VFR) and about 52.2% for a family vacation, followed by 37.1% of people for rest and recreation as compared with any other traditional purpose such as business (only 3.2%). This finding contrasts (Figure 2.3) the Middle Eastern and US markets' outcomes, where business travel is the primary motivation for travel (Choufany, 2020). It is apparent that the compulsory indoor stay and lockdown due to COVID-19-related restrictions motivate citizens and residents to reunite with their family and friends for rest and enjoyment.

Business Corporations in Oman may defer traveling timelines from the Business travel perspective and could have a smaller budget allocation to spend on Business travel.

From the travel decision-making perspective, 71% of the respondents attributed family's opinion as the main motivator to take up travel post-COVID-19 (Figure 2.4). About 32% of the respondents shall consider their Friend's opinion, and 29% of respondents shall rely on social media as a motivating source to travel.

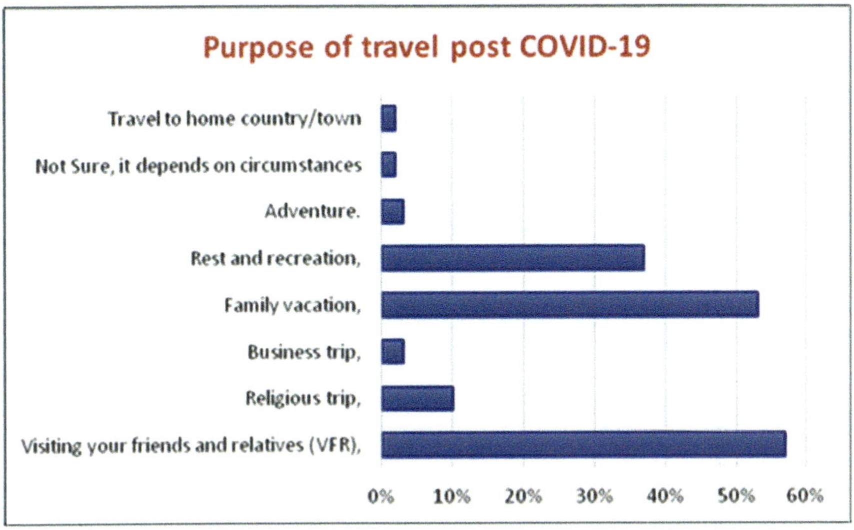

FIGURE 2.3 Purpose of travel post-COVID-19.

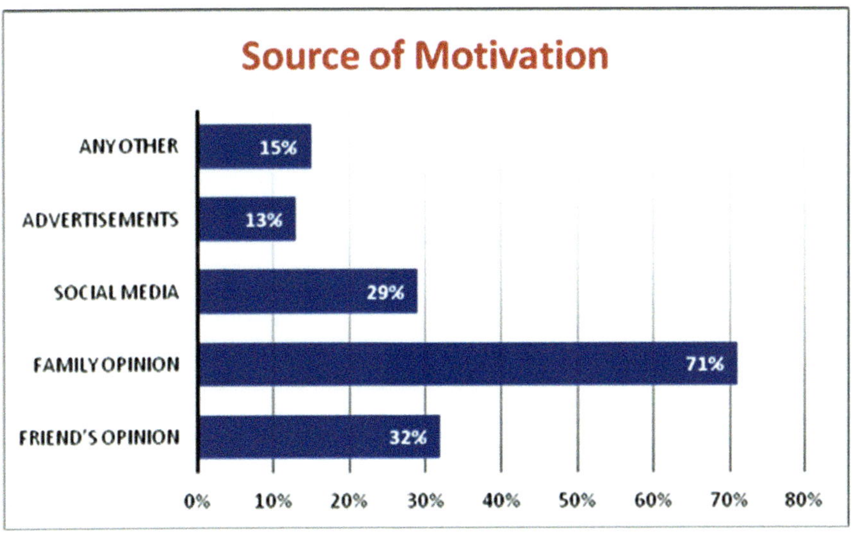

FIGURE 2.4 Source of motivation.

As per the travel style, the citizens and residents specify that their travel companions will be family (76%), single (19%), and few (5%) with a friend's group (Figure 2.5). This finding matches the results of the travelers research study conducted in China (Enger et al., 2020).

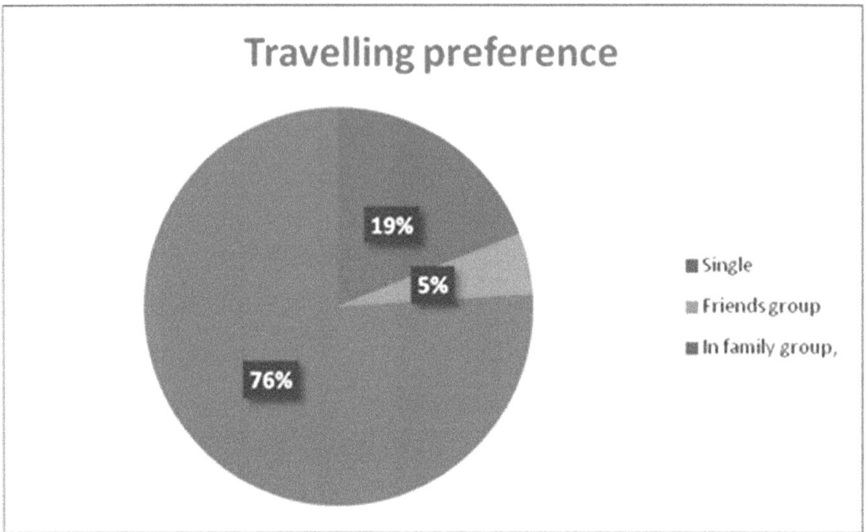

FIGURE 2.5 Traveling preference.

The sales campaign and promotional offers may not be the significant motivator for 40% of citizens and residents to undertake travel post-COVID-19; however, the promotions and offers may motivate approximately 39% of citizens and residents (Figure 2.6). This signifies that the promotional campaign and offers are provided by Tourism organizations, such as hotels and airlines to motivate the travelers to continue traveling.

Approximately, 52% of the citizens and residents made travel plans during the pandemic time, which they would like to take post-COVID-19, and nearly the same 48% of the citizens and residents are still apprehensive about making travel plans (Figure 2.7). This indicates the pressing need to raise and restore the resident's confidence levels regarding traveling. The tourism organizations shall be required to undertake continuous sales and promotion activities to increase trust and confidence to motivate and improve the citizens' and residents' intentions to travel.

The COVID-19 virus will have an immense influence on the travel behavior of Omani citizens and residents and the choice of their holiday destination as well as hotel accommodation. Kourgiantakis et al. (2020) reflect that the COVID-19 pandemic has shifted the preference for domestic trips and localism in tourism consumption. About 61% of citizens and residents would prefer staycations (Figure 2.8).

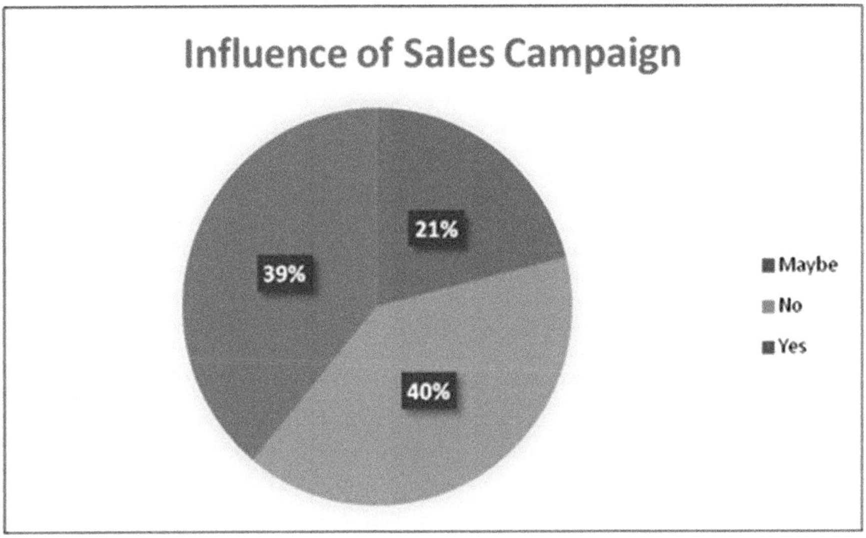

FIGURE 2.6 Influence of sales campaign.

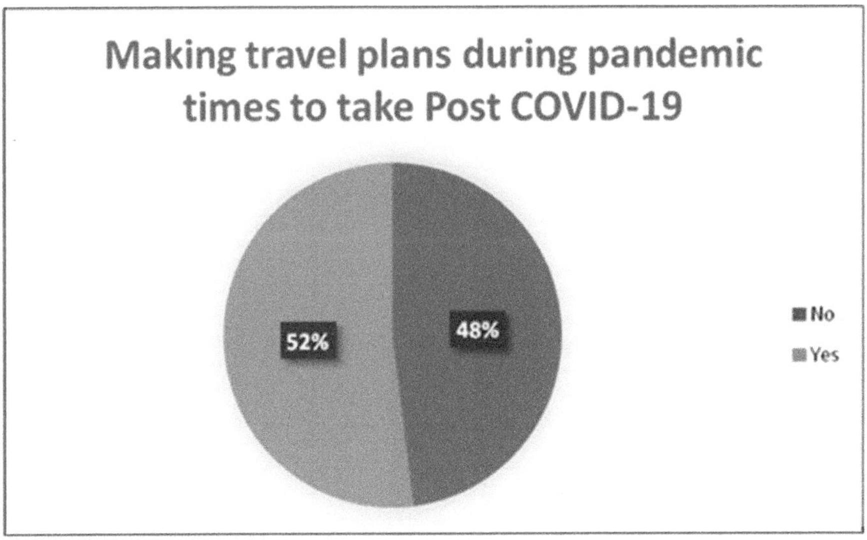

FIGURE 2.7 Making travel plans during the pandemic times to take post-COVID-19.

With reference to the preferred travel experience, the majority of the citizens and resident respondents (55%) do not want to get engaged in common adventure tourism activities in Oman, such as four-wheel drive and dune bashing or trekking. However, the citizens and residents (40%) would prefer to travel to the beaches and get engaged in the sun, sea, and surf activities; 24% citizens and residents would like to visit Oman's popular metro cities, such as Muscat, Sohar, Nizwa, Sur, and Salalah. Approximately, 19% of the citizens and residents would like to travel to the mountains in Oman, such as Jabel Akhdar, Jabel Shams, or Dhofar mountains (Figure 2.9).

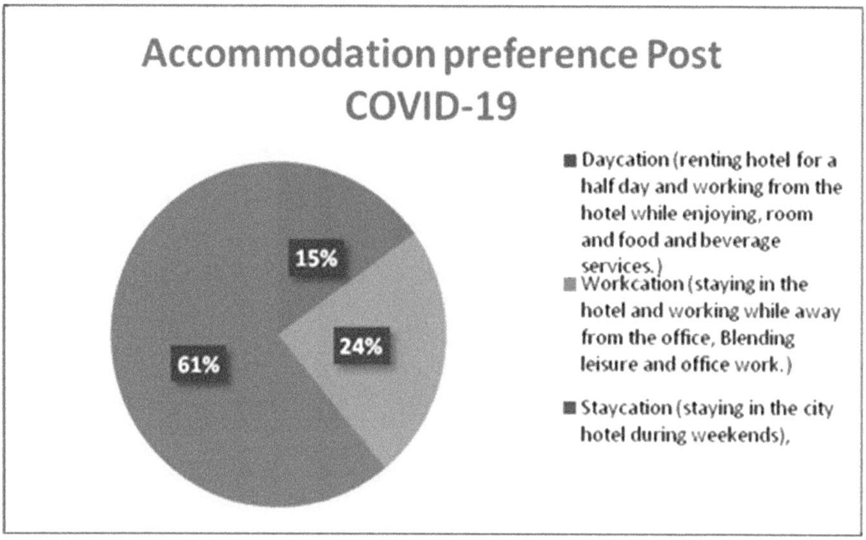

FIGURE 2.8 Accommodation preference post-COVID-19.

The citizens and residents have fragmented opinions about attending popular events, such as the Muscat Festival and Salalah Khareef Festival post-COVID -19. Approximately, 35% of citizens and residents have expressed willingness to attend the event, and nearly the same (34%) have expressed their unwillingness; 31% of citizens and residents are still indecisive about their participation in the events (Figure 2.11). This could perhaps be attributed to the social distancing norms and fear of traveling to the events of the large gathering.

Although cars are the preferred mode of transport in Oman; however, this research's findings indicate the trend toward the increased use of recreational vehicles (RVs) in Oman in the future. Almost half of the citizens and residents (52%) would prefer traveling by camping vehicles or RVs post-COVID-19 (Figure 2.12). According to Vishwanathan, Gautam (2021), OMRAN, the government's tourism investment arm is pushing to grow interest in motorhome travel and provide impetus to tourism at home.

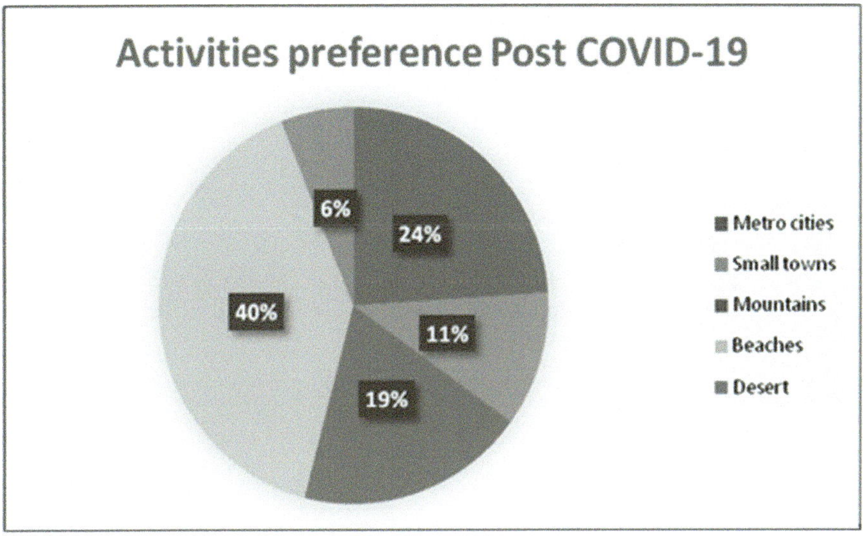

FIGURE 2.9 Activities preference post-COVID-19.

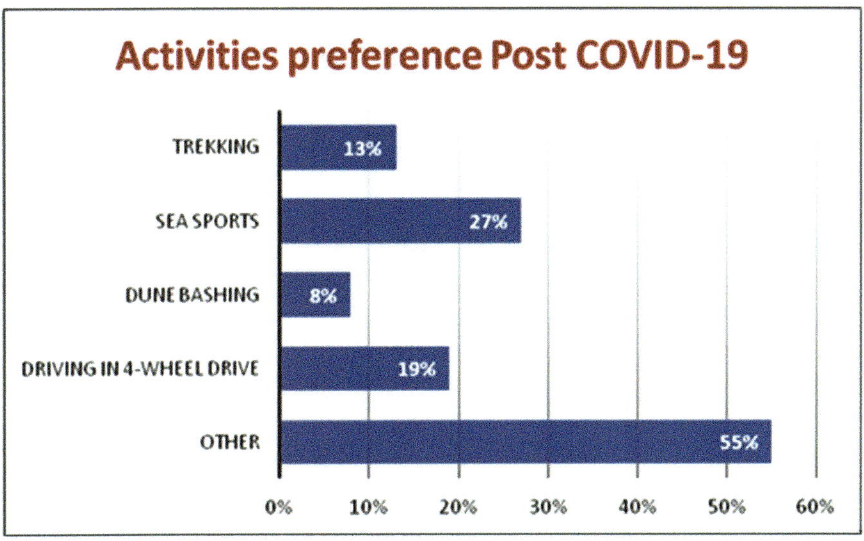

FIGURE 2.10 Activities preference post-COVID-19.

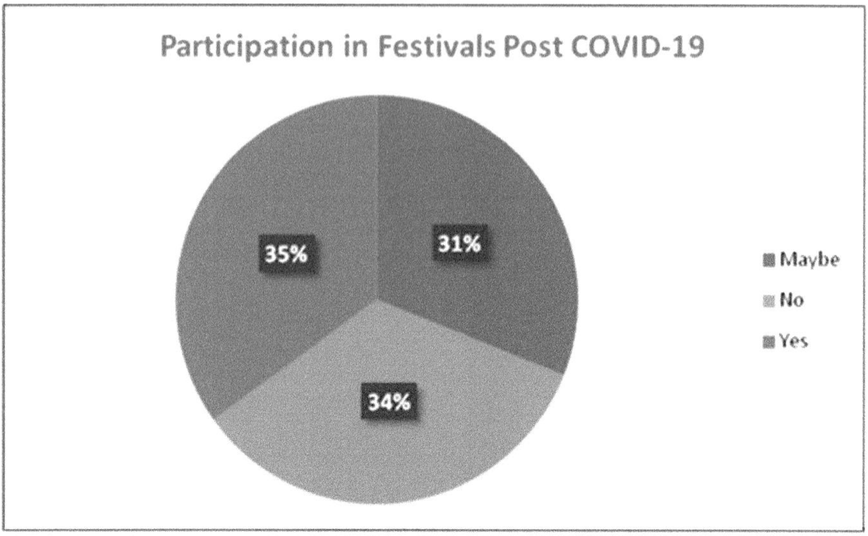

FIGURE 2.11 Participation in festivals post-COVID-19.

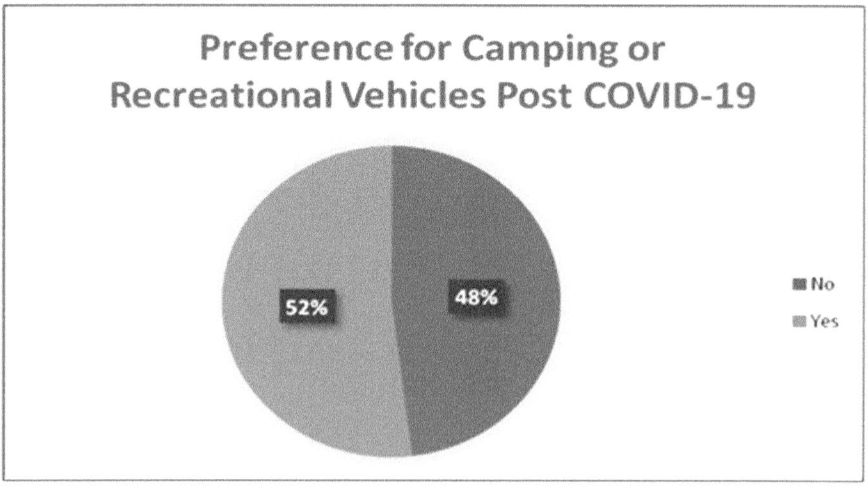

FIGURE 2.12 Preference for camping or recreational vehicles post-COVID-19.

The safety concern regarding travel activities such as eating street food is the cause of concern for 50% of citizens and residents (Figure 2.13). These citizens and residents are majorly concerned while eating in the streets. Safety, sanitation, and hygiene are the most noteworthy new norm of the travel behavior from the time these are highlighted in various other studies also (Kourgiantakis et al., 2020; Wachyuni and Kusumaningrum, 2020).

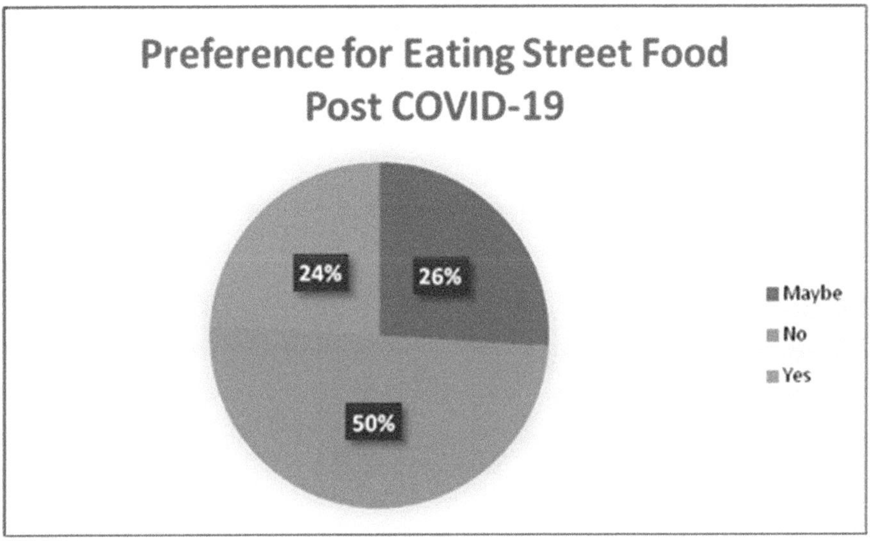

FIGURE 2.13 Preference for eating street food post-COVID-19.

Many international hotel organizations are following stricter health and safety protocols at their premises in order to regain the trust and lost confidence of the guests. The survey revealed that 83.9% of citizens and residents are willing to pay extra money for the more secure and sanitized tourism and hospitality services in Oman (Figure 2.14). The government can play an important role in restoring travelers' safety and confidence and the combined initiatives of the private and public sectors will be required post-COVID-19 to ensure proper addressing of safety, hygiene, and sanitation concerns. Oman's international image as a clean destination will act as an asset in restoring travelers' confidence in visiting Oman.

The citizens and residents were asked about their expectations from the government regarding the management of the tourism and hospitality sector post-COVID-19. The majority of them responded by stating that the role of the government is quite laudable in responding to the management of this pandemic crisis. The majority of the citizens and residents expect that the government should continuously keep on providing them with clear information regarding the pandemic, specify rules, and code of conduct, and update information on the website and apps, such as Tarassud Plus mobile app. According to Kothaneth (2020), this app can alert the user if a person with coronavirus symptoms is present anywhere in the proximity. The medical scout in Oman is available in different languages, such as Arabic, English, Hindi, Urdu, and Bangla. The residents also expect that online tickets and accommodation at reasonable prices should be available to them. The residents also opined that the government should ensure that the tourism and hospitality facilities that are now getting opened for the traveler should strictly adhere to the safety guidelines and sanitation measures.

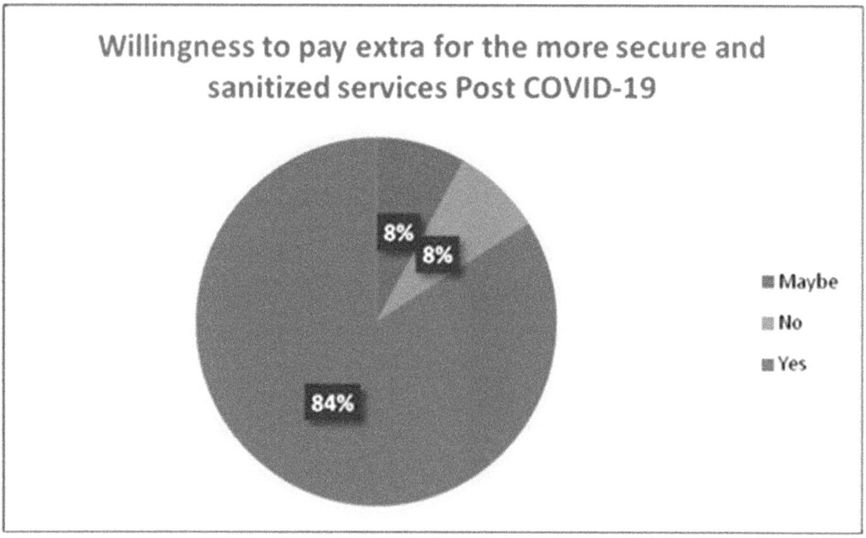

FIGURE 2.14 Willingness to pay extra for the more secure and sanitized services post-COVID-19.

The preferred travel activities of the citizens and residents have altered due to the pandemic. The tourism and hospitality businesses in Oman have to shift their focus to enhanced hygiene, sanitation, and safety guidelines and protocols to enable the residents to stay safe and secure. This new focus will slowly allow the tourism and hospitality industry to come back on track in Oman.

2.5 CONCLUSIONS

The core purpose of this study was to examine the post-COVID-19 travel preferences of Oman's citizens and residents. This research includes the study of citizens' and residents' travel preferences and perceptions regarding motivation for taking up the travel, accommodation preference, traveling style, preferred tourism activities, selected places to visit, sanitation, safety preferences, etc. Due to less travel consumption, many tourism organizations suffered massive losses (Wen et al., 2020); however, this study has provided a ray of hope that citizens and residents will prefer to travel post-COVID-19 pandemic, and this will boost the tourism business to recover the losses. The positive outcomes of this study regarding the inclination of citizens and residents toward resuming traveling activities is a kind of morale booster for the tourism industry in Oman. This study indicates the wide shift in consumption behavior and traveling patterns due to the pandemic. Therefore, this research study can be a reference for tourism and hospitality management professionals to anticipate the citizens and residents preferences toward travel and tourism post-COVID-19 pandemic. From a managerial perspective, the findings implicate that managers must be prepared to adopt changes in the day to day business operations as new regulations, safety and hygiene rules, and social distancing norms will greatly affect the functioning as well as the future of tourism and hospitality businesses in Oman. These organizations cannot survive through this pandemic without the active support and cooperation of the government.

To facilitate the economic revival of the tourism industry, the government of Oman is committed to ensuring a safe travel environment in the country to rebuild citizens' and residents' confidence. The government is helping the tourism sector business to adjust to the new normal by fostering innovation in digital technologies for seamless travel.

As every study has few limitations, similarly, this study also has limitations that must be presented for future studies. Firstly, the data sample size of this study is relatively small and future studies can be carried out with a large sample size for checking the evolutionary insight of travel preferences of residents and citizens in Oman. Secondly, Quantitative methods can also be employed to study the different perspectives on travelers' preferences post-pandemic. Thirdly, this study has discussed general travel preferences, but future studies can emphasize on differentiating between travel preferences of Omani citizens and the expatriate residents in the Sultanate of Oman.

This study attempts to identify the altered travel preferences of the residents and citizens residing in the major towns of Oman and the new emerging pattern of demand post-COVID-19 pandemic. The authors believe that this study will provide much needed insights into new demand patterns, travel preferences, and motivations to the tourism and hospitality businesses and will enable the managers to devise suitable policy and innovative product formulation to ensure a fresh start of the travel and tourism-related business activities post-COVID-19 in Oman.

KEYWORDS

- **post-COVID-19**
- **travel preferences**
- **perception**
- **travel restrictions**
- **hygiene**
- **sanitation**
- **Oman**

REFERENCES

Abdullah, M.; Dias, C.; Muley, D.; Shahin, M. Exploring the Impacts of COVID-19 on Travel Behavior and Mode Preferences. *Transp. Res. Interdiscip. Persp.* **2020,** *8* (October), 100255. https://doi.org/10.1016/j.trip.2020.100255

Al-Hasni, Z. S. The Economic Impact of COVID-19 on the Omani Tourism Sector. *Psychol. Educ. J.* **2021,** *58* (2), 824–830. https://doi.org/10.17762/pae.v58i2.1946

Bhati, A. S.; Mohammadi, Z.; Agarwal, M.; Kamble, Z.; Donough-Tan, G. Motivating or Manipulating: The Influence of Health-Protective Behaviour and Media Engagement on Post-COVID-19 Travel. *Curr. Iss. Tour.* **2020,** *0* (0), 1–5. https://doi.org/10.1080/13683500.2020.1819970

Brouder, P. Reset Redux: Possible Evolutionary Pathways Towards the Transformation of Tourism in a COVID-19 World. *Tour. Geog.* **2020,** *22* (3), 484–490.

Chalk, M.; Seitz, A. R. *For Peer* **2005,** 1–29.

Chebli, A.; Ben Said, F. The Impact of Covid-19 on Tourist Consumption Behaviour : A Perspective Article. *J. Tour. Manage. Res.* **2020,** *7* (2), 196–207. https://doi.org/10.18488/journal.31.2020.72.196.207

Choufany, H. M. HVS COVID-19: Travelled and Hotel Guest Sentiment Findings Middle East. HVS. https://www.hvs.com/article/8766-hvs-covid-19-traveller-and-hotel-guest-sentiment-findings-middle-east.

Claereboudt, M. R. Oman. In *World Seas: An Environmental Evaluation Volume II: The Indian Ocean to the Pacific*, 2nd ed.; Elsevier Ltd. https://doi.org/10.1016/B978-0-08-100853-9.00002-6

Enger, W.; Saxon, S.; Suo, P.; Yu, J. The Way Back: What the World Can Learn from China's Travel Restart After COVID-19. McKinsey & Company. https://www.mckinsey.com/industries/

travel-transport-and-logistics/our-insights/the-way-back-what-the-world-can-learn-from-chinas-travel-restart-after-covid-19.

experienceoman.om. *Sultanate of Oman*, 2021. www.experienceoman.om: https://experienceoman.om/oman/history/

Global Peace Index, I. f. *Measuring Peace in a Complex World, Global Peace Index 2020*; Institute for Economics & Peace: Sydney, 2020.

Gossling, S.; Scott, D.; Hall, C. Pandemics, Tourism, and Global Change: A Rapid Assessment of COVID-19. *J. Sustain. Tour.* **2020,** 1–20.

Gössling, S.; Scott, D.; Michael Hall, C. Pandemics, Tourism and Global Change: A Rapid Assessment of COVID-19. *J. Sustain. Tour.* **2021,** *29* (1), 1–20.

Haarhoff, R. Tourist Perceptions of Factors Influencing Destination Image: A Case Study of Selected Kimberley Resorts. *Afr. J. Hospital. Tour. Leisure* **2018,** *7* (4), 1–21.

Ivanova, M.; Ivanov, I. K.; Ivanov, S. Travel Behaviour After the Pandemic: The Case of Bulgaria. *Anatolia* **2020,** 00 (00), 1–11. https://doi.org/10.1080/13032917.2020.1818267

Jones, P.; Comfort, D. The COVID-19 Crisis, Tourism and Sustainable Development—Research Repository. *Athens J. Tour.* **2020,** *7* (2), 75–86. https://doi.org/10.30958/ajt/v7i2

Kim, H.-Y. Statistical Notes for Clinical Researchers: Assessing Normal Distribution (2) Using Skewness and Kurtosis. *Restorative Dentistry Endodontics* **2013,** *38* (1), 52–54.

Kotler, P. *Manajemen Pemasaran*; Prehalindo: Jakarta, 2000.

Kourgiantakis, M.; Apostolakis, A.; Dimou, I. COVID-19 and Holiday Intentions: The Case of Crete, Greece. *Anatolia* **2020,** 1–4.

Li, J.; Nguyen, T. H. H.; Coca-Stefaniak, J. A. Coronavirus Impacts on Post-Pandemic Planned Travel Behaviours. *Ann. Tour. Res.* **2020,** March, 102964. https://doi.org/10.1016/j.annals.2020.102964

Ministry-of-Tourism. *Sultanate of Oman*, 2021. https://www.omantourism.gov.om/: https://www.omantourism.gov.om/wps/portal/mot/tourism/oman/home/experiences/nature/beaches/!ut/p/a0/04_Sj9CPykssy0xPLMnMz0vMAfGjzOItvc1dg40MzAz8fZzMDTyDQz0Mg92djN2dDPQLsh0VAXtl1rg!/

Molloy, J.; Tchervenkov, C.; Hintermann, B.; Axhausen, K. W. Tracing the SARS-CoV-2 Impact: The First Month in Switzweland. Transport 2020. https://doi.org/10.3929/ethz-b-000414874

Nassri, Z. Tourism Sector Loses Half a Billion Rials Due to Impact of coronavirus. *Oman Observer* Dec 12, 2020. https://www.omanobserver.om/tourism-sector-loses-half-a-billion-rials-due-to-impact-of-coronavirus/

Naumov, N.; Varadzhakova, D.; Naydenov, A. Sanitation and Hygiene as Factors for Choosing a Place to Stay: Perceptions of the Bulgarian tourists. *Anatolia* **2020,** 1–4.

Nguyen, D. Van, Pham, G. H.; Nguyen, D. N. Impact of the Covid-19 Pandemic on Perceptions and Behaviors of University Students in Vietnam. *Data in Brief* **2020,** *31*, 105880. https://doi.org/10.1016/j.dib.2020.105880

Oman-Observer, Sultanate Reports 2 Cases of Coronavirus. Muscat, Muscat, Oman, Feb 24, 2020.

Oman-Observer. Muscat Lockdown from Friday, Schools Suspended Until Further Notice. Muscat, Muscat, Oman, 2020.

Ritchie, B.; Crotts, J.; Zehrer, A.; Volsky, G. Understanding the Effects of a Tourism Crisis. *J. Travel Res.***2014,** *53* (1), 12–25.

Service, T. N. COVID-19 Cases Among Expats Crosses 11,000. Muscat, Muscat, Oman, June 9, 2020.

Sigala, M. Tourism and COVID-19: Impacts and Implications for Advancing and Resetting Industry and Research. *J. Busi. Res.* **2020,** *117*, 312–321. https://doi.org/10.1016/j.jbusres.2020.06.015

TAS-News-Service. *Oman Closes Dhofar Governorate, Masirah, Jabal Shams, Jabal Akhdar and Duqm*. The Arabianstories.com, June 13, 2020. https://www.thearabianstories.com/2020/06/13/oman-closes-dhofar-governorate-masirah-jabal-shams-jabal-akhdar-and-duqm/

Terziyska, I.; Dogramadjieva, E. Should I Stay or Should I Go? Global COVID-19 Pandemic Influence on Travel Intentions of Bulgarian Residents. *SHS Web Conf.* **2021,** *92* (Jan), 01048. https://doi.org/10.1051/shsconf/20219201048

Times-Of-Oman. *Oman's Inbound Tourism Rises to OMR2.3 Billion in Last 4 Years*, 2020. https://www.timesofoman.com: https://www.timesofoman.com/article/omans-inbound-tourism-rises-to-omr23-billion-in-last-4-years

Uğur, N. G.; Akbıyık, A. Impacts of COVID-19 on Global Tourism Industry: A Cross-Regional Comparison. *Tour. Manage. Persp.* **2020,** *36* (September), 100744. https://doi.org/10.1016/j.tmp.2020.100744

UNOCHA. *United Nations Office for the Coordination of Humanitarian Affairs*. www.unocha.org/: https://www.unocha.org/middle-east-and-north-africa-romena/oman#Top

Vishwanathan, G. Motorhome Travel: Your Home on the Road. *The Times of Oman*, Jan 20, 2021. https://timesofoman.com/article/motorhome-travel-your-home-on-the-road

Wachyuni, S. S.; Kusumaningrum, D. A. The Effect of COVID-19 Pandemic: How are the Future Tourist Behavior? *J. Educ. Soc. Behav. Sci.* **2020,** *33* (4), 67–76.

Wen, J.; Kozak, M.; Yang, S.; Liu, F. COVID-19: Potential Effects on Chinese Citizens Lifestyle and Travel. *Tour. Rev* **2020**.

What Is the Oman Climate Like? 2020. www.omanevisa.co.uk: https://www.omanevisa.co.uk/omanclimate/#:~:text=In%20Oman%2C%20the%20climate%20is,northern%20and%20southern%20mountain%20areas.&text=This%20means%20the%20Oman%20climate,up%20to%2050%20degrees%20Celsius

CHAPTER 3

EXPERIENCE DESIGN GUIDE: WINE TOURISM EXPOSURE EXPERIENCE AUDIT FOR REOPENING POST-COVID-19

BYRON MARLOWE

Don Smith Distinguished Professor of Hospitality Management, Director, Wine and Beverage Business, The School of Hospitality Business Management, Carson College of Business Washington State University Tri-Cities Campus, United States

ABSTRACT

The impact of after COVID-19 on wine tourism is very and cautious and critical. In this context, safety and security of visitors should be major consideration of all decisions around the vineyards, wineries environment, and tasting rooms. No doubt, service management, experience design, and creating positive memories are some main visitor intents to revisit and repurchase. Hence, the involved audit process the 4E Model of education, that is, entertainment, esthetic, and escapist experiences have enriched and encouraged the wine visitor experience. Here, this chapter portray a systematically identify components of the 4Ps (Product, Price, Place, and Promotion) within a hospitality service business that enhance the experiential value for their customer with 4E Model of education.

Post-COVID Tourism and Hospitality Dynamics: Recovery, Revival, and Re-Start.
Umendra Narayan Shukla & Sharad Kumar Kulshreshtha (Eds.)

3.1 INTRODUCTION

Service management, experience design, and creating positive memories are at the foundation of visitor intent to revisit and repurchase after the impact of COVID-19 on wine tourism. Safety and security should be considered a component of all decisions around the environment in tasting rooms. The audit process should involve the 4E Model of education, entertainment, esthetic, and escapist experiences (Pine and Gilmore, 1999). This chapter helps to systematically identify components of the 4Ps (Product, Price, Place, and Promotion) within a hospitality service business that enhance the experiential value for their customer. Owners are often too close to their business to accurately assess these experiential elements; so, an internal and external audit is recommended. Therefore, gathering assessments from your staff, customers, or vendors, if possible, on the potential for COVID-19 exposure experiences also your own experiences with COVID-19 in the hospitality service business is valuable and can inform the audit. This collected feedback can provide a richer picture of what you may choose to emphasize or revise in your business's offerings as we reopen wineries for retail and hospitality food and beverage operations.

3.2 EXPERIENCE ECONOMY CONCEPTS

Researchers are active in their efforts to analyze the critical attributes and dynamic business implications of the experience economy. This review targets the tourism, wine tourism, and hospitality fields within this current genre' of research. Studies have defined and elaborated the definition of the "experience economy," which will be referred to as ExEc for the purpose of this review. Pine and Gilmore assert that a consumer experience occurs when a company intentionally uses services as the stage, and goods as props, to engage individual customers in a way that creates a memorable event (Harvard Business Review, 1998). The 4Es, as defined by Pine and Gilmore (1999) conceptualized the multidimensional nature of consumer experiences and defined the components as educational, escapist, esthetic, and entertainment experiences. It has been suggested that the 4Es are appropriate for understanding tourism products, such as hotels and restaurants (Pine and Gilmore, 2002), special events (Pullman and Gross, 2003) and heritage trails (Hayes and Macleod, 2007). In addition, Quadri-Felitti and Fiore (2012) state that the 4Es offer a relevant framework to examine the wine tourism experience. For Pine and Gilmore (2012), before an experience can be staged, it must be successfully themed. Furthermore, theming an experience means scripting a story reliant upon guests' participation. Similarly, effective interpretation at a winery follows an overarching theme unique to the vineyard. According to Pine and Gilmore (1999), the five key principles in successfully theming an experience are (1) an engaging theme must alter a guest's sense of reality; (2) the richest venues possess themes that fully alter one's sense of reality by affecting the experience of space, time, and matter; (3) engaging themes integrate space, time, and matter into a cohesive, realistic whole; (4) themes are strengthened by creating multiple places within a place; (5) a theme should fit the character of the enterprise staging the experience (pp. 49–52). Each of these principles seem to point to the fact that the theme should enhance a sense of place for visitors; also, an important goal for an interpretive theme. Interpretation can serve as a wineries' primary form of communication with their audience

In order to fulfill these purposes, businesses create interpretive plans. Many times, businesses start out solely selling goods or offering services and want to add value by offering distinct customer experiences. The mix of goods and services offered by small firms can range from very focused to diverse. These goods and services may offer cues for the kind of theme or impression around which to build experiences. For instance, a goat farmer who makes Neufchâtel, a cheese originally from the town of the same name in the region of Normandy, France, may find it appropriate to build a Bed and Breakfast or retail store esthetic experience around a French provincial theme (Fiore et al., 2007).

The concept of staging an integrated customer experience has been utilized to evaluate tourism products, such as heritage trails (Hayes and MacLeod, 2007) and special events (Pullman and Gross, 2004). Conversely, scholars have discussed the difficulties of using staging's theatrical metaphors for managing a destination (Morgan et al., 2009). Ali-Knight and Carlsen (2003) called for wineries to pay more attention to staging experiences shared by Pine and Gilmore (1999) in their presentation to Australian wine marketers. The article offered numerous examples associated with staging the experience but fell short of aligning their suggested activities with the 4Es specifically (Ali-Knight and Carlsen, 2003: 5–6). The authors called for the need to empirically examine the role of staged experiences in the tourism setting on wine purchases rather than the tourism experience (Ali-Knight and Carlsen, 2003: 6).

As an experience can engage a person in a number of ways, Pine and Gilmore (1999) discussed the two dimensions of engagement that ultimately create four realms of experiences. The first dimension looks at the level of guest participation: from passive to active participation. Passive participation occurs when the performance is not directly affected or influenced by the customer. Active participation in an experience occurs when the customer does affect or influence the performance or event. Voase (2002) further explored the active and passive visitor experience in the context of cultural and heritage attractions. When the word active is used to describe an activity or event, many people may immediately think of a physical activity. Similarly, a passive activity may bring to mind a contemplative activity. Voase acknowledged that considering active and passive in these terms is misleading. Instead, he suggested viewing active as "mental activity" and passive as "mental passivity" (Voase, 2002). This assertion aligned with Pine and Gilmore's definitions where active = involvement and passive = observation.

The second dimension is used to understand the experience involved in the level of connection or the "environmental relationship" that individuals feel toward an event or performance (Pine and Gilmore, 1999). In this context, absorption in an event or performance occurs when a person's attention is focused on bringing the experience into the mind; immersion in an event or performance occurs when a person becomes physically (or virtually) a part of the actual experience (Pine and Gilmore, 1999). Here, they are treated as opposite ends of a spectrum; in other literature, particularly around the theory of flow, they are treated synonymously (Csikszentmihalyi, 1990). The realms of experience are applicable to a variety of fields, but seem to be especially suited for application within the tourism industry in a variety of settings including wine (Richards, 2001; Oh et al., 2007; Hosany and Witham, 2010; Quadri-Felitti and Fiore, 2012).

3.3 REVIEW OF 4E FRAMEWORK FOR WINE TOURISM

An experiential view of wine tourism has been scant although promising (e.g., Bruwer and Alant, 2009; Galloway et al., 2008; Getz and Carlsen, 2008; Pikkemaat et al., 2009; Quadri-Felitti and Fiore, 2012, 2013), given the hedonic nature of the wine tourism experience. Quadri-Felitti's (2012) dissertation study of rural wine tourism utilized the central constructs of the experience economy model; the 4Es consist of education, esthetics, entertainment, and escapist (Pine and Gilmore,1999) to explain the experiential nature of wine tourism. These studies, along with the fragmentary evidence found in the current wine tourism literature, discussed in the following section, strengthen the call for wine tourism research to explore the applicability of the 4E framework. Quadri-Felitti and Fiore (2013) posited that the 4Es are equally relevant to and useful in conceptualizing the wine tourism experience and that the current wine tourism literature provides evidence of the framework's value on the whole. Descriptions and motivating factors culled from the existing wine tourism research are cataloged according to the constructs of the 4E model. Table 3.1 inventories evidence of individual 4Es selected from the current wine tourism literature. It suggests the need for a more encompassing view of the experiential nature of wine tourism (Adopted from Quadri-Felitti and Fiore 2012). Typical wine tourist activities within the 4E model of the experience economy provide a depiction of typical wine tourist activities positioned within the model, which demonstrates the suitability of this framework in helping to understand the greater wine tourism experience.

3.3.1 EDUCATION

An educational experience could encompass learning in both formal and informal learning environments. When an individual has an educational experience, they are involved in active absorption of the event or performance (Pine and Gilmore, 1999). Pine and Gilmore (1999) believe that in order "to truly inform a person and increase his knowledge and/or skills, educational events must actively engage the mind (for intellectual education) and/or body (for physical training)" (p. 32). Most visitors come to wine destinations with undefined learning goals leading to an informal learning experience. Learning involves multiple processes including perception and memory (Falk and Dierking, 1995; Hooper-Greenhill, 1999). Perception and memory are influenced by prior experience which in turn influences learning. When it comes to experiences, education and entertainment are not mutually exclusive.

Education is repeatedly identified as a motivator of wine tourists (Ali-Knight and Charters, 2001; Charters and Ali-Knight, 2000; Williams and Kelly, 2001), which offers strong evidence that this aspect of the 4E model is appropriate for examining the wine tourism experience. Learning as a consumer motivation consistently materializes in the wine tourism consumer research regardless of the demographics of gender, cohort (e.g., Boomers, Generations X or Y), or life cycle stage (Fountain and Charters, 2010;

Getz and Carlsen, 2008). Some wineries offer home winemaking seminars and partner with chefs and farmers to offer culinary classes. Festivals and galleries, whether retail or exhibition, serve as an educational as well as an entertainment element (Carlsen, 2004; Mitchell and Hall, 2006; Park et al., 2008; Yuan et al., 2005).

Education—Examples of wine tourists experience (Quadri-Felitti and Fiore, 2012).

- Wine tastings and seminars
- Culinary—wine pairing events
- Home winemaking seminars
- Cooking and craft making classes

3.3.2 ENTERTAINMENT

An activity or event as entertainment but most involve some element of fun and enjoyment. In the context of the experience economy, entertainment occurs when someone uses their senses to passively absorb an event or performance (Pine and Gilmore, 1999). There is a connotation to the use of the term entertainment that brings to mind activities that are fun for fun's sake, such as a theme park experience or a Vegas-style show. However, an entertainment experience could also be watching an orchestra performance, reading a book, watching a child play, or laughing at jokes; it is the level of engagement and the emotions evoked from the experience that determines the experience's value as entertainment. The connection of art, music, and heritage as entertainment within wine destinations is well documented (Charters, 2006; Williams and Kelly, 2001). Tours of vineyards and winemaking facilities with informative narration are other examples of both educational and entertainment (Quadri-Felitti, 2012). "Edutainment" is a term also used by Pine and Gilmore (1999) to describe the overlap of education and entertainment in their model (pp. 31–32).

As expressed by Pine and Gilmore (1999), entertainment reflects the passive/absorption dimension of the model wherein performers engage a customer's attention. Whereas agricultural and viticultural activities taking place at a winery, farm, or vineyard may be viewed and absorbed, cultural attractions and events that contribute to wine destinations may also be a considerable draw (Carmichael, 2005; Frochot, 2000; Williams and Kelly, 2001). Tours of vineyards and winemaking facilities with informative narrations are another example of both the educational and entertainment dimensions as tourists learn about viticulture and are engaged by the presenters. Getz and Carlsen (2008) labeled the fun, informative aspects of wine tourism as "edutainment" (p. 262), also a term used by Pine and Gilmore (1999: 31–32) and (Quadri-Felitti, 2013) to describe the synthesis of the educational and entertainment experiences in the experience economy.

Entertainment—Examples of wine tourists' experience (Quadri-Felitti and Fiore, 2012).

- Cellar concerts and music in the vineyard.
- Wine blending demonstration.
- Farm and food demonstrations.

3.3.3 ESCAPISM

In the escapist experience realm, the individual is an active participant in an immersive environment; in essence, the participant directly affects the performance or event. Motion simulator attractions are obvious examples of an escapist experience. However, Pine and Gilmore (1999) make the distinction that an escapist's experience is not just a journey from but also a voyage to some specific place or activity, opening the door for a wider application to leisure pursuits. In leisure studies, escape is one of the dimensions that influence one's leisure behavior (Iso-Ahola, 1982, 1989; Kleiber et al., 2011). The motivation to participate in leisure activities comes from an individual's desire to either seek needs satisfaction or escape their everyday environment (Iso-Ahola, 1982, 1989; Snepenger et al., 2006; Kleiber et al., 2011). The seeking and escaping dimensions of leisure motivation are related to the need for optimal arousal (Dunn and Iso-Ahola, 1991, Snepenger et al., 2006). In particular, individuals seek stimulating activities when at a lower level of arousal and escape over arousing situations by finding less-stimulating activities, thereby achieving their own personal optimal level of arousal. In this new light, we can perhaps view entertainment experiences as those that visitors seek for higher stimulation and escapist experiences can then be viewed as opportunities for individuals to get away from their everyday environment. It is the difference of place (Bruwer, 2003) that draws urbanites to the rural tourism experience and signifies the escapist realm. Various rural wine destinations provide ample activities in which the tourist may be engrossed in activities ranging from hot air ballooning and helicopter rides to grape stomping and harvest volunteering (Quadri-Felitti, 2012).

Research about rural wine tourists' motivations supports the escapist principles within the model. The escapist's experience is highly immersive wherein consumers are engrossed in a different time or place. While tourists may gaze at the wine region's vineyard, they may also immerse themselves by participating in wine country recreation, abundant in the natural, rural setting of wine appellations. Rural wine destinations provide ample sports, recreation, and engaging activities in which the tourist may be engrossed, including hot-air balloon and helicopter rides. The participatory, immersive activities of performing agricultural or viticultural activities, such as taking part in the harvesting or crushing of the grapes, are applicable to this realm. Such activities are consistently rated as important reasons for wine tourism (Fountain and Charters, 2010; Sparks, 2007). This escapist' construct is substantiated in the wine tourism literature with findings that confirm that tourists want more to do than just to visit wineries and taste wine (Beames, 2003; Charters and Ali-Knight, 2002; Cohen and Ben-Nun, 2009).

Escapist—Examples of wine tourists experience (Quadri-Felitti and Fiore, 2012).

- Vineyard hiking and cycling tours
- Hot-air ballooning over vineyards
- Vineyard tour by horse and carriage

3.3.4 ESTHETICS

An esthetic experience occurs when an individual is a passive participant in a fully immersive event or environment. In modern times, esthetics are predominantly associated with art and/or beauty (Diffey, 1986). However, the word esthetic comes from the Greek word for sensation (Cordua, 1986). So, while an esthetic experience could involve taking in a portrait at an art gallery, it could also be a sensory reaction to the physical setting of another leisure activity. Climate and weather are important factors in tourists' decision-making and also influence the successful operation of tourism businesses, climate is defined as the prevailing condition observed as a long-term average in a location, while weather is the manifestation of climate at a specific point in time and destination. Climate and weather play an important role in destination selection because tourists are sensitive to climate and weather change (Maddison, 2001). Understanding how weather can change the aesthetics of a wine destination and the wine-tasting experience is increasingly important as weather patterns fluctuate due to climate change. Climate change may not only impact the coral reefs, glaciers, and polar bears but could also have a significant effect on the U.S. wine industry (Santisi, 2011). While some businesses may engage customers primarily through one realm of experience, many present experiences that encompass multiple realms. An experience that encompasses all four realms is said to hit the "sweet spot" in the middle of the framework. The sweet spot is a "mnemonic place" a place that is distinct from the everyday norm (Pine and Gilmore, 2012).

The esthetic experience entails immersion in a sensual environment. The winescape reflects this opportunity and has proven to be fundamental for wine tourism (Alant and Bruwer, 2004). Places are selected for the pleasure and uniqueness they offer that differs from modern, everyday urban or suburban life. Visitors driving rural roads lined with vineyards and orchards is a typical consumption activity of excursionists spending a day in wine country (Alant and Bruwer, 2004; Alonso et al., 2007; Barber et al., 2008; Getz and Brown, 2006). Williams (2001) found a shift in wine region advertising; more emphasis is being placed on the esthetic dimension of the rural landscape. In addition, some scholars argue that the evaluative consumption of wine, as it takes place in a winery's tasting room, has much in common with consuming music and art (Charters and Pettigrew, 2005). Although wine consumption as an esthetic experience is arguable, enjoying and purchasing wine have been identified as contributing motivators to visit a particular region (Brown and Getz, 2005).

Esthetics—Examples of wine tourists are enriched by sensual environments (Quadri-Felitti and Fiore, 2012).

- Consuming the "winescape."
- Enjoying unique lodging (B&B) and wines.
- Driving rural roads lined with vineyards.
- Art and craft fairs at wineries.

3.4 EXPOSURE EXPERIENCE DESIGN GUIDE

The conceptual framework for this audit was grounded in the concepts set forth in The Experience Economy (Pine and Gilmore, 1999). The authors' major claim was that we are heading into an era with new economic offerings Post-COVID: exposure experiences. These experiences must be rooted in personal safety and security for employees and guests from COVID-19 virus exposure. To further explain this concept, the current layers to our economy were explored. For example, commodities, which were fungible items from the natural world, such as grapes, corn, wheat are the foundational products of an agrarian economy. At the time of the Industrial Revolution, most western societies became industrial economies where tangible products (or goods) were manufactured from commodity items, that is, grapes turned into wine, bottled, and sold. After the end of World War II, the U.S. economy began to shift from an industrial economy to a service economy. Services are the dominant economic offering today. Service economy products are intangible activities undertaken for a client, so a waitress pouring a glass of wine for a customer is a service. Some may argue that experiences are lumped in with services and should continue to be seen in as such. Pine and Gilmore argue that experiences are a different economic offering entirely. Companies who create experiences for their customers use services as "the stage" and goods as "props" in order to engage an individual in a personal way. When you buy an experience, you paid to spend time enjoying a series of memorable events the company has in essence, "staged." Therefore, the value of the experience is derived from the customer's memory of being engaged by the event. No two people can have the exact same experience. Each individual's experience comes from the interaction between the staged event and their prior state of mind but the opportunity cost of their exposure to COVID-19 while visiting a winery is potentially enough to close a winery temporarily or in the worst case permanently. Undoubtedly, POST-COVID -19 experience examples will have virus exposure elements which will need to be addressed redesigning the experience with the expected. Potential virus exposure in the experience offerings of the winery that the audit is performed on will need to be evaluated. After the audit, recreate the experiences while taking away the virus exposure that the previous experience may have provided. Upon conclusion of the audit chart of the winery which has performed the exposure audit, they should be prepared for a POST-COVID experience economy recovery.

Each element should add sensory pleasure, meaning (stories or insights), and personal relevance for the consumer but also be evaluated for potential virus exposure and explored.

3.5 DISCUSSION

In conclusion, a wine tourism post-COVID-19 exposure audit will produce a wide variety of experiences which may or may not have the potential to address your guest's exposure to the COVID-19 virus, it is not a fool-proof exercise but it may mitigate and support your business in reopening and staying open through the pandemic and after its impact. Regardless of the yes or no answers to the above audit, it is important that you've engaged in this exercise. Allow yourself to replicate this investigation at another winery and tasting room and allow others to conduct the audit on your winery as well. Here, are some observations you will want to take note of as well in relationship to your reopening. Effectively describe your new experiences or experience redesign. Commonly, a winery may have shifted to reservations and appointment only to limit crowds and congestion. The wineries' tasting room and building capacity along with occupancy #'s are paramount to controlling crowds, be specific, know your potential for guests based on the legal requirements within the jurisdiction you conduct business. Weekends vs. weekdays, high demand vs. low demand periods, all need to be accounted for when considering your winery tasting room capabilities. The type of operation you choose to conduct and what are the core and supplementary experiences you are providing after the pandemic will speak to your company culture and be recognized by your guests. While conducting the audit, take pictures of the services cape for records and recollection. Create a service blue print for employee execution and a customer journey map for guests to feel safe. Being overly prepared will only reinforce your commitment to the safety and security of your guests and make your tasting room the winery of choice.

TABLE 3.1 4E Exposure Experience Audit Chart (Adapted From Quadri-Felliti and Fiore, 2012)

The 4Ps	**The 4Es of the Experience Economy Exposure Audit**			
Product Price Place (Distribution) Promotion	Educational Experience	Esthetic Experience	Entertainment	Escapist Experience
Traditional Marketing Components	Contributing to the consumer's knowledge, skill or personal development	Immersing the consumer in unique, harmonious, and sensory appealing environment	Experience Absorbing the consumer in a performance	Providing ways for consumers to actively participate in creating a different sense of place or time

Each element should add sensory pleasure, meaning (stories or insights), and personal relevance for the consumer.

Source: Adapted From Quadri-Felliti and Fiore, 2012

To whom are you trying to sell your new (or redesigned) experience? If it is a redesigned experience, do you expect additional or different market segment(s) as a result? Show the size of each of your target markets and

TABLE 3.2 Your Exposure Audit Chart Exercise with Examples (Yes or No For COVID-19 Exposure).

The 4Ps	Educational experience	Esthetic experience	Entertainment	Escapist experience
Traditional Marketing Components	Contributing to the consumer's knowledge, skill or personal development	Immersing the consumer in unique, harmonious, and sensory appealing environment	Experience absorbing the consumer in a performance	Providing ways for consumers to actively participate in creating a different sense of place or time
Product: "The core and augmented experience"	**Each element should add sensory pleasure, meaning (stories or insights), and personal relevance for the consumer**			
Physical Plant: Landscape, parking, entrance ways, architecture, interior design, outdoor seating, company vehicles	Exposure Example #1:	Exposure Example #2:	Exposure Example #3:	Exposure Example #4:
People and Staff: Uniforms, greetings, training, interactions, attention Merchandise: Tangible elements (bedrooms, wine, menu, tableware, etc.), displays events and activities: Tours, special events, parties, packages, partnerships/sponsorship	B&B or winery has heritage rose garden labeled with varieties names and history	Windows of store, shop, or winery frame, an outdoor view, while the ceiling is painted with a celestial pattern	Retail or restaurant has large fish in wall-sized tanks provide activity to watch	Bridal or florist shop windows resembling church stained glass
Traditional Marketing Components	Contributing to the consumer's knowledge, skill or personal development	Immersing the consumer in unique, harmonious, and sensory appealing environment	Experience absorbing the consumer in a performance	Providing ways for consumers to actively participate in creating a different sense of place or time
Price: "An experience has greater economic value"	**Each element should add sensory pleasure, meaning (stories or insights), and personal relevance for the consumer**			
Individuals: Prix fixe Frequency purchases, Gift purchases, Bundled purchases	Exposure Example #1: A class on pairing with tasting menu includes take home materials	Exposure Example #2: Tasting menu includes gift of linens or pre-printed elegant note cards	Exposure Example #3: Private spaces used for folk dance performance	Exposure Example #4: Wines or jams are custom labeled with customers art work

TABLE 3.2 *(Continued)*

The 4Ps	Educational experience	Esthetic experience	Entertainment	Escapist experience
Groups: Added value (bundled) Travel suppliers, preferred vendors (certified, etc.) Customization: Labeling (names, art), blends, limited				
Traditional Marketing Components	Contributing to the consumer's knowledge, skill or personal development	Immersing the consumer in unique, harmonious, and sensory appealing environment	Experience absorbing the consumer in a performance	Providing ways for consumers to actively participate in creating a different sense of place or time
Place: Distribution "Sharing the experience"	**Each element should add sensory pleasure, meaning (stories or insights), and personal relevance for the consumer**			
Unique to tourism and hospitality businesses Geographic: Enhancing natural elements, signage, proximity to cities Digital: Web, social media, customer relationship management Distributors: Placements, retail and wholesale, online, traditional, local, regional, national, consortium, representatives	Exposure Example #1: Winery hosts special tastings for trade professionals, earning them "certification"	Exposure Example #2: Custom labels are created for products for sale at restaurants or special stores	Exposure Example #3: Watching videos of how products are made or how buyers select products while waiting	Exposure Example #4: Apple-scented business cards used by pick your own apple farm.
Traditional Marketing Components	Contributing to the consumer's knowledge, skill or personal development	Immersing the consumer in unique, harmonious, and sensory appealing environment	Experience absorbing the consumer in a performance	Providing ways for consumers to actively participate in creating a different sense of place or time

TABLE 3.2 *(Continued)*

The 4Ps	Educational experience	Esthetic experience	Entertainment	Escapist experience
Promotion: "Invitation to the memorable experience"	**Each element should add sensory pleasure, meaning (stories or insights), and personal relevance for the consumer**			
Web communications: Web site, social media, blogs, video channels, user generated content, connectivity (links to related sites), interactivity, simulate "experience" online Advertising and Print: Tasting notes, educational brochures, cards, business cards Sponsorship, outdoor, travel (meetings) trades Public Relations: "fam" trips for agents, meeting planners; writers; connect to trends, pitches vs. releases, film tourism; event PR	Exposure Example #1: Brochure supplies historical fun facts followed by a quiz for children or adults. Send in quiz for a drawing to win a free gift	Exposure Example #2: Photo contest for visitors select the best to be published in a glossy coffee table book	Exposure Example #3: Product catalog provides funny testimonials from customers	Exposure Example #4: Slide show played on property and web shows customers stomping grapes during harvest festival

potential growth. What are the characteristics of your target market(s)? For example, with a winery, say you are targeting just wine club members and hyper local customers. Provide hard numbers and be SPECIFIC! (e.g., wine club members and hyper local customers population, growth potential, % of male and female, age, life style, and so on).

Identify your existing or potential competitors. Evaluate strengths and weaknesses of your competitors. Describe how you are different from them and what your strengths and weaknesses are. A positioning map can be used in this section. The positioning map must be derived from your target (potential) customers. You can use the survey method to draw an objective positioning map. First, find out (determine) the key attributes of your business (or the industry your business is in). Then list your (main) competitors in the questionnaire and ask respondents to rate all competitors on the attributes (e.g., using a 10-point scale). Lastly, locate your competitors and your business on the map. Make sure you show the scale you used in the survey on the map.

Finally, promotion, including a customer, loyalty program. How are you going to communicate your new (or redesigned) experience to your target market(s)? Think about effective ways and options you have. Are you going to use a local newspaper, magazine, fliers, free coupons, and so forth? What is the cost? Again, be as SPECIFIC as possible. Also, develop and describe a feasible customer loyalty program (frequent user program) for your business. You may want to review relevant, existing loyalty programs before developing your own. Also think about the expense of the program.

Is it worth doing? Yes, this is an exercise of business. Its cyclical in nature, business environments change with time and the experience we create must also change. Meeting the changing expectations of our markets is possible only when we are engaged in our experience offering design.

KEYWORDS

- **experience design**
- **experience audit**
- **4E model**
- **wine tourism**
- **post-COVID-19**

REFERENCES

Alant, K.; Bruwer, J. Wine Tourism Behaviour in the Context of a Motivational Framework for Wine Regions and Cellar Doors. *J. Wine Res.* **2004,** *15* (1), 27–37.

Ali-Knight, J.; Carlsen, J. An Exploration of the Use of 'Extra Ordinary' Experiences in Wine Tourism. In *Proceedings of the International Colloquium in Wine Marketing, Wine Marketing Group*; Lockshin, L., Rungie, C., Eds.; University of South Australia: Adelaide, 2003.

Ali-Knight, J.; Charters, S. The Winery as Educator: Do Wineries Provide What the Tourist Needs? *Aust. NZ Wine Ind. J.* **2001,** *16* (6), 79–86.

Alonso, A.; Fraser, R. A.; Cohen, D. A. Investigating Differences Between Domestic and International Winery Visitors in New Zealand. *Int. J. Wine Busi. Res.* **2007,** *19* (2), 114–126.

Barber, N. A.; Donovan, J. R.; Dodd, T. H. Differences in Tourism Marketing Strategies Between Wineries Based on Size or Location. *J. Travel Tour. Market.* **2008,** *25* (1), 43–57.

Beames, G. The Rock, the Reef and the Grape: The Challenges of Developing Wine Tourism in Regional Australia. *J. Vacation Market.* **2003,** *9* (3), 205–212.

Brown, G. P.; Getz, D. Linking Wine Preferences to the Choice of Wine Tourism Destinations. *J. Travel Res.* **2005,** *43* (3), 266.

Bruwer, J. South African Wine Routes: Some Perspectives on the Wine Tourism Industry's Structural Dimensions and Wine Tourism Product. *Tour. Manage.* **2003,** *24* (4), 423–435.

Bruwer, J.; Alant, K. The Hedonic Nature of Wine Tourism Consumption: An Experiential View. *Int. J. Wine Busi. Res.* **2009,** *21* (3), 235–248.

Carlsen, P. J. A Review of Global Wine Tourism Research. *J. Wine Res.* **2004,** *15* (1), 5–13.

Carmichael, B. Understanding the Wine Tourism Experience for Winery Visitors in the Niagara Region, Ontario, Canada. *Tour. Geog.* **2005,** *7* (2), 185–204.

Charters, S. *Wine & Society: The Social and Cultural Context of a Drink*; Elsevier Butterworth-Heinemann: Oxford, 2006.

Charters, S.; Ali-Knight, J. Who Is the Wine Tourist? *Tour. Manag.* **2002,** *23* (3), 311–319.

Charters, S.; Ali-Knight, J. Wine Tourism—a Thirst for Knowledge? *Int. J Wine Market.* **2000,** *12* (3), 70–81.

Charters, S.; Pettigrew, S. Is Wine Consumption an Aesthetic Experience? *J. Wine Res.* **2005,** *16* (2), 121–136.

Cohen, E. A Phenomenology of Tourist Experiences. *Sociology* **1979,** *13* (2), 179–201.

Cohen, E.; Ben-Nun, L. The Important Dimensions of Wine Tourism Experience from Potential Visitors' Perception. *Tour. Hospital. Res.* **2009,** *9* (1), 20–31.

Cordua, C. A Critique of Esthetics. In *Possibility of the Aesthetic Experience*; Mitias, M., Ed.; Martinus Nijhoff Publishers: Dordrecht, 1986.

Csikszentmihalyi, M. *Flow: The Psychology of Optimal Experience*; Harper & Row, Publishers: New York, 1990.

Diffey, T. J. The Idea of Aesthetic Experience. In *Possibility of the Aesthetic Experience*; Springer: The Netherlands, 1986; pp 3–12.

Dolcos, F.; Cabeza, R. Event-Related Potentials of Emotional Memory: Encoding Pleasant, Unpleasant, and Neutral Pictures. *Cogn. Affect. Behav. Neurosci.* **2002,** *2* (3), 252–263.

Dunn Ross, E. L.; Iso-Ahola, S. E. Sightseeing Tourists' Motivation and Satisfaction. *Ann. Tour. Res.* **1991,** 18 (2), 226–237.

Falk, J. H.; Dierking, L. D. Learning from Museums: Visitor Experiences and the Making of Meaning; Altamira Press. 2000.

Fountain, J.; Charters, S. Generation Y as Wine Tourists: Their Expectations and Experiences at the Winery—Cellar Door. In. *Tourism and Generation Y*; Benckendorff, P., Moscardo, G., Pendergast, D., Eds.; CAB International: Wallingford, 2010; pp 47–57.

Frochot, I. Wine Tourism in France. Wine Tourism Around the World: Development, Management and Markets **2000,** 67.

Galloway, G.; Mitchell, R.; Getz, D.; Crouch, G.; Ong, B. Sensation Seeking and the Prediction of Attitudes and Behaviours of Wine Tourists. *Tour. Manage.* **2008,** *29* (5), 950–966. DOI: 10.1016/j.tourman.2007.11.006

Getz, D.; Carlsen, J. Wine Tourism Among Generations X and Y. *Turizam: znanstveno-stručni časopis* **2008,** *56* (3), 257–269.

Hayes, D.; MacLeod, N. Packaging Places: Designing Heritage Trails Using an Experience Economy Perspective to Maximize Visitor Engagement. *J. Vacation Market.* **2007,** *13* (1), 45–58.

Hooper-Greenhill, E. Education, Communication and Interpretation: Towards a Critical Pedagogy in Museums. *Educ. Role Museum* **1999,** *2*, 3–27.

Hosany, S.; Witham, M. Dimensions of Cruisers' Experiences, Satisfaction, and Intention to Recommend. *J. Travel Res.* **2010,** *49* (3), 351–364.

Iso-Ahola, S. *Social Psychological Perspectives on Leisure and Recreation*; Charles C. Thomas Publisher: Springfield, 1989.

Kleiber, D.; Walker, G.; Mannell, R. A *Social Psychology of Leisure*, 2nd ed.; Venture Publishing, Inc.: State College, 2011; pp 51–179.

Maddison, D. In Search of Warmer Climates? The Impact of Climate Change on Flows of British Tourists. *Clim. Change* **2001,** *49* (1–2), 193–208.

Mannell, R. C.; Iso-Ahola, S. E. Psychological Nature of Leisure and Tourism Experience. *Ann. Tour. Res.* **1987,** *14* (3), 314–331.

Mitchell, R.; Hall, C. M. Wine Tourism Research: The State of Play. *Tour. Rev. Int.* **2006,** *9* (4), 307–332.

Morgan, M.; Elbe, J.; Curiel, J. Has the Experience Economy Arrived? The Views of Destination Managers in Three Visitor—Dependent Areas. *Int. J. Tour. Res.* **2009,** *11* (2), 201–218.

Neulinger, J. *The Psychology of Leisure: Research Approaches to the Study of Leisure*, Vol. 111; Springfield, 1974.

Oh, H.; Fiore, A. M.; Jeoung, M. Measuring Experience Economy Concepts: Tourism Applications. *J. Travel Res.* **2007,** *46* (2), 119–132.

Otto, J. E.; Ritchie, J. R. The Service Experience in Tourism. *Tour. Manage.* **1996,** *17* (3), 165–174.

Park, K. S.; Reisinger, Y.; Kang, H. J. Visitors' Motivation for Attending the South Beach Wine and Food Festival, Miami Beach, Florida. *J. Travel Tour. Market.* **2008,** *25* (2), 161–181.

Pikkemaat, B.; Peters, M.; Boksberger, P.; Secco, M. The Staging of Experiences in Wine Tourism. *J. Hospital. Market. Manage.* **2009,** *18* (2–3), 237–253.

Pine, B. J.; Gilmore, J. H. *The Experience Economy: Work Is Theatre & Every Business a Stage*; Harvard Business Press, 1999.

Prentice, R. C.; Witt, S. F.; Hamer, C. Tourism as Experience: The Case of Heritage Parks. *Ann. Tour. Res.* **1998,** *25* (1), 1–24.

Pullman, M. E.; Gross, M. A. Ability of Experience Design Elements to Elicit Emotions and Loyalty Behaviors. *Decis. Sci.* **2004,** *35* (3), 551–578.

Pullman, M. E.; Gross, M. A. Welcome to Your Experience: Where You Can Check Out Anytime You'd Like, But You Can Never Leave. *J. Busi. Manage.* **2003,** *9* (3).

Quadri-Felitti, D.; Fiore, A. M. Experience Economy Constructs as a Framework for Understanding Wine Tourism. *J. Vacation Market.* **2012,** *18* (1), 3–15.

Richards, G. The Experience Industry and the Creation of Attractions. *Cult. Attract. Eur. Tour.* **2001,** 55–69.

Santisi, J. Warming Up the Wine Industry. *E Wk* October 31, **2011,** *22* (6), 9.

Snepenger, D.; King, J.; Marshall, E.; Uysal, M. Modeling Iso-Ahola's Motivation Theory in the Tourism Context. *J. Travel Res.* **2006,** *45* (2), 140–149.

Sparks, B. Planning a Wine Tourism Vacation? Factors That Help to Predict Tourist Behavioural Intentions. *Tour. Manage.* **2007,** *28* (5), 1180–1192.

Voase, R. Rediscovering the Imagination: Investigating Active and Passive Visitor Experience in the 21st Century. *Int. J. Tour. Res.* **2002,** *4* (5), 391–399.

Williams, P. W. Positioning Wine Tourism Destinations: An Image Analysis. *Int. J. Wine Market.* **2001,** *13* (3), 42.

Williams, P. W.; Kelly, J. Cultural Wine Tourists: Product Development Considerations for British Columbia's Resident Wine Tourism Market. *Int. J. Wine Market.* **2001,** *13* (3), 59–76.

Yuan, J. J.; Cai, L. A.; Morrison, A. M.; Linton, S. An Analysis of Wine Festival Attendees' Motivations: A Synergy of Wine, Travel and Special Events? *J. Vacation Market.* **2005,** *11* (1), 41–58.

CHAPTER 4

MAINTAINING HOTELS' COVID-19 PROTOCOLS POST-PANDEMIC TO ENHANCE SERVICE FOR GUESTS WITH AUTISM: AN OPPORTUNITY FOR INCREASED ACCESSIBLE TOURISM

WAYNE W. SMITH[1*], KAREN PEESKER[1], DANIEL GUTTENTAG[2], and JULIE KELLERSHOHN[1]

[1]*Ted Rogers School of Hospitality and Tourism Management, Ryerson University, Toronto, Ontario, Canada*

[2]*Hospitality and Tourism Management, College of Charleston, Charleston, South Carolina, USA*

ABSTRACT

Autism spectrum disorder (ASD) is an increasingly prevalent neurodevelopmental condition, and individuals with ASD and their families face myriad barriers in their lives. Research has categorized such barriers as relating to social safety and physical safety in a leisure setting, and barriers of both types are common in hotels. Nevertheless, many of the protocols that hotels introduced to promote health and safety during the COVID-19 pandemic (e.g., online check-in, reduced capacity public spaces, and wayfinding signage) have inadvertently created a hotel experience that mitigates many of the challenges that hotels pose for guests with ASD and their families. This chapter details how various protocol modifications have fostered a hotel design that is more ASD-friendly in terms of both social and physical safety. It is suggested that hotels take advantage of such circumstances by

Post-COVID Tourism and Hospitality Dynamics: Recovery, Revival, and Re-Start.
Umendra Narayan Shukla & Sharad Kumar Kulshreshtha (Eds.)

finding ways to maintain these protocols postpandemic to facilitate tourism opportunities for individuals with ASD and their families, giving the hotels a potentially lucrative competitive advantage.

4.1 INTRODUCTION

The COVID-19 pandemic pummeled the hotel industry. According to the American Hotel and Lodging Association (2021a), the effects exceeded 9/11 by at least 9 times. Hotels represented one of the first industries to be impacted by the pandemic and are expected to be one of the last to recover (AHLA, 2021a). During the pandemic, hotels quickly implemented various protocols to remain open and provide a safe environment for both employees and guests (AHLA, 2021a). These protocols generally entailed new cleaning and disinfecting regimens, and guest-service modifications to ensure physical distancing (AHLA, 2021b). The increased safety protocols increased operational and infrastructure costs at a time when the industry was enduring historically low occupancy rates. Nevertheless, these challenges have given rise to a potential unintended positive consequence—many of the newly introduced protocols can now be retained or modified to enhance business operations postpandemic, particularly regarding accessible tourism.

Accessible tourism refers to offering hospitality-related products that are functionally usable by those with mobility, vision, hearing, cognitive, and developmental disabilities (Darcy, 2010). While the hotel industry, in particular, has successfully become increasingly accessible for those with physical disabilities (Perry and Kozub, 2011) there has been little work done on creating accessible operations for those with developmental disorders (Freund et al., 2019). Autism spectrum disorder (ASD) is one developmental disorder that the tourism sector has largely overlooked. While there are some examples of improvements in this area—for instance, Hartsfield–Jackson Atlanta International Airport introduced a multisensory room as a calming oasis for individuals with ASD (LaGrave, 2016)—there remains much opportunity for progress in better-supporting travelers with ASD and their families. In light of the COVID-19 pandemic, there has been a call to foster a more inclusive environment for those with ASD (Spain et al., 2021). Spain et al. (2021) argued that because protocol-related services and business models are already being altered to create a new standard for what is considered "normal," now is the ideal time to reshape service provisions with adaptions that support those with ASD.

ASD is a neurodevelopmental disorder characterized by the following two criteria: (1) A persistent deficit in social communication and interactions across multiple environments and (2) restricted, repetitive patterns of behaviors, interests, or activities (Hodges et al., 2020). Those who have ASD typically demonstrate deficits in social-emotional reciprocity. They also often have difficulty with non-verbal communicative behaviors and may have challenges maintaining and understanding human relationships. Some people with ASD may display repetitive motor movements, engage in repetitive use of objects or speech, be ritualized in their routines, or become fixated on abnormal interests in their intensity or focus. Further, those with ASD may experience hyper- or hyporeactivity to sensory input or unusual interest in an environment's sensory aspects (Hodges et al., 2020). ASD diagnoses in the USA have been increasing in recent decades (Maenner et al., 2020). In 2000, one in 150 children was diagnosed with ASD, but by 2016 this had increased to one in 54 children (Maenner et al., 2020). This growth has created an increased need for hotels and other hospitality suppliers to enhance their capacity to provide satisfactory services to this growing population of individuals and their families.

The traditional pre-pandemic hotel experience often would pose challenges for those with ASD, as long wait times, high sensory environments, and obligatory complex interactions with staff members often fostered an uncomfortable environment for such individuals. The introduction of COVID-19 protocols has led to new environments and alternate approaches to staff interactions that have unintendedly addressed many of these challenges. Given the significant disruption caused by COVID-19 and the hotel industry's adoption of myriad new protocols in response, now is an opportune time for these properties to develop more accessible tourism options providing high-quality services for those with ASD. This chapter will examine this possibility, and revolves specifically around the work of Freund et al. (2019), Li et al. (2019), and Sedgley et al. (2017), who examined barriers to participation in services for those with ASD and their families. These studies indicated the following two areas in which those with ASD face perceived barriers: (1) Access to a socially safe environment and (2) access to a physically safe environment. The rest of this chapter will explore how recently implemented COVID-19 protocols help to address these two significant types of barriers.

4.2 UNDERSTANDING BARRIERS FOR THOSE WITH ASD

This chapter adopts a universal design model in which all society members can access services with limited barriers (Michopoulou et al., 2015). Adopting the universal design model calls for businesses to introduce practices whereby people of all abilities can access and successfully use the services. Given the general lack of research and guidance on providing hotel services to those with ASD, it is expected that pre-pandemic hotel design was largely inaccessible for people with ASD.

Most of the research into people with ASD has centered on either medical-based services (Spain et al., 2021) or educational services (Kinnealy et al., 2012; Mcallister and Maguire, 2012; Mostafa, 2008; Vogel, 2008). While a few studies have explored how a physical environment can positively affect children with ASD, these studies have typically focused on building and classroom design (Kinnealy et al., 2012; Mcallister and Maguire, 2012; Mostafa, 2008; Vogel, 2008). In addition, a few studies examined playground or recess-based treatments (Menear et al., 2006), demonstrating that outdoor play-based treatments have been beneficial to those with ASD (Licciardello et al., 2008; Mason et al., 2014). There is, however, evidence that there remain invisible barriers inhibiting those with disabilities from participating in the same activities as their peers (Dorsch et al., 2016). Spain et al. (2021) noted that, due to COVID-19 restrictions, barriers to service provision in the medical field have further increased. In general, while there has been limited research on how to cater to those with ASD outside of the healthcare or educational environments (Li et al., 2019), there is strong evidence that barriers limit participation in services of all types.

Three studies that examined barriers to participation in leisure-based activities for those with ASD were Li et al. (2019), who looked at park access, Freund et al. (2019), and Sedley et al. (2017), who examined hospitality customer service. These studies identified several potential barriers to service provision. Li et al. (2019) stated that while the benefits of park access included sensory, motor, emotional, and social development, there were also significant barriers to access to parks for families with children with ASD. For example, some parents held internal fears of their children having inappropriate social interactions; for example, their children may throw tantrums or display odd behavior or echolalia (repetition of another person's speech). Likewise, some parents expressed safety concerns relating to traffic, wandering, tripping and falling, bullying, and stranger danger. There were also concerns about their children's phobias, intensely visual and acoustic inputs, small or dark spaces, and crowded or noisy environments. One of the biggest concerns raised in the study focused on issues with the public realm, specifically regarding social inclusion and others' judgment. Finally, other barriers, such as financial and time constraints, were mentioned as hindrances to park usage. This study's findings indicated that counteracting barriers would precipitate greater park usage by those with ASD.

A second study by Freund et al. (2019) found that perceptions of accessibility were critical to travel choices. Like Li et al.'s (2019) study, there was a fear among parents of children with ASD that their presence would disturb other patrons. This fear is exacerbated by the perception that hotel staff are untrained to handle service encounters with those who have ASD. Freund et al. (2019) also found that environmental constraints played a significant role in perceptions of accessibility. Specifically, they argued that environmental constraints such as wait times during check-in and in restaurants, a limited menu or an unwillingness to adapt menu options, the loudness of the lobby and restaurant areas, and room placement/design (near elevators and loud spaces) could have a significant impact on a family's perceptions of usability.

Like both Freund et al. (2019) and Li et al. (2019), Sedgley et al. (2017) found that one of the most significant challenges to hospitality settings is the "gaze of discipline and surveillance" (p.20), whereby parents are put into an unwelcoming and uncomfortable position due to being surrounded by people who do not understand their children's behavior. The research indicated that guests with ASD found the need for flexibility to be particularly challenging (e.g., waiting until 5:30 p.m. for a reservation made for 5:00 p.m.). If food cannot be adapted in addition to an unexpected wait time, this can create a challenge of compounding issues for those with ASD. When accommodations are requested, staff often do not have the skills to address the stressful encounter deftly. Like Freund et al. (2019), Sedgley et al. (2017) found that hospitality locales' sensory environments can be overstimulating, with few places to escape. These compounding elements led Sedgley et al. (2017) to conclude that, for parents of children with ASD, all too often, "A holiday is never relaxed" (p.21). The three studies highlight the previously described barriers to having a socially safe and physically safe environment. Interestingly, many of these barriers have unintentionally been addressed by the protocols many hotels have instituted during the COVID-19 pandemic.

4.3 CREATING A SOCIALLY SAFE ENVIRONMENT

A socially safe environment includes certainty, comfort concerning routines, and safe social interactions between staff and guests. In terms of COVID-19 protocols, many hotels have now shifted to online, touchless check-in and check-out procedures (CDC 2020). The online check-in procedure assists those with ASD in many ways. The first is that it allows the person with ASD to have certainty, particularly if rooms can be pre-selected (especially if there are room pictures available), and the check-in procedure is communicated before arrival. Providing additional information about the room environment can provide further certainties, such as the floor, distance to the elevator, and the window's view. By having the online service available, queueing to check in (another high-stress area) is also alleviated. An online check-in process also could provide an opportunity for families to communicate information about a guest with ASD, so that hotel staff are aware of the individual who is found wandering around the hotel alone. As a result of COVID-19 protocols, many hotel restaurants likewise have adopted an online ordering system. An online restaurant ordering system creates certainty by allowing the client to preview food choices and communicate needed adaptions. These systems also allow the individual to pre-order and pay for his or her meal, reducing miscommunication potential and adding a timing element for when the food will be ready. Implementing this system furthers the environment of certainty. Pre-ordering when dining in creates a better experience for those with ASD and is additionally favorable to many other patrons.

Beyond online communications, due to COVID-19 protocols, many staff are being taught to recognize when patrons are ill (CDC, 2020). This training should extend to looking for people who are demonstrating traits and behaviors associated with ASD. Staff members should look for signs that guests are feeling distressed and then offer assistance. As part of the COVID-19 protocols, staff should communicate in a succinct, clear, and concise manner to reduce contact time, and that practice should continue (CDC, 2020). For instance, at check-in, interactions with the front desk agent are limited to necessary information only, reducing the likelihood of misunderstanding. Social distancing guidelines that encourage people to space themselves from one another should also continue postpandemic, thereby avoiding perceived crowding. Social distancing can be included at check-in, in the elevators, and in the lobby. As lobbies return to being hubs of activity postpandemic, one idea would be to adopt a dedicated 60-min window in which a quiet check-in period can occur. The quiet check-in period could entail reduced auditory and visual stimuli (e.g., lowered the music and adjusted lighting). This period also could provide additional staffing to assist those in need; for example, the concierge, bell staff, and operations managers could be trained to assist those who may require a more patient approach to checking in. This period additionally could involve special access to elevators for newly checked-in guests; for example, having one elevator dedicated to check-ins with express service to room floors.

These socially safe changes would create an improved environment for those with ASD, who often have difficulty with non-verbal communicative behaviors. The new environment could provide a streamlined hotel experience with the bonus of improving services for all. Additionally, providing these socially safe services would allow people with ASD to become more comfortable with the hotel experience by having it feel ritualized, which will lower the stress and the stress of those who accompany them. Finally, by continuing many of these practices that had already started due to COVID-19 protocols, those with ASD would experience reduced sensory input, providing additional comfort to them and their travel companions. These socially safe services would not just enhance the experience for those with ASD; they would increase the accessibility of the services to a broader range of people and would be welcomed by those who are neurodivergent, have physical disabilities, or require additional assistance.

4.4 CREATING A PHYSICALLY SAFE ENVIRONMENT

During the COVID-19 pandemic, hotels in the United States have often adopted a series of alterations to their physical and superstructural environment to protect guest health (CDC, 2020). One such modification is the increased usage of wayfinding signage. Wayfinding signage helps to alleviate stress related to unknown environments. As demonstrated during the COVID-19 pandemic, wayfinding signage is also a helpful tool to create physical distancing by guiding people's flow through space and reducing people bumping into one another. The continued adoption of wayfinding signage would create a more comfortable environment for those with ASD by assuring direction and space. In addition to wayfinding signage, it is recommended that hotels provide signage in the lobby/check-in area stating that they provide accessibility support, which will help to alleviate stress for ASD customers. Furthermore, continued usage of ground markings indicating spacing between guests waiting in line also would enhance the experience for those with ASD. While this practice requires additional space, it has an established history of reducing customer stress, reducing queue jumping, and discouraging those who do not need to be in a queue from joining it (Mann, 1969; Weng et al., 2017).

Other areas significantly altered by COVID-19 protocols are hotel lobbies, restaurants, and public spaces. Social distancing regulations required furniture to be eliminated or moved, pathways between public spaces to be decluttered, and limitations placed on the number of guests in hotels' public spaces. While postpandemic, these public spaces likely will be returned to their pre-pandemic usages; there have been benefits derived from the social distancing, and if some of this distancing were maintained, it would be of great assistance for those with ASD. For instance, in restaurants, keeping a percentage of socially distanced tables away from televisions and other stimuli would benefit guests with ASD, in addition to many others (e.g., guests with hearing loss). Also, if reservations are made and managed using an online system through which individuals are texted when their table is ready, the space requirements for waiting areas outside of restaurant spaces can be reduced.

In terms of lobby space, COVID-19 protocols dictated that many hotels shift towards best practices in which the design should be a functional, clean, and well-maintained space for both work and social purposes (Bennan, 2011). According to Brennan (2011), music should be low to allow for conversation and/or concentration, and visual distractions should be minimized. These design elements postpandemic should incorporate a quiet space located adjacent to the main lobby area that could be used for those working, those who want a quiet meeting space, or those who need to disconnect from the lobby's commotion to reorient themselves. Like the previously mentioned Atlanta airport, hotels could create a small multisensory room for ASD customers or provide a small tactile activity gift that can provide a calming effect as the ASD individual checks in and adjusts to the new hotel environment. During COVID-19 protocols, lobby space should incorporate design elements that direct and maintain the guests' flow. In postpandemic times, thoughtfully guiding the guest flow should remain, with large gathering spaces placed in areas where flow-through traffic is unnecessary.

Due to the COVID-19 protocols, many room features also were modified to maintain cleanliness. The practice whereby key elements are inspected and sealed following cleaning should be maintained, as the sealing of the room gives a sense of greater safety. The policy of allowing for non-contact service, whether it be room service or daily servicing, should remain an option for those who wish it. This option would allow those who may have challenges with human interactions to feel more comfortable knowing that their room will not be entered without their knowledge or permission. Lastly, there should be an option for those with ASD to choose rooms located in a hotel's quiet areas, away from elevators, and away from large or busy public spaces (such as next to the ice machine).

Many of the physical design changes due to COVID protocols create an opportunity to make the hotel increasingly accessible for those who have ASD. The design elements required by COVID-19 protocols increased the usage of wayfinding and other directional signage to facilitate traffic flows. The queuing policies moved the hotels to implement best queue design practices, increasing guest comfort while maintaining function and safety. The public areas were reconfigured to increase the spaces' functionality, allowing for greater usage and easier flow-through. The cleaning and non-contact service options give a greater sense of personal security and safety. While all of these changes were for COVID-19-related purposes, they also created greater accessibility for those with ASD.

4.5 SUSTAINABLE DEVELOPMENT AND ACCESSIBLE TOURISM

If hotels were to maintain specific COVID protocols to make their properties more suitable for guests with ASD, it would facilitate tourism by individuals with ASD and their families, which is in line with sustainability goals. The benefits that tourism offers, such as joy, bonding, relaxation, education, and personal growth, should be accessible. Indeed, Article 7 of the World Tourism Organization (2021) states, "The prospect of direct and personal access to the discovery and enjoyment of the planet's resources constitutes a right equally open to all the world's inhabitants," and later on the article explicitly posits, "[T]ourism for people with disabilities [should] be encouraged and facilitated." Unfortunately, however, as was previously described, the tourism sector requires significant progress in accessibility for individuals with ASD. According to research undertaken by the Family Travel Association (FTA) and the International Board of Credentialing and Continuing Education Standards (IBCCES) (2021), 87% of parents who have children with ASD do not take family vacations, which is likely because 89% are not satisfied with the current travel options available for children with ASD. These findings highlight how the immense barriers to travel faced by ASD individuals and their families are impeding such families from enjoying the benefits of travel, which is especially unfortunate due to the increased stress that parents of children of ASD often experience (Schieve et al., 2007). Therefore, hotels' modified protocols during the COVID-19 pandemic present a remarkably straightforward and effortless opportunity to help increase tourism's accessibility.

Nevertheless, these benefits should not be contextualized in purely altruistic terms, as hotels that become more ASD-friendly will enjoy a competitive advantage. As has been mentioned, ASD is a widespread condition, and families of children with ASD often avoid travel due to the barriers they face. In strategic terms, this segment, therefore, represents a largely untapped and potentially lucrative market. The previously mentioned research by the FTA and IBCCES (2021) found that 93% of parents of children with autism would be more inclined to travel if autism-certified options were available, while also noting that US$262 billion is spent annually on autism-related services in the USA. ASD-friendly hotel design likely would represent a dominant value proposition for prospective travel parties with an ASD guest, thereby granting the hotels greater pricing power, eliminating the impact of other possible shortcomings, and permitting significant differentiation.

Further, many of the recommendations made here would reduce the hotel's environmental impact by reducing energy usage (by turning down lights and music) and reducing food wastage (with online ordering systems put into place). Moreover, the protocols would not be costly to introduce, given that they already have been implemented during the pandemic, and the protocols often would appeal to many individuals who do not have ASD. In total, hotels are becoming more accessible to enhance their ability to traverse the triple bottom line successfully.

4.6 CONCLUSIONS

Sedgley et al. (2017), Li et al. (2019), and Freund et al. (2019) explored barriers to leisure and hospitality participation for those with ASD. All three studies identified how stressful it was for parents and individuals with ASD to participate in leisure or hospitality-related activities due to social and physical barriers. COVID-19 required hotel management to implement thoughtful systemic and structural changes that were often challenging and costly to implement and required guests and staff members to alter their experience expectations. However, inadvertently, these changes created new efficiencies, opened new avenues to problem-solving, and created a more adaptable hotel industry for what the new normal postpandemic could be. What that new normal entails postpandemic requires thoughtfulness as to which protocols to keep, adapt, and discard. As demonstrated in this chapter, many of the protocols adopted during COVID-19 could now, in the postpandemic environment, make hotels more accessible to those with ASD. As has been described, these protocols could contribute directly towards producing a hotel experience whereby guests with ASD will feel socially and physically safer. By creating a safer environment, those with ASD will be able to engage in the hotel experience more successfully, thus alleviating many of the fears expressed by parents in the research by Sedgley et al. (2017) and Freund et al. (2019). Further, as discussed throughout the chapter, these changes are built on a universal design philosophy in which it is not only those with ASD who derive the benefits but also many other customers who would be attracted to the new standard practices.

KEYWORDS

- **autism**
- **COVID-19**
- **hotel**
- **barriers**
- **accessible tourism**

REFERENCES

AHLA. AHLA's State of the Hotel Industry 2021a. *AHLA*. https://www.ahla.com/sites/default/files/2021_state_of_the_industry_0.pdf

AHLA. Enhanced Industry-Wide Hotel Cleaning and Safety Guidelines in Response to COVID-19. *AHLA*, 2021b. https://www.ahla.com/sites/default/files/SafeStayGuidelinesV5_031221_FINAL.pdf

Baio, J.; Wiggins, L.; Christensen, D. L. et al. Prevalence of Autism Spectrum Disorder Among Children Aged 8 Years—Autism and Developmental Disabilities Monitoring Network, 11 Sites, United States, 2014. *MMWR Surveillance Summaries* 2018, *67* (No. SS-6), 1–23.

Brennan, S. N. The Positive Physical Environment in Hotel Lobby Design; Doctoral Dissertation, San Francisco State University, 2011.

Centers for Disease Control and Prevention. COVID-19 Employer Information for Hotels, Resorts and Lodges. https://www.cdc.gov/coronavirus/2019-ncov/community/organizations/hotel-employers.html (accessed on Mar 16, 2021).

Chikuta, O.; du Plessis, E.; Saayman, M. Accessibility Expectations of Tourists with Disabilities in National Parks. *Tour. Plan. Dev.* **2019,** *16* (1), 75–92.

Darcy, S. Inherent Complexity: Disability, Accessible Tourism and Accommodation Information Preferences. *Tour. Manage.* **2010,** *31* (6), 816–826.

Dorsch, T. E.; Richards, K. A. R.; Swain, J.; Maxey, M. The Effect of an Outdoor Recreation Program on Individuals with Disabilities and Their Family Members: A Case Study. *Therap. Recreat. J.* **2016,** *50* (2), 155.

Family Travel Association & The International Board of Credentialing and Continuing Education Standards Family Travel Association (FTA) and IBCCES have teamed up to ensure travel agents from all over the globe have access to training and certification in the special needs travel market, 2021. https://ibcces.org/fta/ (accessed Mar 16, 2021).

Fasihi, H. Urban Parks and Their Accessibility in Tehran, Iran. *Environ. Justice* **2019,** *12* (6), 242–249.

Freund, D.; Cerdan Chiscano, M.; Hernandez-Maskivker, G.; Guix, M.; Iñesta, A.; Castelló, M. Enhancing the Hospitality Customer Experience of Families with Children on the Autism Spectrum Disorder. *Int. J. Tour. Res.* **2019,** *21* (5), 606–614.

Hodges, H.; Fealko, C.; Soares, N. Autism Spectrum Disorder: Definition, Epidemiology, Causes, and Clinical Evaluation. *Transl. Pediatr.* **2020,** *9* (Suppl 1), S55.

Kokaridas, D.; Demerouti, I.; Margariti, P.; Krommidas, C. The Effect of an Indoor Climbing Program on Improving Handgrip Strength and Traverse Speed of Children with and Without Autism Spectrum Disorder. *Palaestra* **2018,** *32* (3).

LaGrave, K. This Airport Launched a New Room for Fliers with Autism. *Condé Nast Traveler.* https://www.cntraveler.com/stories/2016-04-21/this-airport-launched-a-new-room-for-autistic-fliers.

Li, D.; Larsen, L.; Yang, Y.; Wang, L.; Zhai, Y.; Sullivan, W. C. Exposure to Nature for Children with Autism Spectrum Disorder: Benefits, Caveats, and Barriers. *Health & Place* **2019,** *55*, 71–79.

Li, L.; Du, Q.; Ren, F.; Ma, X. Assessing Spatial Accessibility to Hierarchical Urban Parks by Multi-Types of Travel Distance in Shenzhen, China. *Int. J. Environ. Res. Public Health* **2019,** *16* (6), 1038.

Mann, L. Queue Culture: The Waiting Line as a Social System. *Am. J. Sociol.* **1969,** *75* (3), 340–354.

Odom, M. The Impact of Socioeconomic Factors on Park Quality and Accessibility Within the Kalamazoo City Park System; Doctoral Dissertation, Kalamazoo College, 2019.

Maenner, M. J.; Shaw, K. A.; Baio, J. Prevalence of Autism Spectrum Disorder Among Children Aged 8 Years—Autism and Developmental Disabilities Monitoring Network, 11 Sites, United States, 2016. *MMWR Surveillance Summaries* **2020,** *69* (4), 1.

Michopoulou, E.; Darcy, S.; Ambrose, I.; Buhalis, D. Accessible Tourism Futures: The World We Dream to Live in and the Opportunities We Hope to Have. *J. Tour. Futures* **2015,** *1* (3), 179–188.

Perry, H. B.; Kozub, M. Family Travel Experiences When One Member Has a Developmental Disability. *Proceedings of the 2011 Northeastern Recreation Research Symposium*, 2011.

Schieve, L. A.; Blumberg, S. J.; Rice, C.; Visser, S. N.; Boyle, C. The Relationship Between Autism and Parenting Stress. *Pediatrics* **2007,** *119* (Suppl 1), S114–S121.

Singer, J. 'Why Can't You Be Normal for Once in Your Life?' From a 'Problem with No Name' to the Emergence of a New Category of Difference. In *Disability Discourse*; Corker, M., French, S. Eds.; McGraw-Hill Education, 1999; pp 59–67.

Spain, D.; Mason, D.; Capp, S. J.; Stoppelbein, L.; White, S. W.; Happé, F. "This May Be a Really Good Opportunity to Make the World a More Autism Friendly Place": Professionals' Perspectives on the Effects of COVID-19 on Autistic Individuals. *Res. Autism Spectr. Disorders* **2021,** *83*, 101747.

Weng, S. J.; Gotcher, D.; Kuo, C. F. Lining Up for Quick Service—The Business Impact of Express Lines on Fast-Food Restaurant Operations. *J. Foodservice Busi. Res.* **2017,** *20* (1), 65–81.

World Tourism Organization. Global Code of Ethics for Tourism, 2021. https://www.unwto.org/global-code-of-ethics-for-tourism (accessed Mar 18, 2021).

CHAPTER 5

SUSTAINING DUBAI'S LUXURY IMAGE: THE POST-PANDEMIC ROAD MAP

SANJAY NADKARNI[1] and ANDY HEYES[2]

[1]*Emirates Academy of Hospitality Management, Dubai, UAE*

[2]*Stenden Hotel Management School, Leeuwarden, The Netherlands*

ABSTRACT

This chapter looks at the resilience of Dubai's policy prerogatives and the attendant course of action in the immediate aftermath of COVID-19 to revive the tourism and hospitality sectors. Discussions include that of the impacts of COVID-19 on Dubai's hospitality and tourism markets including that of the now postponed Expo 2020 Dubai, closures of hotels, and the loss of jobs.

The methodology adopted is a synthesis of secondary data analysis with a review of academic literature and public policy publications at the destination level. This in turn will look to outline the current problems/predicaments in Dubai as well as suggest possible methods to move forward when it comes to managing the crisis on part of government authorities. Using a wide range of hotel and tourism data, the authors have attempted to gauge a viable roadmap in the planning for the future of Dubai's revitalization post-COVID-19.

5.1 INTRODUCTION

The year 2020 was one which is likely to live on in the minds for decades for nearly "*everyone old enough to count to ten.*" Within the space of 12 months,

Post-COVID Tourism and Hospitality Dynamics: Recovery, Revival, and Re-Start.
Umendra Narayan Shukla & Sharad Kumar Kulshreshtha (Eds.)

the way in which humanity lived their lives changed dramatically as severe acute respiratory syndrome coronavirus 2 (SARS-CoV-2) as known as coronavirus which as a result was simplified to COVID-19 has swept across the globe, creating what can only be described as a worldwide pandemic. Aspects such as social distancing (2 m), excessive hand washing, the wearing of face masks, the amount of people who can meet in a household at any one time and in some countries, the forbiddance of paying with cash are just some of the aspects which have affected the way people are surviving across the world and have continued to survive in some cases.

While media reports have categorically claimed that the virus had been transmitted from China—moreover a place called Wuhan (Brewster, 2020)—discussions and arguments remain about how the virus came about (i.e., was it man made in a laboratory or something which god meticulously created?). Regardless, it is fair to say that the pandemic has caused havoc on a social, political, economic, and psychological scale both at a national and international level.

The impacts on the hospitality and tourism industries worldwide have been exponential. That is to say and to be argued, no one singular country has not been affected in some way due to the unprecedented emergence of COVID-19 as it enters across worldwide borders.

5.1.1 ABOUT DUBAI

The United Arab Emirates (UAE) is no exception. At the time of writing this chapter, within the country, an estimated 600,000 people of the country's estimated 9.7 million population had been clinically confirmed to have contracted coronavirus, with the death toll standing at close to 1800—equating to a fatality rate of just over 2.5% (MOHAP, 2021). There is no data on the nationalities of the deceased available in the public domain. To which degree and to which nationalities the deceased belonged is unknown.

Since 2016, in a series of research articles, the authors of this chapter have extensively discussed the current state of the luxury hotel and tourism industries of Dubai (Nadkarni and Heyes, 2016; Heyes and Nadkarni, 2020).

Nevertheless, the impact of COVID-19 on the UAE and more specifically on Dubai is something that is still to be explored or discussed in its entirety. This chapter will provide a zeitgeist, examining the impacts of COVID-19 on the Emirate of Dubai, an analysis of the crisis management measures and possible ways to deal with the impact, and furthermore, will look to suggest future planning possibilities and scenarios heading into the future.

5.2 WHY AND HOW WAS DUBAI HIT SO HARD BY COVID-19

Reasons why Dubai's hospitality and tourism markets were affected by the pandemic are vastly complicated to which multiple reasons could apply. First, discussing the positioning of Dubai regarding its geographical location within the globe, Dubai had made significant investments to transform Dubai International Airport (DXB) into what is now one of the busiest airports in the world based on passengers numbers transiting through its three terminals. Alongside this Dubai's full-service airline (Emirates) has been collecting multiple accolades, including that being named the world's best airline, the world's most valuable airline ($3.7 billion), and having a route map covering over 165 different destinations (Emirates.com, 2020). Dubai has vastly become an international hub connecting East to West and North to South. While there is no conclusive scientific evidence, there is speculation that major hubs across the globe including Dubai may have inadvertently contributed to the spread of the COVID-19 virus, given that so many different nationalities from a variety of locations use transit hubs.

As a result, Dubai's government was quick to act by introducing the best practices in COVID-19 management protocols leading to a gradual easing of restrictions to open up the economy and minimize inconvenience to the residents (Gulf Today, 2020). Reasons for such measures could have been for multiple reasons, not just for controlling the worldwide spread of the virus based on the protection of human life, but also protecting the image, reputation, and long-term strategic direction which was officially set out many decades ago—for Dubai to become one of the leading hospitality, tourism, and financial centers of the world (DMCC, 2021).

Apart from suspending flights from countries deemed high risk depending on the prevailing pandemic ground realities, Dubai has continued to welcome visitors across the board. This strategy has largely paid off as travelers keen to escape lockdowns in their places of domicile have opted to choose Dubai, one of the few destinations that are open. A case in point is the influencers who lost many of their followers for what was perceived by the fans as being reckless behavior during a pandemic. The UK for instance ended up banning inbound flights from UAE and after the easing of the ban, has maintained the UAE on the red list meaning inbound passengers have to undergo institutional quarantine at their own expense (Hymas, 2021).

5.2.1 DUBAI'S STAKEHOLDERS AND THE NEED FOR CRISIS MANAGEMENT

Systematically, crisis management can be defined in multiple ways with no singular theory seen to be in existence. While some could argue that a crisis is a means for destruction, others could see this as an opportunity for better development opportunities in the future. As per Fearne–Banks (2002: 2), crisis management can be defined as "*a major occurrence with a potentially negative outcome affecting organizations, company's, individuals, industry's/ sectors and even nations as a whole*". While the success of Dubai has been well documented, others could argue that there is still room for improvement—whether that be from forming/providing a greater variety of hospitality and tourism outlets or improving and updating the current inventory. There is therefore a need for anticipation, planning, preparation, and sufficient training to meet the needs of the market. In regards to the crisis of 2020—that of COVID-19—no one could really anticipate how much the current situation was going to affect the market before its so-called introduction.

There is however a significant number of heavyweight stakeholders in Dubai, ranging from entities with a global corporate footprint to those directly owned or controlled by the UAE government entities or rulers' representatives. Thus, there is an overarching drive toward generating returns on investments and ensuring growth. Real estate and service sector verticals (banking and hospitality) have traditionally been the growth drivers of Dubai's economy.

5.2.2 STOCK MARKET

While Dubai is famous for its touristic and hospitable nature, Sheikh Maktoum's vision was also to transform Dubai into a world-leading business and tourism center (DIFC, 2020). The introduction of Dubai's financial center led many of the global corporate giants and investors to park funds into the Dubai market while also supporting other medium-sized enterprises. These somewhat risky investments it could be argued have come at a cost in line with the requirement for overseas businesses to tie up with a local partner in order to operate in the UAE; however, as part of the government's robust response to the pandemic situation, this requirement has been relaxed for what are considered to be priority sectors (DMCC, 2021).

5.2.3 EXPO 2020 DUBAI

Following the successful bid by the Government of UAE to host EXPO 2020, large tranche of investments has been realized and channeled into transforming 438 ha (1080 acres) of desert land into what can now only be described as a city within a city. Initial costs of supporting facilities alone (hospitality, residential, commercial, transportation, leisure, etc.) were estimated to be US$33 billion worth of investment to support incoming visitors with both corporate and leisure facilities (Gulf News, 2017).

The awarding of the EXPO 2020 to Dubai in 2013 was said to enable Dubai to build on its established image for excellent infrastructure, attractions, hotels, affordable amenities, easy accessibility, and highly developed air and road transportation systems.

The impact of COVID-19 on the EXPO 2020 expedition has been substantial. What was originally planned to run from 20 October 2020 to 10 April 2021 has had to be postponed to 1 October 2021–31 March 2022. Some could, therefore, suggest that EXPO 2020 has now—in fact—become EXPO 2021. However, is this the case and to what extent are these pushbacks likely to affect Dubai?

No sector within the UAE has felt the brunt of the impact more than the hospitality and tourism industry, which had been preparing for the Expo with full zeal ever since Dubai won the bid to host the event in 2013. The COVID-19 crisis has no doubt hit the core of Dubai's hospitality and tourism industries significantly meaning there is a means by which things need to be completed. With travel restrictions being enforced in the main source markets, the revenue streams from tourism have significantly dipped.

The building of the EXPO 2020 infrastructure along with further supporting facilities such as hotels had as a result created a substantial number of jobs. Since then, the city has steadily added massive capacity to its hotel room inventory, spruced up the transportation infrastructure including a dedicated metro line, and helped to provide its residents with support heading forward (DMCC, 2021). The introduction of local as well as international hotel brands has significantly enhanced the city's room inventory. Similarly, further activities/facilities have also been added or revitalized to assist with a larger destination mix. (See Image Below).

Mega projects such as the Marsa Al Arab are in construction. While other cities globally were pushing back against Airbnb and other players in the shared economy space partly egged on by their respective hotel lobbies, Dubai pragmatically entered into an agreement with Airbnb to legitimize the operations despite earlier conflicts of regulations based on taxes and fees. This is not the case however based on Uber. This served two purposes- to boost the real estate sector which was reeling from oversupply and add further capacity to accommodate the visitors to the Expo 2020 Dubai event. For a complete analysis see Nadkarni (2019). Though no reliable estimates are available, the postponement of the Expo 2020 Dubai by a year has upset the planning, execution, and operations of several projects in the public and private sectors. Mega projects such as the hitherto mentioned Masra Al Arab have been impacted because of the current COVID-19 crisis and the same is applicable to the investment pipeline that was set up prior to the crisis. Despite such setbacks, the events industry has been gaining traction—for example, the annual Omega Dubai Desert Classic at Dubai Creek golf course is 2021 in itself in its 32nd year and was successfully concluded at the beginning of the year.

On the flip side, projects where time overruns which would have otherwise found it challenging to meet the October 2020 deadline, ended up with breathing space to ensure the infrastructure was in place as planned. With sustainability being one of the guiding themes of Expo 2020 Dubai, the purpose-built Expo site has been developed in a manner that can be repurposed to a mixed-use integrated ecosystem known as District 2020—or otherwise DISTRICT 2021 depending on how one looks at it from here on in considering the 2021 re-engagement. Contingency plans are, however, expected based on the current predicaments in the representing nations that have invested significant money and efforts in custom-built country pavilions to showcase the cultural, economic, and technological achievements.

5.2.3.1 SUSTAINING DUBAI'S LUXURY IMAGE

Whilst there is no universally accepted definition of luxury, there are several markers of luxury that have remained consistent across demographics (Maman Larraufie and Kourdoughli, 2014). Dubai checks on a host of these markers and have been making a concerted effort at positioning itself as a luxury destination to overcome what Henderson (2006) has underscored is a perceived lack of conventional tourist attractions. For instance, in a study on tourism shopping and urban entertainment, Zaidan (2016) infers that Dubai is perceived as a luxury shopping destination which gives it a competitive edge. The sheer volume of luxury transactions Dubai has spawned has led Stephenson (2014) to coin the term "massclusivity"—an oxymoronic combination of mass and exclusivity which pretty much sums up the image Dubai has cultivated over the years. The predominance of luxury hotel rooms across home has grown and international hotel brands as well as the positioning of Dubai's full-service carrier—Emirates in the premium segment (Kotsi and Michael, 2015)—are other markers that buttress the luxury image. More recently, since Airbnb was allowed to legitimize operations in Dubai by the authorities, a casual search of the listings points to the disproportionately higher number of luxury accommodation options in comparison to other destinations. Dubai's Road Transport Authority has introduced premium vehicles including Tesla in its taxi fleet. The city's ever-expanding glitzy skyline further adds to the luxury story.

Thus, to summarize, the long journey toward building an image of luxury has stood Dubai in good stead in reaping economic dividends across key touristic markers, as illustrated in Figure 5.1.

FIGURE 5.1 Luxury destination's touristic markers.

From a theoretical perspective, Ko et al. (2019) have provided an extensive review of scholarly work in understanding luxury and have identified quality, authenticity, prestige (image), price premium, and resonance as the key criteria for measuring luxury. What needs to be highlighted at this juncture is the dearth of extensive studies in Dubai to gauge the tourists' responses to these measures which are critical to explaining the destination image perception using models such as Brand Luxury Index (BLI) (Vigneron and Johnson, 2004).

It needs to be underscored that Dubai's luxury tag has come to be associated with ostentation and conspicuous consumption- a phenomenon validated by Bearden and Etzel's (1982) work which demonstrated that luxury goods consumed in public were more likely to be conspicuous goods. Such hyper-consumerism runs were contrary to sustainability best practices. Hence, there is a conscious pivot toward sustainable luxury which is also evidenced in sustainability being one of the driving themes of the Expo 2020 Dubai. Whether this perception of responsible and sustainable luxury catches on among tourists is debatable. For instance, Athwal et al. (2019) have identified three themes—consumer practices and concerns, organizational practices, and cross-cultural issues—which need to be factored by the stakeholders of Dubai's hospitality and tourism industry. Adding a theoretical dimension to this point, the theory of uniqueness (Stokburger–Sauer and Teichmann, 2013) underscores the need to differentiate from others when there is too much similarity in the socio-economic milieu. A move in this direction is the repurposing of the EXPO 2020 site post-event for residential and commercial use- unique, high quality, and yet sustainable offering.

5.2.4 HOSPITALITY AND TOURISM DOWNSIZING IN DUBAI

Within Dubai, local hotel asset owners were forced to close some of the hotel properties as a response to the pandemic and focus on monetizing their key (Diaa, 2020). Local hotel brands such as The Address (owned by Emaar Hospitality) were one of the first hotel brands to close establishments, including that of the infamous The Address Downtown—with many of the buildings repurposed into offices/apartments. These transformations it could be argued are vital for the emergence of the sector as a result of the continuous developments throughout the world.

As a result, redundancies and a loss of jobs were inevitable. This somewhat of a contingency plan in itself is a representation of the contemporary version of the life cycle/destination life cycle model (Butler, 2006)—that is to say when the markets (tourism and hospitality) start to "dry up," there is always a need for buildings to become revitalized whether that be through stabilizing the industry/market or converting the said buildings (as mentioned) into similar/different facilities altogether (including that of residential buildings).

However, to do this is a very difficult scenario. Furthermore, it could be argued to be a means to meet the needs and wishes of industry stakeholders—with many highly invested in real estate and infrastructure—therefore the "filling" of these buildings makes it substantial for the economy to build itself here-on-in.

Similarly, Dubai's flagship carrier Emirates publicly admitted the need to have made a cut in the headcount to meet the decrease in international travel and tourism numbers not only in the city of Dubai but also who use the airlines' services worldwide (Nair, 2020). This decrease can also be said to assist in the management of tourists in and out of Dubai on a regular basis—controlling incoming and outgoing flights on a continuous basis and also restricting the number of visitors to the Emirate. Most closures have been planned on a temporary basis and the impact was most acute starting in March 2020 when the WHO declared COVID-19 to be a worldwide pandemic. This pandemic as a result is still ongoing, though on account of the rollout of vaccinations, the situation has been easing somewhat. The situation subsequently continued to worsen globally. Smith Travel Research (STR) data suggests that the average daily rates (ADR)* fell by 20% year-on-year (y-o-y) and the impact on revenue per available room (RevPAR) was even more pronounced, dropping by 43% y-o-y. The trends continued until June 2020. However, since July 2020, Dubai allowed a gradual re-opening of its hospitality and tourism-related businesses and started welcoming overseas visitors once again from all over the world. This coupled with the staycation market gave a great boost to the sector, however, questions are asked about how far the so-called staycation market can go in relation to the inclusion of concepts such as Airbnb banned from certain areas of the city (outlawhotels.com, 2019).

From a crisis management perspective, Dubai has handled the COVID-19 situation with guarded optimism and pragmatism. The contingency plans that the authorities set out are quite clearly one of strategic direction and can infiltrate a wide range of stakeholders on a regular basis. Whilst jurisdictions across geographies were imposing successive lockdowns, the only notable restriction on public movement in Dubai lasted for 2 weeks, beginning March end until mid-April. Even during this period, an efficient digital platform for getting approval from authorities to visit grocery stores or supermarkets (the only businesses allowed to stay open) was put into place, including police cordons stretched for multiple miles. Although businesses in general suffered, particularly hospitality and allied verticals, the residents were not put through inconvenience in accessing daily essentials including food, etc. There was a gradual opening up of the economy thereafter. The proactive decision making despite the prevailing uncertainty has served Dubai in a positive frame and is in alignment with what Laws et al. (2007: 3) recommend in terms of a baseline response strategy- immediate action, empathy for those impacted and minimizing the impact of adverse publicity. The emphasis on restoring normality in the aftermath of a crisis has been put forth by Scott et al. (2008); Dubai has stretched this concept by attempting to restore normality, during the crisis, and not in the aftermath of unknown reasons. Whether it has achieved total success is still to be seen and discussed. In doing so, to complement the short-term tactical responses that have been contingent on the evolving pandemic situation, together with the other Emirates that constitute the UAE, Dubai has accelerated strategic policy changes that were already in the pipeline pre-COVID-19, in a way leapfrogged into attaining many a new normal; some examples are discussed in the following.

- The Abraham accords which established formal diplomatic ties between the UAE and Israel were signed in August 2020. Among the industries to reap early harvest benefits was the hospitality and tourism sector; by conservative estimates, by December 2020, over 50,000 Israelis visited Dubai (Hendrix, 2020), bringing much-needed cheer to the sector.
- In November 2020, laws were amended to allow 100% foreign ownership of businesses (Copper-Ind, 2020). Until then, foreign shareholders could not own over 49% of a locally registered company. The amended laws are expected to provide a much-needed fillip to inbound investments and business activities, including hospitality and tourism.
- Despite a burgeoning expatriate workforce, none of the countries of Gulf Cooperation Council have ever offered a route to citizenship. In January 2021, the UAE became the first country among member states to offer citizenship to a restricted category of expatriates, paving way for a more inclusive, versatile, and talented citizen base bringing tangible benefits to the society at large (Magdy, 2021).

These examples reflect a desire on part of the government to ace the charts in the league tables of nation building, that too in the midst of a raging pandemic. The ability to make an opportunity out of a crisis through swift decision making has been honed over the years since the global financial crisis of 2008 which had a detrimental impact on Dubai's economy (Renaud, 2012). With reference to the hitherto mentioned tactical response, a case in point is the rapid mobilization for COVID-19 testing, setting up of field hospitals, and the ongoing vaccination campaign. On each of these fronts, the UAE has been in the top league on a per capita basis all this while allowing businesses including hospitality and tourism to operate with minimal restrictions in line with COVID-19 safety protocols.

From a theoretical perspective, the tactical and strategic responses hitherto discussed can be explained with reference to the qualitative attributes of Chaos theory, which are sensitive to initial conditions, discreteness of change, attraction to specific configurations, structural invariance at different scales, and irreversibility (Thietart and Forgues, 1995: 19). Dubai's crisis management strategy resonates with Keown–McMullan's (1997: 9) observation that entities undergo significant change (long-term strategic policy changes- or the new normal) even when they are successful in mitigating the crisis with rapid action tactical responses. An alternative framework to understand the overall crisis management response offered by Dubai is the single loop/double loop approach with reference to community adjustment capabilities proposed by Richardson (1994). It fits into the double loop approach which challenges the traditional linear mindset, as opposed to the single loop approach which is status quo.

5.3 DUBAI'S HOTEL AND TOURISM MARKETS

There is a considerable body of literature on Dubai's luxury hotel and tourism sectors discussing a wide range of factors including room rate transparency, visitor numbers in the Emirates, and the number of hotels classified under the luxury classification (Nadkarni and Heyes, 2016; Heyes and Nadkarni, 2019, 2020). While the perceptions of Dubai as a luxury destination are still strong, the authors have brought this into question on the grounds that the abundance of luxury prevailing in Dubai dilutes its "exclusivity" value by making it widely accessible (Heyes and Nadkarni, 2016)—accessible to a magnitude of other people. However, the impact and emergence of COVID-19 are likely to have changed this—as previously suggested with diminishing inbound arrivals on account of COVID-19 restrictions (Hymas, 2021).

Another case in point is the issue of the oversupply of hotel rooms at the luxury end of the spectrum. With reference to Figure 5.2, data coming out of STR suggests that over the past 5 years, the supply in the luxury, upper upscale, and upscale segments which have already been high historically, continued on an upward trajectory in the run-up to Expo2020, despite a sluggish demand. Zooming in on the last 2 years of data as shown in Table 5.2, this growth trend actually turned negative as a result of COVID-19. The difference in magnitude of the negative year-on-year percentage change is the key metric shown in Table 5.1 points to attempts at keeping occupancy higher by providing and maintaining attractive room rates, which has seen some success, given the combined demand of the staycation market and tourist. Again, to what extent this combines Airbnb is unknown.

TABLE 5.1 Key Hotel Metrics—Year-on-Year Change.

Occ %		ADR		RevPAR		Percent Change from YTD 2019					
2020	2019	2020	2019	2020	2019	Occ	ADR	Rev PAR	Room Rev	Room Avail	Room Sold
48.7	74.4	125.80	148.79	61.33	110.72	−34.5	−15.4	−44.6	−46.7	−3.8	−37.0

Source: STR.

TABLE 5.2 Hotel Rooms by Class—Year-on-Year Change.

Class	2019	2020	Difference
Luxury	17,928	17,710	(218)
Upper Upscale	23,613	23,410	(203)
Upscale	14,284	15,348	1,064
Upper Midscale	9,652	9,752	100
Midscale	5,906	4,699	(1,207)
Economy	1,271	1,284	13

Source: STR.

In addition to the physical constraints imposed by a successive lockdown in source markets, the contraction in supply (flight capacity and hotel rooms) is also a result of the subdued sentiment on the demand side. As Dragouni et al. (2016: 80) have underscored, "*sentiment and mood indicators are net transmitters of spillover shocks to outbound tourism demand*" and are directly correlated to consumer behavior and spending decisions (Bryant and Macri, 2005).

5.4 WHAT DOES THE FUTURE OF DUBAI HOLD?

Nevertheless, it is anticipated that Dubai's tourism sector will bounce back quicker and stronger as a result of the authorities' determination to support the sector which is a key driver of the economy (Rahal, 2020). In regards to 2021, Dubai's Destination Management Organization (DMO) "Visit Dubai" has been quick to act in regards to re-assuring visitors to the Emirate of its health and safety regulations/procedures in accordance with tackling the COVID-19 situation (visitdubai.com, 2021).

Upon Dubai's emergence, there is little doubt that His Highness Sheikh Mohammad bin Rashid Al Maktoum envisioned his Emirate to be one that would bring prosperity to his people but also one which would be able to overcome problematic situations. As Zwaya (2020) quoted his highness based on his Twitter account "*The year 2020 taught us a lot: we learned that our power is in unity, happiness is with the family, we learned that being better is in giving and that health comes before politics and economy.*"

Nevertheless, with decades of work and investments into the Emirate, there is undoubtedly still a need for a revitalization plan to be put into action in order to sustain Dubai's current and future presence as a leading destination globally. The impact of COVID-19 is likely to cast its shadow, albeit temporarily.

As such, it is likely that a strategy is being or already has been planned which will assist in the re-emergence of Dubai's hospitality and tourism industries as well as the rejuvenation of its financial center. Nevertheless, as discussed in the articles by Nadkarni and Heyes (2016), Heyes and Nadkarni (2019, 2020), it is possible to suggest future plans which may be able to assist and be possible.

The completion of mega projects currently in the pipeline will expand the inventory of hotels, and the offerings for touristic activities and increase real estate opportunities, providing tourists and future expatriates with greater options from which to choose.

Based on earlier studies by Heyes and Nadkarni (2016, 2019), it is moot whether these additions would contribute any further in solidifying Dubai's position as a luxury destination, given the potential dilution of exclusivity and rarity which are proxies of luxury (Frank, 1999; Yeoman and McMahon-Beattie, 2006). Simultaneously, to sustain and grow the economy, there is a need to forge a plan to restore the demographics diluted because of the pandemic on account of job losses.

A sustained and naturally made target market can look to revolutionize itself into a new era. The model illustrated in Figure 5.2 shows the drivers that could potentially sustain the luxury element in Dubai's destination perception.

5.4.1 HISTORY AUTHENTICITY UNIQUENESS PRICE

Despite its spectacular growth into a sprawling cosmopolitan metropolis, Dubai retains history and authenticity which is richer than what meets the eye. There are several characteristics that make Dubai stand apart- its demographics, seamless infrastructure, economic opportunities, and governance to name a few. From a price elasticity perspective, while the "value for money" tag has gradually eroded over the years to be replaced by "overpriced," in the post-COVID-19 scenario, value-driven pricing seems to be making a comeback.

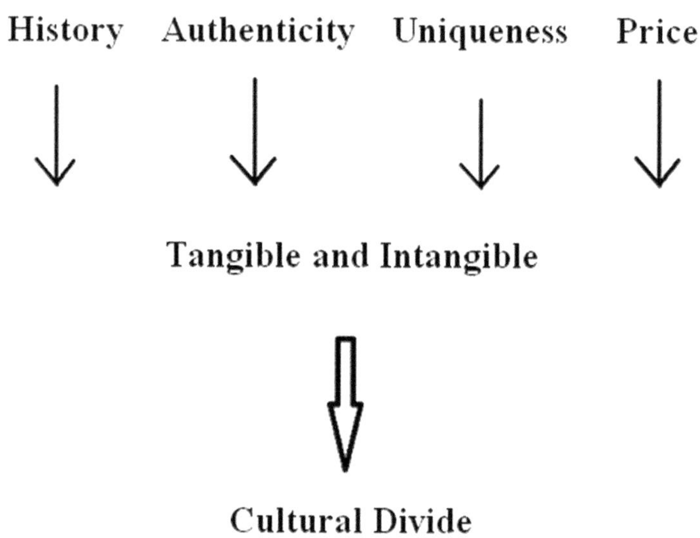

FIGURE 5.2 Drivers for sustaining Dubai's luxury destination identity.

In conclusion, from a destination planning perspective, it is inevitable that Dubai pivots toward the triple bottom line approach (Elkington, 1998) to ensure long-term economic, environmental and social viability as a destination of choice. Toward this end, as travelers grow more conscious of the sustainability aspect, the perception of ostentation and conspicuous consumption will be a hindrance to drawing in the growing breed of environmentally and socially sensitive tourists across source markets.

As hitherto discussed, sustainable luxury is a natural policy choice for Dubai to be able to retain its 'luxury destination' tag. Decoupling ostentation from rarity is a long-drawn process and can be achieved through alternative paradigms. A case in point is the usage of electric vehicles in public transportation which Dubai has already demonstrated by incorporating Tesla cars in its taxi fleet- an example of achieving sustainability without compromising on luxury.

The already fast pace of adoption of Industry 4.0 technologies in Dubai has only accelerated in the post-COVID-19 context, thus making stronger, the dimension of uniqueness and rarity. Driverless metro networks, blockchain-based contracts, and 5G coverage are some examples. The advantage of using Industry 4.0 technologies such as AI, blockchain, IoT, and robotics is that they are also the very drivers for sustainability. Thus, adoption at scale would not only help Dubai achieve its sustainability targets aligned with the UN's Sustainable Development Goals (SDGs) but also help retain the luxury destination tag, albeit with the adjective "sustainable" preceding luxury.

KEYWORDS

- **destination recovery**
- **Dubai**
- **luxury**
- **policy and planning**

REFERENCES

Anthonisz, A.; Mason, G. Reinventing Tourism: The Dubai Phenomenon. *Worldwide Hospital. Tour. Themes* **2019,** *11* (3), 279–286. DOI: 10.1108/WHATT-01–2019–0004

Athwal, N.; Wells, V. K.; Carrigan, M.; Henninger, C. E. Sustainable Luxury Marketing: A Synthesis and Research Agenda. *Int. J. Manage. Rev.* **2019,** *21* (4), 405–426.

Bearden, W. O.; Etzel, M. J. Reference Group Influence on Product and Brand Purchase Decisions. *J. Consum. Res.* **1982,** *9* (2), 183–194.

Brewster, J. A Timeline of the COVID-19 WUHAN Lab Origin Theory. *Forbes*, May 24, 2020. https://www.forbes.com/sites/jackbrewster/2020/05/10/a-timeline-of-the-covid-19-wuhan-lab-origin-theory/?sh=21b3e36e5aba

Bryant, A. W. D.; Macri, J. Does Sentiment Explain Consumption? *J. Econ. Finance* **2005,** *29* (1), 97–110.

Butler, R., Ed. *The Tourism Area Life Cycle*, Vol. 1; Channel View Publications, 2006.

Copper-Ind, C. UAE to allow 100% foreign ownership of businesses. *Int. Invest.* **2020**. UAE to allow 100% foreign ownership of businesses (internationalinvestment.net).

Diaa, S. Emaar Hospitality Suspends Bookings at Three Hotels in Dubai. *Gulf News* March 10, 2020. https://gulfnews.com/business/property/emaar-hospitality-suspends-bookings-at-three-hotels-in-dubai-1.70280320

DIFC. A Leading International Financial Hub in the Middle East, Africa and South Asia (MEASA) Region, 2020. https://www.difc.ae/

DMCC Discover DMCC, 2021. https://www.dmcc.ae/

Dragouni, M.; Filis, G.; Gavriilidis, K.; Santamaria, D. Sentiment, Mood and Outbound Tourism Demand. *Ann. Tour. Res.* **2016,** *60*, 80–96. DOI: 10.1016/j.annals.2016.06.004

Dubai Experience Marsa Al Arab—These Two Islands Are to Enclose the Burj Al Arab, 2020. https://dubai-experience.com/marsa-al-arab-project/

Dubai to Stop Temperature Checks and Thermal Scanners from Jan. 1. *Gulf Today* Dec 30, 2020.

Elkington, J. Accounting for the Triple Bottom Line. *Measur. Busi. Excellence* **1998**.

Emirates.com Home, 2020. https://www.emirates.com/ (accessed Dec 31, 2020).

Fearne-Banks, K. *Crisis Communications: A Casebook Approach*; Lawrence Erlbaum Associates: New Jersey, 2002.

Frank, R. *Luxury Fever: Why Money Never Fails to Satisfy in an Era of Excess*; The Free Press: New York, 1999.

Henderson, J. C. Tourism in Dubai: Overcoming Barriers to Destination Development. *Int. J. Tour. Res.* **2006,** *8* (2), 87–99.

Hendrix, S. Package-Tour Diplomacy: Thousands of Israeli Tourists Flock to Dubai After Peace Deal. *The Washington Post* Dec 14, 2020. https://www.washingtonpost.com/world/middle_east/israel-dubai-uae-tourists-peace/2020/12/14/0d492838–3973–11eb-aad9–8959227280c4_story.html

Heyes, A. What Is Luxury Hospitality? A Need to Move Towards a Scientific Understanding. *Res. Hospital. Manage.* **2021,** *11* (2), 67–69. DOI: 10.1080/22243534.2021.1917919

Heyes, A.; Nadkarni, S. Luxury Consumption in Tourism: Is Dubai Losing Its Luxury *Appeal. Research in Hospitality Management* **2016,** 6 (2), 213–218. DOI: 10.1080/22243534.2016.1253293

Heyes, A.; Nadkarni, S. Luxury Consumption in Tourism: The Case of Dubai Part II. *Res. Hospital. Manage.* **2020,** *10* (1), 51–54. DOI: 10.1080/22243534.2020.1790209

Heyes, A.; Nadkarni, S.; Sustaining a Luxury Destination: The Case of Dubai. In *Sustainable Destination Branding and Marketing: Strategies for Tourism Development*; Sharma, A., Fernandez J-I., Hazzan, A., Eds.; CABI: Oxford, 2019.

Hymas, C. Social Influencers Banned from Holidaying in Dubai: All 80 UK Flights a Day from/to the United Arab Emirates Will Be Grounded from 1pm on Friday. *The Telegraph*, Jan 28, 2021. https://www.telegraph.co.uk/news/2021/01/28/social-influencers-banned-holidaying-dubai/

Keown-McMullan, C. Crisis: When does a mole-hill become a mountain? *Disaster Prevent. Manage.* **1997,** *6* (1), 4–10. DOI: 10.1108/09653569710162406

Ko, E.; Costello, J. P.; Taylor, C. R. What Is a Luxury Brand? A New Definition and Review of the Literature. *J. Busi. Res.* **2019,** *99*, 405–413.

Kotsi, F.; Michael, I. Planning and Developing 'Destination Dubai' in the Context of the United Arab Emirates (UAE). In *Planning for Tourism: Towards a Sustainable Future*; CABI: Oxfordshire, Boston, 2015; pp 149–168.

Laws, E.; Prideaux, B.; Chon, K. S., Eds. *Crisis Management in Tourism*; CABI: London, 2007.

Magdi, M. UAE Opens Citizenship to Select Foreigners to Boost Economy. *Bloomberg* Jan 30, 2021. https://www.bloomberg.com/news/articles/2021–01–30/uae-plans-to-offer-citizenship-to-select-group-of-foreigners.

Maman Larraufie, A. F.; Kourdoughli, A. The E-Semiotics of Luxury. *J. Global Fashion Market.* **2014,** *5* (3), 197–208.

MOHAP. COVID 19 Information Center, 2021. https://www.mohap.gov.ae/en/Awareness Center/Pages/COVID19-Information-Center.aspx

Nadkarni, S. The Pre-and Post-Event Planning Roadmap for Expo 2020: Dubai's Coming-Out-Party. *Worldwide Hospital. Tour. Themes* **2019,** *1* (3), 259–265.

Nair, D. (**2020,** June 9). Dubai's Emirates Airline Confirms More Job Redundancies: Carrier Says It's Become 'Impossible to Sustain Excess Resources. *Gulf News* June 9, 2020. https://gulfnews.com/business/aviation/dubais-emirates-airline-confirms-more-job-redundancies-1.1591707058775

Outlawhotels.com *AirBnB Banned from Downtown Dubai* 2019. https://outlawhotels.com/breaking/airbnb-banned-from-downtown-dubai/

PrivateIslands. The World Islands—Dubai—United Arab Emirates, Asia—Private Islands for Sale, 2020. privateislandsonline.com

Projects Linked to EXPO 2020 Worth $33 Billion. *Gulf News* Oct 19, 2017. https://gulfnews.com/expo-2020/news/projects-linked-to-expo-2020-worth-33-billion-1.2109430

Rahal, N. Dubai's Hospitality Sector Eyes Strong Finish to 2020 after months of Uncertainty. *Arabian Business* Dec 31, 2020.

Renaud, B. Real Estate Bubble and Financial Crisis in Dubai: Dynamics and Policy Responses. *J. Real Estate Literature* **2012,** *20* (1), 51–77. DOI: 10.1080/10835547.2012.12090313

Richardson, B. Crisis Management and Management Strategy-Time to "Loop the Loop"? *Disaster Prevent. Manage.* **1994,** *3* (3), 59–80. DOI: 10.1108/09653569410795632

Scott, N.; Laws, E.; Prideaux, B. Tourism Crises and Marketing Recovery Strategies. *J. Travel Tour. Market.* **2008,** *23* (2–4), 1–13. DOI: 10.1300/J073v23n02_01

Sheikh Mohammed's New Year Post Recaps How UAE Beat COVID-19 Challenges. *Khaleej Times* Dec 31, 2020. https://www.khaleejtimes.com/news/video-sheikh-mohammeds-inspiring-new-year-post-recaps-how-uae-beat-covid-challenges

Stephenson, M. L. Tourism, Development and 'Destination Dubai': Cultural Dilemmas and Future Challenges. *Curr. Iss. Tour.* **2014,** *17* (8), 723–738.

Stokburger-Sauer, N. E.; Teichmann, K. Is Luxury Just a Female Thing? The Role of Gender in Luxury Brand Consumption. *J. Busi. Res.* **2013,** *66* (7), 889–896.

STR. (n.d.). The Global Hotel Industry Standard for Improving Performance and Measuring Success | STR. https://str.com/

Thietart, R. A.; Forgues, B. Chaos Theory and Organization. *Org. Sci.* **1995,** *6* (1), 19–31. DOI: 10.1287/orsc.6.1.19

Vigneron, F.; Johnson, L. W. Measuring Perceptions of Brand Luxury. *J. Brand Manage.* **2004,** *11* (6), 484–506.

Visitdubai.com Travel to Dubai Covid 19 Advisory Services, 2021. https://www.visitdubai.com/en/coronavirus-advisory

Yeoman, I.; McMahon-Beattie, U. Luxury Markets and Premium Pricing. *J. Revenue Pricing Manage.* **2006,** *4* (4), 319–328. DOI: 10.1057/palgrave.rpm.5170155

Zaidan, E. A. Tourism Shopping and New Urban Entertainment: A Case Study of Dubai. *J. Vacation Market.* **2016,** *22* (1), 29–41.

CHAPTER 6

STRATEGIZING THE POST-COVID TOURISM SECTOR RECOVERY

HARSH VARMA

United Nations World Tourism Organization, Madrid, Spain

ABSTRACT

Of all the economic sectors hit by Covid19, tourism continues to be hardest hit with tourist arrivals going down between 70% and 80% in 2020 with resultant losses in employment and incomes. The impact continues in the first half of 2021 with cautious optimism for 2022, particularly in developing countries. A number of governments and their private sectors have taken strategic measures to restart the sector by adopting flexible policies and strategies, restoring tourists' confidence, introducing new health regimens, undertaking product diversification, working out new business models, reducing or deferring taxation, providing fiscal and monetary incentives, improving the physical condition of tourism resources, and motivating upskilling and re-skilling of personnel. Many governments have seen Covid19 as an opportunity to realign and redesign tourism for the future which would pay rich economic, social and environmental dividends, and ensure the sector's expeditious revival.

6.1 INTRODUCTION

Tourism continues to be one of the sectors hardest hit by the Covid-19 and, as of today, the outlook remains grim and highly uncertain. International

Post-COVID Tourism and Hospitality Dynamics: Recovery, Revival, and Re-Start.
Umendra Narayan Shukla & Sharad Kumar Kulshreshtha (Eds.)

organizations and major tourist receiving countries have estimated that the sector went down between 70% and 80% in 2020. The devastating impact has continued into the first quarter of 2021, and the optimism of slow recovery in the year has now been pushed to 2022, the major contributing factors being continuing mandatory testing, quarantine regulations, complete closure of borders, and the below-expectation vaccination rate. Even outlook for 2022 remains cautious at the moment, though bordering on hope and optimism.

According to the World Tourism Organization (UNWTO), 2020 resulted in one billion fewer arrivals, a loss of USD 1.3 trillion in export revenues, and 100–160 million jobs being put at risk with at least 50 million deemed as permanently lost. The loss of arrivals was more or less uniform with Asia and the Pacific at −96%, Europe and Africa at −85%, Middle East at −84%, and Americas at −79%. It is estimated that 32% of the world's popular destinations remain closed (69 destinations), out of which 30 are in the Asia-Pacific, 15 in Europe, 11 in Africa, 10 in the Americas and 3 in the Middle East. Destinations that are heavily dependent on international, business, and events tourism have suffered particularly, with many coastal, mountaineous, and rural areas faring better than urban areas. The news of Covid vaccine is indeed encouraging and offers a glimmer of hope for recovery, though challenges still persist, with the sector expected to remain in survival mode till at least the beginning of 2022. While domestic tourism in countries like China and Russia and some other destinations has helped to a considerable extent in mitigating the impacts on incomes, employment, and businesses, experts believe that the real recovery would take place only when international tourism rebounces.

While many governments and their private sectors realized the gravity of the situation during the first quarter of 2020 itself and swung into action like a cat on a hot tin roof, others, regrettably, decided to wait and watch. Those who woke up early have taken tangible actions to offset the blow, yet more needs to be done and in a much more coordinated way with the private operational sector and local authorities.

6.2 POLICY AND STRATEGIC MEASURES FOR TOURISM RECOVERY

With innovation, imagination, and improvization as *sine qua non*, and with the aim of regaining value and protecting jobs and incomes, some of the policy and strategic measures taken to restart the sector *inter alia* include:

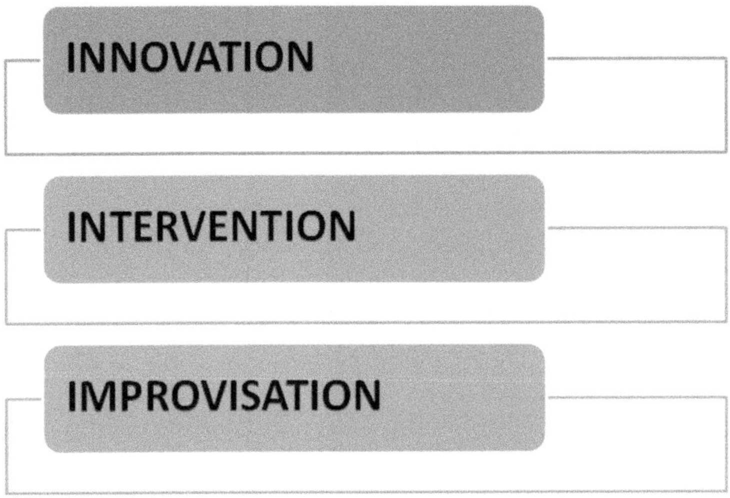

FIGURE 6.1 POST-COVID tourism revival model.

Restoring tourists' confidence:

*restoring tourists' confidence by introducing mandatory health and safety operating procedures, and providing them with all essential real-time information thus limiting uncertainty. For this purpose, dedicated interactive websites were created combining information from public and private sector, and destinations.

Product diversification, new product development:

*based on primary research in established and potential markets and segments within these markets, the private sector was quick to adopt a strategy of product diversification, new product development, and enhancement of the appeal of the existing products.

New business models:

*this strategy was backed up by new business models based on financial and institutional restructuring, for example, lower margins of profit, lower overheads, and lower deployment of personnel. Marketing programmes were minimized, agreements with marketing agencies suspended, and the print production programme was put on hold. The focus of marketing operations shifted to social media and digital marketing.

Focus on micro-level marketing:

*based on the refined data on new consumer preferences provided by NTOs, an intensive product diversification exercise was undertaken. The refined data helped the private sector to focus on selected markets and segments within these markets to direct their marketing efforts. A major decision taken was to avoid macrolevel marketing as it will not yield desired results. Instead, undertake microlevel marketing directed at specific market segments, the underlying dictum being "the narrower your focus, the broader your appeal."

Open-air and activity-oriented holidays:

*a major conclusion of the research was that tourists who are prepared to travel would like to avoid crowded urban areas and use of public transport. They prefer open-air and activity-oriented holidays in rural areas and closer to home destinations. Another preference was for national parks and nature reserves. Thus, immediate action was initiated to upgrade homestays in rural areas and also create new accommodation capacity by putting up prefabricated structures and luxury tents. The upper limit of a cluster was fixed at 30 pax. New packages were created with local elements such as wine tasting, cuisine, folklore, arts and crafts while activities added included cycling, fishing, nature walks, dancing with the locals, and star gazing. Catchy bylines were created to support electronic and print brochures, an example being: "Far from the madding crowd, to peace and serenity, where nature is still nature – unhurried, uncommon, unspoilt."

Self-driving motoring holidays:

*self-driving motoring holidays were another strong attraction for those who did not want to fly. Tour operators did detailed route planning, suggested attractive breaks in between, tied up with service stations to meet emergencies, and provided other logistics.

Positive encouragement:

*encouraged by the positive market response received during the second half of 2020, the travel trade, hoping that the current tourist preferences would stay, has adopted a policy of product diversification whereby these new products would find a permanent place in their overall product chain.

Value addition services:

*value addition was another trump card. With new, diversified itineraries, goodies thrown in included day trips to other rural areas and reverse excursions, that is, from rural areas to urban areas to visit museums, historic buildings, and other attractions. Free guide services and free transport were added. These provided a different and unique experience to tourists.

Strict health and sanitary protocols:

*the hospitality sector also played its part with the theme "something for everyone." In addition to observing strict health and sanitary protocols, hotels and other establishments added open-air dining opportunities by increasing street-level terrace spaces with support from the local authorities. Where weather conditions did not permit open-air dining, gazebos, pergolas, and canopies were used to create a cosy atmosphere.

Self-catering options:

*fine dining was replaced by moonlit dining, fixed menus were introduced in place of *a la carte* menus, and where possible, self-catering options were offered with foodkits.

Inclusion of value-addition activities

*value was added to the overall product by adding a number of extras which included free use of watersports equipment in resorts, free use of gymnasia and health centers and lower rates for spas. In addition, free half-day tours including transport and guide services were offered.

Complementary offerings:

*several hoteliers went a step further by offering free breakfast, 30% discount on food and beverages, and free extra beds in rooms.

Re-skilling:

*both travel trade and hospitality sector undertook massive upskilling and re-skilling progammes for their personnel–upskilling in new safety protocols and re-skilling for those who wanted to change their career paths.

6.3 STRATEGIC ROLE OF GOVERNMENT FOR TOURISM REVIVAL AND RE-START

Governments also placed their role through their tourism boards and public sector corporations. In their efforts to mitigate the impacts of the Corona19 crisis, some of the salient actions taken include:

* governments have reduced VAT on services, reduced/deferred taxation, provided lower energy rates, waived off local taxation, written off property tax till the end of 2021, and extended mortgage periods without any additional charge or penalty to provide relief to enterprises. Other relief measures relate to higher fiscal and monetary benefits to new investors, lower or no licence fees, and minimal duty on capital goods imports.
* framework of principles has been prescribed to help resume transport services with focus on targeted measures instead of general prohibitions, lifting restrictions on individualized transport; and ramping up safer collective transport options. These principles ensure that measures taken are compatible, coordinated and mutually accepted across destinations. Again, this provides tourists more freedom, safety, and certainty to travel.
* since more than 78% services to the tourism sector are provided by SMEs, immediate steps were taken to provide them liquidity to cover labor costs, materials and operational inputs, inventories and overheads, rent, and utilities. Special medium-term assistance was provided for redesigning, repackaging, and repricing their products and services to match the new consumption patterns. This has considerably assisted in implementing new smart solutions for recovery, creating a new value chain of innovative, responsible and economically viable and sustainable tourism.
* with times changing very fast, it is but natural that travel habits, patterns, and preferences have also undergone a radical change. The consequence has been that consumption habits of tourists have also changed drastically. Therefore, through effective communications and periodical webinars, suppliers have been kept regularly informed of the new consumption habits of tourists, and to redesign their products accordingly to meet the new demand. Thus, a number of enterprises in the supply chain, particularly SMEs, were saved from the ordeal of ending up with outdated products and services.

* governments saw an opportunity in the crisis to improve the physical condition of tourism resources as part of their product enhancement policy. Tangible steps were taken to spruce up historic monuments, clean beaches, create new hiking and trekking trails with associated facilities, put up signage, and relocate facilities in national parks and nature reserves to distribute tourists keeping social distancing in mind. Similarly, projects were launched to upgrade roads leading to destinations, install street lights, and create new public facilities such as toilets and car parks. In short, the endeavor was to develop the infrastructure for the future.
* governments provided financial assistance and acted as facilitators through their overseas tourist offices, embassies, and marketing agencies for the private sector to carry out extensive primary research in existing and potential markets to gauge the mood of tourists. NTOs deployed their human resources to analyze the raw data and provided refined data to the private sector.

6.4 RETHINK, REORGANISE, REALIGN, AND REDESIGN TOURISM FOR THE FUTURE

The public and private sector realized that the Covid-19 crisis is an opportunity to reconsider, reform, rethink, reorganize, realign, and redesign tourism for the future. Undoubtedly, tourism is at crucial crossroads and it is well-acknowledged that actions taken today will shape the sector of tomorrow and beyond. Governments and the private sector need to take into account the longer-term implications of the crisis while putting in place flexible policies and strategies, capitalizing on digitalization, and at the same time, encouraging the restructuring of the sector to ensure a kind of development which is economically viable, financially profitable, environmentally sustainable, and socioculturally acceptable.

6.5 VACCINATION PASSPORT

The European Union Covid-19 passport/certificate has come into force across the 27-country block on 01 July 2021, with the objective of standardizing Covid related documentation such as proof of vaccination or

PCR test results. This move was initiated after a number of EU countries announced that they would issue such certificates to not only facilitate free movement of travelers but also to end fake PCR tests and vaccine certificates by creating a homogenized system by which the holder's data can be kept updated and verified. The certificate will enable the EU citizens and legal residents to travel within the EU countries without the condition of mandatory quarantines—assuming that the current situation does not worsen. The certificate seeks to standardize the documents issued by each country so that they can be recognized and accepted by each member state. The certificate will guarantee the use of national certificates to prove that the holder has been vaccinated against Covid-19, has had a negative PCR or antigen test, or has recovered from the illness. Thus, the holder will enjoy a period of immunity from reinfection. The certificate would be available to all citizens and legal residents of a member country irrespective of their nationality. The use of the certificate is not confined to free movement of travelers across the countries but will also be used for social and other events. For example, Austria would like to use the certificate for entry into hotels, restaurants, and cultural evenings such as musical performances and operas.

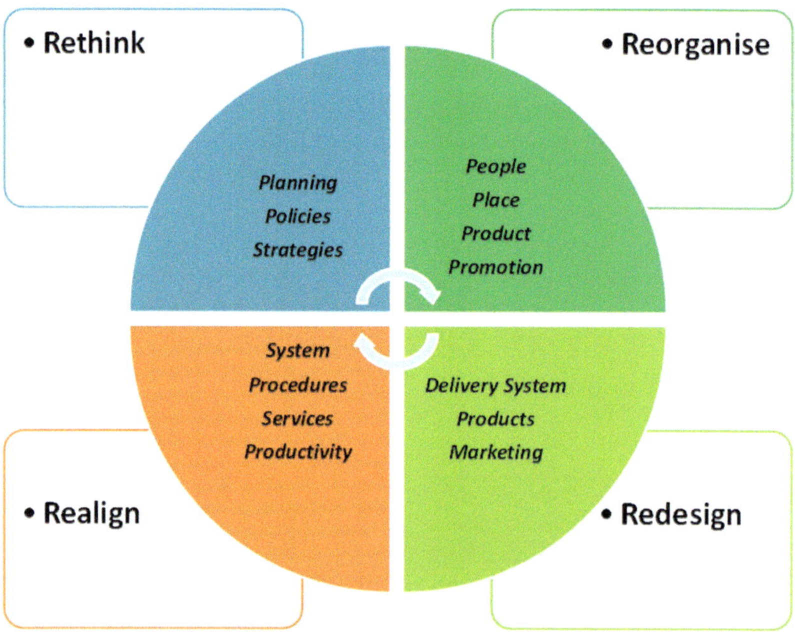

FIGURE 6.2 POST COVID tourism revival approach.

The data included in the certificate would facilitate cross-EU functionality so that a country can accept a document that has been issued by another member state. In particular, the certificate would include a barcode or QR code that would permit the verification of the authenticity and validity of the document, amongst other essentials information. The certificate would be issued in the official language of the issuing country and in English. Another document will include the details of the Covid-19 vaccine the holder has received, the result of a PCR test or any other information that provides evidence that the holder has overcome the virus should he/she has contracted in previously. The certificate would be available digitally and in hard form with a validity of 1 year.

6.6 CONCLUSIONS

For the countries that are dependent on tourism, the Covid19 pandemic has had a devastating effect on their tourism sector, particularly on their economies, with resultant losses in earnings, employment, and government revenues. Since March 2020, the global tourism industry has come to a grinding halt, and in spite of governments' efforts to protect their population through lockdowns, quarantines, and restrictions on international and domestic movements, these measures have only marginally improved the situation. Encouraged by the high rate of vaccinations, strict adherence to Covid-19 Standard Operating Procedures (SOP), and restoration of consumer confidence through vigorous assurance campaigns, a few developed countries, particularly in Europe and North America, have gradually started to open up their tourism and hospitality industry by June 2021 in the hope that by the end of the first quarter of 2022, the industry may come back to the 2019 level, provided no new variants surface and the pandemic situation does not worsen. As contrast to this, the beginning of 2021 has been far worse for several developing countries with arrivals going down by as high as 85%–90% when compared with prepandemic level. The ripple or cascading effects of this decline are far more serious: skilled and unskilled human resources, capital, experience, and expertise remain unutilized; and lack of demand for goods and services has a negative impact on the sectors which provide backward and forward linkages to tourism. Overall, the outlook for the future still looks grim, and according to UNWTO (2021), due to prevalent restriction on international and regional movements, slow rate of vaccination, low consumer confidence, and widespread negative economic environment, recovery may only be possible by 2023 and may even spill over to 2024. Therefore, countries and their tourism sector have to be prepared for the long-haul when it comes to the restoration of prepandemic level.

With every possible innovation and improvization, many countries have taken tangible steps to amend their policies, planning approach, and strategies to meet the Covid19 challenge, but the real need of the hour is to adopt a flexible approach–an approach that is dynamic in nature, and can be easily and quickly adapted to the ever-changing pandemic scenario. Many countries have constituted working groups and task forces comprising public and private sector to review and monitor the situation on a monthly basis and provide their recommendations for modifications. Many countries have adopted a three-phased strategy: the first is damage control; the second is consolidation of minor/initial gains; and, the third is to perform and endeavor to achieve the prepandemic level. Efforts to restore consumer confidence would be an integral and essential component of all the three phases.

It is the unanimous view of experts based on the experience of the last more than eighteen months that in the postpandemic world, times will change very fast. The new norm will witness new travel patterns and habits. With these, the consumer preferences will also change. The entire value chain, comprising consumption on one hand, and production and supply on the other, will undergo a sea of change. Therefore, countries will have to adopt a two-pronged strategy to meet this change: conduct regular and intensive research to identify the new consumption patterns, and then to ensure that production and supply appropriately match the new demands.

With new consumption patterns will also emerge new demands for destinations, services, and facilities. Again, continuous research will have to be conducted on the identification and recognition of new demand patterns including markets and specific segments within the identified markets. And this will bring into play the entire process of further product development and product diversification in consultation and coordination with the travel trade. The new demand patterns which have emerged so far clearly indicate that consumers would like to avoid run-of-the-mill products, destinations that are located in crowded urban areas, locations that involve mass transportation and closed door products such as museums, art galleries, and shopping malls. They would definitely like to patronize open-air destinations which have the possibilities of activity-oriented holidays with families, community-based tourism experience, and visits to wildlife parks and nature reserves. The product diversification policy of countries should revolve around these essentials elements. Thus, rural or community-based enterprises will have to be provided with all-possible support including microcredits and soft loans, and assistance for business development, marketing, and human resource development.

However, the urban destinations or established/traditional products cannot be neglected while keeping new products in the forefront. The only way to keep them "alive" and make their optimum utilization would be through consumer reassurance by way of strict observance of SOPs in accommodation premises, mass transportation, museums and art galleries, and fully vaccinated and adequately trained personnel in new norms. It goes without saying that a series of value-added elements will have to be thrown in to make these products more attractive and appealing.

Governments have played their part in providing much-needed relief to various sub-sectors of their tourism industries by way of cash subsidies, unemployment allowance, tax deferment, additional monetary and fiscal incentives, and financial support to MSMEs. This crucial support must continue or even further enhanced till the sector is back on the rails.

To conclude, a Herculean effort would be required by public and private sector alike to revert the sector to a prepandemic level. In this whole process, what would really hold the key are expeditious vaccination, observance of health and hygiene SOPs, focus on digital marketing, optimum utilization of social media, and further product development and diversification. No doubt, a number of initiatives have been launched and a lot has been achieved in these areas but a lot more still remains to be done.

KEYWORDS

- **new policies and strategies**
- **new normal**
- **new opportunities**
- **new business models**
- **COVID 19**

REFERENCES

United Nations Conference on Trade and Development. 100% of Global Destinations Now Have Covid19 Travel Restriction, UNWTO Reports, 2020a.

United Nations Conference on Trade and Development. Impact of Covid19 on Tourism in Small Island Developing States, 2020b.

United Nations Conference on Trade and Development. Covid19 and Tourism: Assessing the Economic Consequences, 2020c.

United Nations World Tourism Organization. Tourism Dashboard, 2021.

United Nations World Tourism World Tourism Barometer January 2021.

United Nations World Tourism World Tourism Barometer March 2021.

United Nations World Tourism Travel Restrictions Report 2021, 9th ed.

CHAPTER 7

STRATEGIES FOR THE RECOVERY OF THE TOURISM INDUSTRY IN THE NEW NORMAL ERA AFTER THE COVID-19 PANDEMIC

MEHMET ERTAŞ

Pamukkale University, Faculty of Tourism, Denizli, Turkey

ABSTRACT

Tourism is one of the sectors most affected by the COVID-19 pandemic that emerged in the People's Republic of China. International travel bans restricted tourists from traveling. Therefore, the tourism industry has come to a standstill. The normalization movement that started in June 2020 has revived tourism, albeit a little. However, with the transition of the northern hemisphere to the winter season, cases have increased again all over the world in October and November. For this reason, the tourism industry must take necessary caution and continue its activities with the pandemic. Based on these points, the purpose of this study is to offer strategies for the recovery of the tourism industry in the new normal era after the COVID-19 pandemic. The study also provides advice to accommodation facilities, transportation facilities, tour operators and travel agents, governments, and provides predictions about how tourist behavior will change.

7.1 INTRODUCTION

The COVID-19 first appeared on December 31, 2019, in Wuhan, Hubei Province of the People's Republic of China. Later, cases were detected all

Post-COVID Tourism and Hospitality Dynamics: Recovery, Revival, and Re-Start.
Umendra Narayan Shukla & Sharad Kumar Kulshreshtha (Eds.)

over the world in the first 3 months of 2020. In March 2020, the World Health Organization (WHO) declared the disease, which is seen all over the world and has no effective solution, as a pandemic. After the announcement of the pandemic, many countries closed their borders and took some measures to prevent its spread. These measures have considerably affected the tourism industry. Practices such as travel bans and social distance have caused tourists to travel less and cancel their travels. Also, many tourism enterprises had to stop their activities for a certain period (Hao et al., 2020). However, as of June, many countries made normalization statements, gradually opening their borders, and removing the restrictions within the country.

The COVID-19 crisis is the most important crisis experienced by the tourism industry. No previous crisis has caused a decrease in tourist numbers by over 50%. It is estimated that 50 million jobs worldwide are lost in tourism (WTTC, World Travel and Tourism Council, 2020). United Nations World Tourism Organization (UNWTO) (2020a) announced a 65% decrease in tourist arrivals in the first 6 months of 2020 compared with the previous year. The decrease in the number of tourists in April and May is approximately 90%. However, with some tourist activity in July and August, the decrease in the number of tourists is around 80%. While some markets, such as Germany, showed some improvement in international travel demand in June, major outbound markets such as the United States and the People's Republic of China remain in a standstill position (UNWTO, 2020a).

The COVID-19 pandemic is defined as the major global crisis for the first time after World War II. Although governments have taken some measures to stop the viral spread, no country has survived the pandemic. When and how the pandemic will end remains uncertain (Koh, 2020). With September to October 2020, the second wave of the COVID-19 pandemic occurred in many countries and the rate of spread increased. In September 2020, the highest number of cases announced in one day in the world was reached (WHO, 2020). Besides, pandemics, earthquakes cause longer-term damage, unlike natural disasters such as floods and fires (Bae and Chang, 2020). The virus's circulation is unlikely to cease neither in 2020 nor in 2021. Vaccination studies have indeed reached a certain stage, but it will take a long time to produce and distribute enough vaccines for all humanity. Therefore, as humanity, we must live with the COVID-19 pandemic. Health should always be at the forefront; however, tourism enterprises need to start their activities to survive. The purpose of this study is to provide strategies for how the tourism industry will protect itself against this great crisis and how to recover in the new normal era after the COVID-19 pandemic. In this direction, the study offers suggestions to enterprises such as accommodation facilities, transportation facilities, tour operators, and travel agents, makes predictions about how tourist behavior will change and suggests strategies to be made to the government officials in the recovery of the tourism industry. The study presents these strategies and recommendations by compiling previous studies on the COVID-19 pandemic.

7.2 ACCOMMODATION FACILITIES

According to STR, the hotel industry continued to experience double-digit decreases in three metrics—revenue per room available, average daily rate, and occupancy, with low levels of performance across all world regions in July 2020. Hotel room occupancy reached record low levels in July with 17% in Africa, 19% in Central and South America, 27% in Europe, 35% in the Middle East, and 46% in Asia and the Pacific, and 47% in the United States (UNWTO, 2020a). As can be seen from the figures, accommodation facilities continue to be affected by the pandemic crisis.

Since hoteliers thought it was a short-term crisis in the early days of the pandemic, they pursued a policy of price reduction to increase demand, however, this policy did not work due to the spread of the pandemic all over the world and the spread could not be stopped. Hoteliers have taken some precautions after realizing that the pandemic will take a long time. They prefer to reduce human resources costs (unpaid leave, dismissal), operating costs (office and system), and marketing costs (advertising and promotion) to reduce their general costs (Lai and Wong, 2020). However, as hoteliers consider their costs, they should look for ways how their enterprises will host tourists in this period. UNWTO (2020b) has made some suggestions for accommodation facilities as they continue their activities during the pandemic. These include suggestions such as increasing cleanliness in public areas and contact surfaces, the possibility of contactless check-in and check-out, obtaining detailed information about guests, having spare rooms, suitability for physical distance, the flexibility of operations, partnering with healthcare providers, and partnering with local providers. In these rules, hygiene, sanitation, and social distance are at the forefront. Also, in many news, it is said that the virus is transmitted by contact with surfaces. Therefore, the hotel room must be sufficiently disinfected and cleaned before using a new customer (Menegaki, 2020). Besides, it is recommended that accommodation businesses not operate at full occupancy rates as the pandemic is continuing. In this way, it will be easier to provide social distance within the hotel. Hotel facilities should emphasize that they disinfect all parts of hotel buildings at certain times of the day and provide social distance at the hotel in their advertising and marketing.

Another important point after the pandemic will be the safety of tourists. Ensuring that tourists feel safe and rely on the services offered by tourism providers will be essential in the next new era (Bhattacharya, 2020; Roy et al., 2020). Therefore, in the new order, hotel facilities must make some renewals in line with the new health norms. With new building designs, the hotels can be adapted to the pandemic. For instance, a glass strip can be placed in front of the reception to protect employees, or activities for guests can be carried out in open-air environments (Assaf and Scuderi, 2020). Also, these redesigns can alter tourism experiences by removing the sensory boundaries of service areas. For instance, the smell of cleanliness, social distancing, and the number of customers' coexistence in restaurants, festivals, and other tourism settings will affect new standards of psychological comfort and acceptable levels of the perceived crowd (Sigala, 2020).

Hao et al. (2020) gathered the strategies that hotels will make during the pandemic under the headings of leadership and communication, human resources, service provision, financial, disaster management, and social responsibility. In *leadership and communication*, hotel facilities should create disaster management plans for emergencies that may occur and prioritize them in their sales and marketing strategies. In *human resources*, hotels should flexibly reduce nonessential labor costs by socializing service production, sharing labor with the relevant service industry, adopting smart devices to replace manual labor, and firing workers with low-performance efficiency and poor working behavior. In *service provision*, hotel facilities should develop pricing and sales strategies such as pre-sale and package sales, especially with remote online applications. *Financially*, hotels must reduce their non-essential costs to stimulate the reduced cash flow. As *disaster management*, it should continuously monitor the pandemic situation, monitor the measures taken, and implement the most appropriate pandemic practices for the hotel. As a *social responsibility*, hotels should open their beds for health purposes when the number of hospital beds is decreasing. The number of hospital beds has decreased considerably during the peak of the pandemic. Places with hospital extension were needed for people who were infected and has a milder disease. Many hospitality facilities opened their doors as an example of solidarity when the Wuhan government had to arrange shelter for the COVID-19 medical staff posted in the state and construction workers involved in the construction of the COVID-19 hospitals (Hao et al., 2020).

7.3 TRANSPORTATION FACILITIES

IATA (2020) announced that international air passenger demand fell 67% from the first 7 months of 2020 compared with the previous year. In the latest data on international air passenger demand, there was a 92% decrease in July. However, considering that all international flights are closed in April 2020, it shows that the number of passengers has improved. July is also in a better condition than June when a 97% decrease was recorded. Despite this, it is a fact that the demand for international air passengers has decreased considerably. Besides, many countries have started to close their country borders again in October 2020.

The COVID-19 pandemic also affected passenger traffic planned globally for 2020, resulting in a 50% to 51% reduction in the number of seats offered by airlines. In the first 7 months of 2020, the number of available seats decreased by 59% compared with the previous year. Compared to the baseline originally planned, the number of passengers will decrease by between 2858 and 2891 million, and the airlines' gross operating revenues will be lost between approximately USD 387 and 391 billion. Total air passenger traffic (international and domestic) will decrease by 59% to 60% in 2020 compared with 2019 (ICAO, International Civil Aviation Organization, 2020).

Transportation facilities continue to be affected during the pandemic, despite the transition to the new normal era. The distress of normal life with the COVID-19 has caused tourists to avoid traveling to destinations with severe pandemic. For this reason, making passengers feel safe throughout the journey will play an essential role in future travel planning. Obligation to wear a mask during travels, seat filling suitable for social distance, measurement of body temperature, and similar practices will give passengers a safe feeling (Menegaki, 2020). In addition to popularizing such practices, it should be ensured that the passengers who will use the transportation facilities make their travels without forming crowds. For instance, practices such as ensuring that passengers get on the plane or bus without coinciding with any employee, providing social distance in the cabin, taking a certain amount of in-cabin baggage will reduce the crowd. Besides, technological applications should be transitioned gradually to limit contact. For instance, check-in kiosks should be used more widely, and payments should be received by bank transfer (Assaf and Scuderi, 2020). Similarly, UNWTO (2020b) makes recommendations such as health checks for safe air transport, ensuring aircraft and cabin hygiene, providing masks for passengers and officers, and limiting movement in the cabin, paying attention to social distance, providing disinfectants for passengers, reducing the number of items in the cabin, and reducing costs at airports. All these recommendations will help the passengers to travel safely as well as help the transportation facilities develop a positive image.

Safety is important when people travel outside of the places, they live in. Also, it is thought that the first travels during and after the pandemic will be to nearby places in the country. For this reason, transportation facilities need to make local-focused campaigns. Price policies should also be adapted for domestic tourists. Besides, since long-term travels will be replaced by short-term trips, it will be advantageous for countries marketing to nearby destinations (Menegaki, 2020). However, Chebli and Said (2020) concluded in their studies that 76% of them would prefer more distant destinations rather than prefer a destination close to their place of residence in their holiday choices. This shows that when travel restrictions are lifted and the pandemic is eliminated, tourists will continue their travels as before.

7.4 TOURISTS

Travel elements during the pandemic period for tourists consist of four components (Wilson and Chen, 2020). These risk classification of the person, travel elements, health insurance, and policies in the destination country. The first component includes the risk factors and risk threshold (the age of the tourist, the diseases the tourist has, the risk tolerance of the person, and the nature of the trip) the tourist has for the COVID-19 disease. The second component deals with the details of the travel. A special vehicle that is carefully disinfected among tourists is safer than a crowded bus, train, or metro system. The location, size, traffic volume, and time spent at the airports or bus/train stations there will affect the potential exposure level due to travelers coming together during the journey. It also contributes to possible exposures related to the risk level of the destination, its socioeconomic situation, its capacity to detect and stop the spread, and measures to disinfect accommodation, limit the concentration of guests, and promote physical distance. The third component deals with health insurance and government regulations that are not under the tourist's control. Tourists planning travel should be aware of the health insurance policy on coverage if they become seriously ill with the COVID-19 abroad. The fourth component is the decision regarding the opening of borders, the resumption of flights, and the resumption of travel. These decisions are also not under the control of tourists. Restarting safe travel assumes that tests to detect acute infection in symptomatic persons and screening of close contacts are widely available (Wilson and Chen, 2020).

International experts argue that post-pandemic tourism will reach old travel figures even if it takes a long time. Certainly, these travels will not be the same in the new normal era after the pandemic (Ayad et al., 2020). Crowded transportation and accommodation facilities for mass tourism will be less preferred as it is thought that they will have difficulty in providing social distance (Menegaki, 2020). It is thought that tourists will stay away from the mass tourism type, especially since mass tourism creates large groups due to its nature (Aydın and Doğan, 2020). Similarly, Chebli and Said (2020) concluded that many tourists will choose less crowded destinations for their next travel. This may cause some types of tourism to lose their popularity and others to become more popular. In their research, Wachyuni and Kusumaningrum (2020) concluded that after the end of the pandemic, participants plan when and where they will travel within 6 months and that nature tourism will be the most popular attraction with important aspects of safety, cleanliness, and beauty to meet the tourist demand. Besides, holiday types that are not commonly experienced such as camping, caravan, seasonal housing rental, and yacht rental, which are far from the crowd and more suitable for social distance, will be preferred more in the new normal era (Aydın and Doğan, 2020).

In the new normal era after the COVID-19, it is predicted that human behavior will change greatly (Ma et al., 2020). Practices such as frequent hand washing and a healthier lifestyle and social distance that have entered our lives with the pandemic will most likely be the new norm in human behavior (Baraero-Era and Del Rosario, 2020; Brouder, 2020; Chebli and Said, 2020; Sigala, 2020). Besides, transparency and accurate information will play an essential role in tourists' new travel decisions. For instance, tourists will make travel decisions based on governments' travel policies and people's travel behavior related to health and safety (Baraero-Era and Del Rosario, 2020). Similarly, Wachyuni and Kusumaningrum (2020) states that the number of patients in a country will affect the number of tourist arrivals. Therefore, health safety is an important factor affecting the travel choice of tourists to the destinations they intend to visit. At this point, it is a fact that states that struggle with the pandemic correctly and produce solutions to stop its spread will be one step ahead.

7.5 TOUR OPERATORS AND TRAVEL AGENTS

The change in tourist behavior in the new period after the COVID-19 pandemic will affect the business processes of tour operators and travel agencies. In the short period after the pandemic, tourists will tend to travel to less crowded destinations. For this reason, tour operators and travel agencies should design tour plans according to these requests and expectations of tourists. For instance, short-distance, flexible-priced, cancellable tours can be planned with small groups. Ayad et al. (2020) also predicted changes in tourists' behavior in their study on Egyptian tour operators. The most important changes for tourists; they want to book well-known and trusted brands; they chose hotels that emphasize hygiene and sanitation; looking for hotels and restaurants with minimum contact and social distance; they do not prefer long queues during check-in and check-out. Similarly, UNWTO (2020b) makes recommendations for tour operators and travel agencies, such as the implementation of health and safety protocols, digital communication, developing sustainable products, creating new tourism experiences, package tours involving smaller groups, encouraging domestic tourism and nearby destinations, expanding the content of tour packages, designing flexible tour packages, coordinating with insurance companies. Considering that tour operators and travel agencies play a key role in the advertising and marketing of the tourism industry, these suggestions should be highlighted in promotional activities.

In the new normal era, it is thought that package tours will be replaced by individual tours. Tourists who do not want to interfere with the crowd, especially due to social distance and hygiene issues, will continue to travel alone or in small groups (Menegaki, 2020; Sigala, 2020). In their study, Chebli and Said (2020) concluded that tourists choose unknown places choice of destination, avoid group transportation, pay attention to the quality of the health conditions of the destination to be traveled, and choose travel insurance carefully. In line with these recommendations, in the new normal era, tour operators and travel agencies should make the tour planning by considering the whole process from the beginning to the end of the travel.

7.6 GOVERNMENTS

Government policies will play a key role in the continuation of tourism facilities in the new normal era as in the first destructive times of the COVID-19 pandemic. Governments should expand the details of their support packages as they are one of the building blocks and key stakeholders of the tourism industry. For instance, interest-free loans should cover the whole sector, provide funds for the promotion of destinations, visa restrictions should be lifted or mitigated in countries where the epidemic is relatively less, local government practices should be encouraged, tourism businesses should be tax exempted (Assaf and Scuderi, 2020). Governments must also be prepared and foresighted by the fact that other pandemics may occur in the future. In this direction, sustainable plans should be prepared, and it should be noted that these plans include adaptable, inclusive, and innovative strategies. Besides, governments should work in coordination with all stakeholders of tourism to find solutions to the negative effects of the pandemic (Baraero-Era and Del Rosario, 2020).

The tourism industry also needs government assistance to support local economies, build market confidence, and reduce the risk of the spread of the pandemic. Governments should work harder to take credible measures, assist tourism facilities in sustainability, and encourage innovation (Assaf and Scuderi, 2020; Carr, 2020). The media and political actors' strategy for the COVID-19 and the tourism industry is to return to normalcy as soon as possible. However, the COVID-19 pandemic is an opportunity to assess the tourism industry's way of managing tourism in reducing negative impacts on the natural environment and communities (Prayag, 2020). Lapointe (2020) states that as the tourism market will shrink in the new normal era after the pandemic, tourism enterprises will have to turn to local consumers and their needs. At this point, sustainability will be at the forefront thanks to the pandemic. The pandemic has enabled environmental changes to be made to make more sustainable tourism (Jones and Comfort, 2020). For this reason, both national and local governments should develop strategies that will highlight the sustainability of tourism in the new normal era.

7.7 RESULT AND DISCUSSION

Tourism is a labor-intensive sector, and the tourism ecosystem includes the airline industry, tour operators, travel agencies, hotels, restaurants, transportation services, car services, and all other ancillary services related to travel and entertainment (Tyler and Dangerfield, 2020). In many sectors other than tourism, the effects of the COVID-19 pandemic have been tried to be minimized by using technology. For instance, businesses have been able to perform some of their jobs using remote working methods. However, it is a sector in which remote methods do not work since touristic products and services must be purchased where they are produced. Therefore, it would not be wrong to say that tourism is one of the sectors most affected by the COVID-19 pandemic. Besides, it will take longer to return to normal with the pandemic. It may also take a long time for the tourism industry to recover. For this reason, the tourism industry should anticipate and prepare for the short and long-term effects of the pandemic through a multi-level (global, national, and community levels), sustainable, inclusive, adaptable, and innovative tourism recovery plan with different stakeholders (Baraero-Era and Del Rosario, 2020).

At the beginning of the pandemic, academic studies stated that tourism will start with domestic tourism first (Baum and Hai, 2020; Brouder, 2020; Prideaux et al., 2020; Thomas, 2020; Wen et al., 2020). As stated by the researchers, domestic tourism has increased especially after the normalization period. While the recovery of international tourism is slow, demand for domestic tourism is increasing in many major markets such as China. Air capacity in Russia is also supported by increased domestic travel (UNWTO, 2020a). However, the contraction in international demand and the decrease in international tourism expenditures continue. While some markets, such as France and Germany, showed some improvement in international travel demand in June, major outbound tourism markets such as the United States and China remain in a standstill position (UNWTO, 2020a).

Peter et al. (2020) states that VFR (Visit Friends and Relatives) tourism will come to an important position when travel begins due to the isolation caused by travel and contact restrictions. Also, these travels will be less critical than others. Similarly, Senbeto and Hon (2018) supports this view by stating that VFR tourism is more resilient when a crisis occurs and that there were increases in both short-distance and long-distance VFR visits during the 2003 SARS epidemic, the 2008 global financial crisis, and the 2015 bird flu epidemic. Therefore, it is recommended to encourage VFR tourism. On the other hand, it seems that conference tourism will be organized online for a certain period. This type of tourism, where the technological infrastructure is used in the best and correct way, can continue to be performed online until the pandemic is fully solved (Menegaki, 2020).

Ayad et al. (2020) states that travel insurance will come to the fore in the new normal era and that it will not be taken without a mask on tours and museum entrances. Also, tourists will decide to travel when they are sure of the safety of transportation and the hotels they will stay in (Menegaki, 2020). It will become inevitable to choose boutique facilities where personalized services are offered, individual or small group holiday, and travel services instead of more crowded tourism facilities, transportation, restaurants, and entertainment venues, tour programs, and transfers (Demir et al., 2020). Similarly, Chebli and Said (2020) concluded in their study that 61% of the participants suspended group travel. Social distancing concerns are effective in tourists making this decision. Therefore, decision-makers in the tourism industry must assure tourists that these conditions are fulfilled. Besides, it should emphasize that these conditions are fulfilled with advertisements and promotions.

The recommendations of the study for the recovery of the tourism industry in the new era of the COVID-19 pandemic as follow.

- Domestic tourism and local products should be encouraged, costs should be reduced, flexible reservation conditions should be established, high value-added products should be highlighted in promotion, all kinds of expenditures that would not make a profit should be avoided, social distance should be provided, package service applications should be expanded, governments should provide financial support to all tourism facilities, intensify promotions to certain target groups (Roy et al., 2020).
- Tourism facilities should emphasize and encourage social distance, cleanliness, and sanitation (Assaf and Scuderi, 2020).
- All sector employees must be persuaded to meet high health and hygiene standards to revive the tourism industry. In this context, training and guidance should be given to all employees. Training should include pandemic communication guidelines, standard precautions, infection control, and hygiene techniques. Providing training will help reduce the likelihood of employees and guests getting the virus, while also increasing the hotel's reputation (Lai and Wong, 2020; Ulak, 2020).
- Room service, food service, and field visits should be minimized until a permanent solution is found on the COVID-19 pandemic (Ranasinghe et al., 2020).
- The period of uncertainty and risk, tourism facilities should turn to social media marketing. It should be emphasized that profit is not at the forefront and human health is more important than anything else in this social media marketing. In this way, the tourism facilities will provide a business image that gives importance to safety in the eyes of the guests (Menegaki, 2020). The word-of-mouth suggestions that tourists traveling during the pandemic will make in their immediate surroundings will also provide a strong image of the destination (Abbaspour et al., 2020).
- In the new normal era, digital health passports and certificates, digital identity applications, technologies that provide social distance, new

hygiene standards, contactless trips will be at the forefront. Such practices should be widespread in the business processes of tourism facilities (Sigala, 2020).

- In restaurants, attention should be paid to practices such as hand hygiene, masks and gloves for employees, social distance, the limited number of customers, continuous disinfection, temperature measurement, and disposable material use (Gürsoy et al., 2020; Wen et al., 2020).
- Positively guided policy responses should be created to mitigate the potentially shocking effects of the pandemic on the well-being of vulnerable economic activities. Grants and wage subsidies should be provided to these sectors to minimize layoffs. Necessary supports for micro and small enterprises such as tax exemption, delay or waiver of taxes, low-interest loans, grants for small companies should be expanded (Ranasinghe et al., 2020).
- Governments should take care of tourism facilities and continue with a slow economy, rather than go bankrupt and shut down. As long as facilities continue their lives, albeit lame, unemployment will be low. Besides, closing enterprises or firing employees can have devastating consequences for the entire country. For this reason, governments should put forward policies for the continuation of activities even if it is a slow economy (Yeh, 2020).

Given all this, we will not be able to travel as we have done before. The tourism industry will also not be as it was before the pandemic (Renaud, 2020). Although the pandemic had less impact in the summer of 2020 in the northern hemisphere, it is observed that the number of cases in many countries increased by October. Many European countries (France, Greece, Germany, United Kingdom) reverted to restrictions in November 2020. Therefore, it is still uncertain when the COVID-19 pandemic will end and what process it will follow. The expanded scenarios of UNWTO (2020a) for 2021–2024 predict a strong recovery in 2021 based on the assumption of a reversal in the evolution of the pandemic, a significant improvement in passenger confidence, and a major lift of travel restrictions by mid-year. However, in terms of international arrivals, returning to 2019 levels will take between 2½ and 4 years (UNWTO, 2020a). For this reason, it is essential to emphasize sustainability in crisis management policies of tourism facilities and to produce proactive solutions.

KEYWORDS

- **COVID-19 pandemic**
- **new normal era**
- **tourism industry**
- **strategies**
- **recovery**

REFERENCES

Abbaspour, F.; Soltani, S.; Tham, A. Medical Tourism for COVID-19 Post-Crisis Recovery? *Anatolia* **2020,** 1–4. DOI: 10.1080/13032917.2020.1815067.

Assaf, A.; Scuderi, R. COVID-19 and the Recovery of the Tourism Industry. *Tour. Econ.* **2020,** *26* (5), 731–733. DOI: 10.1177/1354816620933712.

Ayad, T.; Moustafa, M. A.; Khan, M. The Future of Tourism Post Covid-19 Pandemic: From the Egyptian Tour Operators' Perspectives. *J. Assoc. Arab Univ. Tour. Hospital.* **2020,** *18* (3), 50–64.

Aydın, B.; Doğan, M. Yeni Koronavirüs (COVID-19) pandemisinin turistik tüketici davranışları ve Türkiye turizmi üzerindeki etkilerinin değerlendirilmesi (Evaluation of Effects of the COVID-19 Pandemic on Touristic Consumption Behavior and Tourism in Turkey). *Pazarlama Teorisi ve Uygulamaları Dergisi* **2020,** *6* (1), 93–115.

Bae, S. Y.; Chang, P. J. The Effect of Coronavirus Disease-19 (COVID-19) Risk Perception on Behavioural Intention Towards 'Untact' Tourism in South Korea During the First Wave of the Pandemic (March 2020). *Curr. Iss. Tour.* **2020,** 1–19. DOI: 10.1080/13683500.2020.1798895.

Baraero-Era, J. S.; Del Rosario, J. M. Examining Tourism Resilience Practices as Basis for a Post-Covid 19 Recovery in the Philippines. *ASEAN J. Hospital. Tour.* **2002,** *18* (1), 1–12. DOI: 10.5614/ajht.2020.18.1.01.

Baum, T.; Hai, N. T. T. Hospitality, Tourism, Human Rights and the Impact of COVID-19. *Int. J. Contemp. Hospital. Manage.* **2020,** 1–11. DOI: 10.1108/IJCHM-03–2020–0242.

Bhattacharya, S. Challenges and Opportunities of the Tourism Industry Amidst COVID-19 Pandemic. In *Hospitality and Tourism Industry Post Pandemic Challenges and Opportunities*; Khanna, P. P., Ganguly, S., Mukherjee, M., Mukherjee, S., Eds.; Bharti Publications: New Delhi, 2020; pp 26–36.

Brouder, P. Reset Redux: Possible Evolutionary Pathways Towards the Transformation of Tourism in a COVID-19 World. *Tour. Geog.* **2020,** 1–7. DOI: 10.1080/14616688.2020.1760928.

Carr, A. COVID-19, Indigenous Peoples and Tourism: A View from New Zealand. *Tour. Geog.* **2020,** 1–12. DOI: 10.1080/14616688.2020.1768433.

Chebli, A.; Said, F. B. Thei of COVID-19 on Tourist Consumption Behaviour: A Perspective Article. *J. Tour. Manage. Res.* **2020,** *7* (2), 196–207. DOI: 0.18488/journal.31.2020.72.196.207.

Demir, M.; Günaydın, Y.; Şen-Demir, Ş. Ş. Koronavirüs (Covid-19) salgınının Türkiye'de turizm üzerindeki öncülleri, etkileri ve sonuçlarının değerlendirilmesi (The Evaluation of the Antecedents, Effects, and Consequences of Coronavirus (Covid-19) Pandemic on Tourism in Turkey). *Int. J. Soc. Sci. Educ. Res.* **2015,** *6* (1), 80–107. DOI: 10.24289/ijsser.734263.

Gursoy, D.; Chi, C. & Hengxuan-Chi. *Restaurant and Hotel Customers' Sentiment Analysis: Would They Come Back? If They Would, WHEN? COVID-19 Report 2: Restaurant and Hotel Industry*; Pullman, 2020.

Hao, F.; Xiao, Q.; Chon, K. COVID-19 and China's Hotel Industry: Impacts, a Disaster Management Framework, and Post-Pandemic Agenda. *Int. J. Hospital. Manage.* **2020,** *90*, 1–11. DOI: 10.1016/j.ijhm.2020.102636.

IATA, International Air Transport Association. *COVID-19 June Data and Revised Air Travel Outlook*, 2020. https://www.iata.org/en/iata-repository/publications/economic-reports/june-data-and-revised-air-travel-outlook/ (Access Date: 03.11.2020).

ICAO, International Civil Aviation Organisation, *Effects of Novel Coronavirus (COVID-19) on Civil Aviation: Economic Impact Analysis*, International Civil Aviation Organization, Air Transport Bureau, Montréal, Canada, 2020. https://www.icao.int/sustainability/Documents/COVID-19/ICAO_Coronavirus_Econ_Impact.pdf. (accessed on Nov 3, 2020).

Jones, P.; Comfort, D. The COVID-19 Crisis, Tourism and Sustainable Development. *Athens J. Tour.* **2020,** *7* (2), 75–86. DOI: 10.30958/ajt.7–2-1.

Koh, E. The End of Over-Tourism? Opportunities in a Post-Covid-19 World. *Int. J. Tour. Cities* **2020,** 1–9. DOI: 10.1108/IJTC-04–2020–0080.

Lai, I. K. W.; Wong, J. W. C. Comparing Crisis Management Practices in the Hotel Industry Between Initial and Pandemic Stages of COVID-19. *Int. J. Contemp. Hospital. Manage.* **2020,** 1–22. DOI: 10.1108/IJCHM-04–2020–0325.

Lapointe, D. Reconnecting Tourism After COVID-19: The Paradox of Alterity in Tourism Areas. *Tour. Geog.* **2020,** 1–6. DOI: 10.1080/14616688.2020.1762115.

Ma, S.; Zhao, X.; Gong, Y.; Wengel, Y. Proposing "Healing Tourism" as a Post-COVID-19 Tourism Product. *Anatolia* **2020,** 1–4. DOI: 10.1080/13032917.2020.1808490.

Menegaki, A. N. Hedging feasibility perspectives against the COVID-19 for the international tourism sector. *Preprints* **2020,** 1–18. DOI: 10.20944/preprints202004.0536.v1.

Peter, K.; Peters, J.; Peters, N. *Visit People Tourism Recovery after disaster*. KPPM Strategy, 2020.

Prayag, G. Time for Reset? COVID-19 and Tourism Resilience. *Tour. Rev. Int.* **2020,** *24* (2–3), 179–184. DOI: 10.3727/154427220X15926147793595.

Prideaux, B.; Thompson, M.; Pabel, A. Lessons from COVID-19 Can Prepare Global Tourism for the Economic Transformation Needed to Combat Climate Change. *Tour. Geog.* **2020,** 1–12. DOI: 10.1080/14616688.2020.1762117.

Ranasinghe, R.; Karunarathna, C.; Pradeepamali, J. After Corona (Covid-19) Impacts on Global Poverty and Recovery of Tourism Based Service Economies: An Appraisal. *J. Manage. Tour. Res.* **2020,** *3* (1), 5–19. DOI: 10.2139/ssrn.3591259.

Renaud, L. Reconsidering Global Mobility–Distancing from Mass Cruise Tourism in the Aftermath of COVID-19. *Tour. Geog.* **2020,** 1–11. DOI: 10.1080/14616688.2020.1762116.

Roy, A.; Suresh, P.; Nirmal, S. S. *Impact of the Covd-19 Pandemic on Travel and Tourism Industry and Future Revival Strategy with Special Reference to Indian Tourism Sector*. In *Hospitality and Tourism Industry Post Pandemic Challenges and Opportunities*; Khanna, P. P., Ganguly, S., Mukherjee, M., Mukherjee, S., Eds.; Bharti Publications: New Delhi, 2020; pp 18–25.

Senbeto, D. L.; Hon, A. H. The Impacts of Social and Economic Crises on Tourist Behaviour and Expenditure: An Evolutionary Approach. *Curr. Iss. Tour.* **2020,** *23* (6), 740–755. DOI: 10.1080/13683500.2018.1546674.

Sigala, M. Tourism and COVID-19: Impacts and Implications for Advancing and Resetting Industry and Research. *J. Busi. Res.* **2020,** *117*, 312–321. DOI: 10.1016/j.jbusres.2020.06.015.

Thomas, F. *Traveling After COVID.* The Responsible Tourism Advisor. Solutions and Action to Build Responsible Enterprises, 2020.

Tyler, D.; Dangerfield, J. M. Ecosystem tourism: A resource based philosophy for ecotourism. *Journal of Sustainable Tourism* **1999,** *7* (2), 146–158, Doi: 10.1080/09669589908667332.

Ulak, N. COVID-19 Pandemic and Its Impact on Tourism Industry in Nepal. *J. Tour. Adventure* **2020,** *3* (1), 50–75.

UNWTO, United Nations World Tourism Organization. *World Tourism Barometer* August/September 2020a, *18* (5), 1–36.

UNWTO, United Nations World Tourism Organization. *Global Guidelines to Restart Tourism. Tourism and COVID-19 Priorities for Tourism Recovery Global Guidelines* 2020b. https://webunwto.s3.eu-west-1.amazonaws.com/s3fs-public/2020–05/UNWTO-Global-Guidelines-to-Restart-Tourism.pdf (accessed on Nov 10, 2020).

Wachyuni, S. S.; Kusumaningrum, D. A. The Effect of COVID-19 Pandemic: How Are the Future Tourist Behavior? *J. Educ. Soc. Behav. Sci.* **2020,** *33* (4), 67–76. DOI: 10.9734/JESBS/2020/v33i430219

Wen, J.; Kozak, M.; Yang, S.; Liu, F. COVID-19: Potential Effects on Chinese Citizens' Lifestyle and Travel. *Tour. Rev.* **2020,** 1–14. DOI: 10.1108/TR-03–2020–0110.

WHO, World Health Organization. WHO Coronavirus Disease (COVID-19) Dashboard, 2020. https://covid19.who.int/. (Access Date: 10.11.2020).

Wilson, M. E.; Chen, L. H. Re-Starting Travel in the Era of COVID-19: Preparing Anew. *J. Travel Med.* **2020,** *27* (5), 1–5. DOI: 0.1093/jtm/taaa108.

WTTC, World Tourism and Travel Council. Coronavirus Puts Up to 50 Million Travel and Tourism Jobs at Risk, 2020. https://skift.com/2020/03/13/coronavirus-puts-50-million-tourism-jobs-at-risk-says-wttc-report/ (accessed on Nov 10, 2020).

Yeh, S. S. Tourism Recovery Strategy Against COVID-19 Pandemic. *Tour. Recreat. Res.* **2020,** 1–7. DOI: 10.1080/02508281.2020.1805933.

CHAPTER 8

THE FUTURE OF TOURISM IN A WORLD WITHOUT TOURISTS? RESILIENCY, ADAPTATION, AND REBIRTH

MAXIMILIANO E. KORSTANJE

Department of Economics, University of Palermo, Buenos Aires, Argentina

ABSTRACT

After the pandemic originated by the outbreak of a new virus in Wuhan, China, the future of tourism remains uncertain and grim worldwide. Basically, this happens because the steps and standard operating procedures adopted by different governments led to a strict lockdown which implied the closure of airspaces, borders, and free transit prohibition measures. What seems to be more important, the global economy of the world, at least as we know it has been fully paralyzed. In the mid of this mayhem, scholars adopted two positions. Some voices claimed for the rebirth of a new tourism industry, a more ecofriendly version of tourism which helps to mitigate the negative impacts of climate change. Others, rather, emphasized on the urgency to re-elaborate new methods to study tourism in a world characterized by low mobilities. This present chapter deals with the challenges and future of tourism research as well as the opportunities of scholarship to be finely integrated to shape more resilient destinations.

Post-COVID Tourism and Hospitality Dynamics: Recovery, Revival, and Re-Start.
Umendra Narayan Shukla & Sharad Kumar Kulshreshtha (Eds.)

8.1 INTRODUCTION

Paragraphing Adrian Franklin (2007), we must accept that tourism studies are mainly based on what he dubbed as "tourist-centricity" which means an overvalorization for considering tourists as the only reliable source of applied research. It is unfortunate that tourist centricity not only led the epistemology in tourism fields into a gridlock but also affected negatively the credibility of tourism research before other disciplines. Over the recent years, some critical voices have alerted on the fact that the economic-based paradigm has monopolized a whole portion of knowledge production relegating other voices and knowledge toward a peripheral position (Hall, 2004; Coles et al., 2006; Pritchard and Morgan, 2007; Harris et al., 2007). More interested in business and profits—for protecting the organic image of tourist destinations—than in expanding the nature of tourism, these "pseudo-scientific works" have ushered the discipline into a state of fragmentation and indiscipline (Tribe, 1997, 2007, 2010). Having said this, the recent outbreak of COVID-19 (also known as COSARSv2) has invariably placed the current tourism research between the wall and the deep blue sea. Basically, this new virus not only paralyzed the world and global commerce, even the tourism industry, but also posited a great epistemological dilemma for tourism research: how might professional researchers study tourism in a world without tourists?

Starting from the premise that tourism research is now centered on the belief, which punctuates tourists are reliable grounds to validate their operational hypotheses, the cancellation of flights, the closure of borders, and even the airspace make this task almost impossible to perform. What is more important, COVID-19 is wreaking havoc in the tourism and hospitality industries as well as other service sectors compromising seriously their evolution and future in the short run. Are we witnessing the end of tourism research—at least as we know it?

This chapter is also divided into two parts. The first one signals to a discussion on the economic as well as sociocultural effects of COVID-19 in the Western imaginaries and the tourism industry. To the economic crisis the virus prompts, experts agree there are some unexpected risks associated with the re-emergence of anti-eastern sentiments of xenophobia and racism. Chinese and other Asian tourists are portrayed as the carriers of a "covert" enemy which should be stopped. The second section, complementarily, explores the opportunities for applied research to survive in a world without tourists. The economic-based paradigm has enthusiastically enthralled the voice of tourists as the only valid source of information for applied research. In consequence, other valid knowledge, voices, and methodologies were pushed to the periphery. Nowadays, the economic-centered theory has many problems to overcome the obstacles of what Adrian Franklin named as "tourist-centricity." Besides, the section discusses some practical lesson to design more resilient destinations.

8.2 COVID-19 AND ITS IMPACT IN THE TOURISM INDUSTRY

The tourism industry has experienced countless global threats as well as economic downturn from its inception (Bianchi, 2006), but no one characterized a terminal crisis we are facing today like coronavirus disease (Williams and Balaz, 2015; Becken and Huguey, 2013). To some extent, risk perception theory debates the hypothesis that tourists avoid dangerous destination while selecting safer choices for their holidays (Rack et al., 2005; Kozak et al., 2007). The turn of the century, experts reach some consensus, has brought many unparalleled risks ranging from global terrorism operating in the European cities to natural disasters and virus outbreaks (McKercher and Chon, 2004; Laws et al., 2007; Saha and Gap, 2014). In the two last decades, the industry went through serious crises originated by the outbreaks of lethal viruses like SARS, Ebola, H1N1, and MERS, only to name a few. One might speculate that policymakers had the opportunity to develop more resilient destinations to placate the negative consequences of virus outbreaks, which paradoxically facilitated the technological breakthroughs posed to the stimulation of travels and connectivity, but COVID-19 proved precisely the opposite. Some studies have enthusiastically vaccinated that the inevitable economic crisis will allow capitalism to develop more sustainable forms of consumptions, as well as more ecofriendly destinations. For these scholars, the current crisis very well gives opportunities for a rebirth which denotes an economic slowdown that may reorganize human relations (Higgins-Desbiolles, 2020; Sigala, 2020; Gossling et al., 2020), while others imagine the chances for a restarting process in the tourism epistemology (Wen et al., 2020). Wen et al. (2020) coordinate efforts in developing new methodologies to understand the activity in a post-COVID-19 world. Per their viewpoint, interdisciplinary research plays a leading role in the configuration of reliable instruments helpful to overcome the obstacles of the current health crisis. Some approaches indicated the urgency for governments to bail out service industries in the pandemics to avoid a situation of extreme instability. In a postindustrial ethos based on services, mobilities, and mass consumption, tourism occupies a central position in the layout of the capitalist system (Prideaux et al., 2020; Yeh, 2020). In this vein, Ioannides and Gymothy (2020) argue convincingly that neoliberalism amply showed no few problems to deal with ecological issues and climate change. Other viruses have interrogated the material asymmetries of capitalist societies, above all the problems in the health system, but COVID seems to be a foundational event invigorating for the mainstream cultural values of neoliberal societies to be radically shifted. The strict lockdowns associated with the needs of adopting measures to keep the "social distance" lead the industry toward the brink of collapse. In this token, Gossling et al. (2020) proffer an excellent analysis of the pandemic in different countries and subsectors alerting on its devastating effects in densely overcrowded cities. As they hold: the COVID-19 crisis should thus be seen as an opportunity to critically reconsider tourism's growth trajectory, and to question the logic of more arrivals implying greater benefits. This may be with a review of the positive outcomes of the COVID-19 pandemic (Gossling et al., 2020: 13–14).

The crisis of tourism has certainly awakened some existential anxieties as never before. In view of this, some evolutionary protective mechanism as ethnocentrism and the hostility to over (if not) mass tourism surfaced. In other cases, it represents an opportunity for some ethnic minorities—like aborigines—which were historically relegated to live in a peripheral position, to gain further privileges in a post-pandemic world (Carr, 2020).

Per our viewpoint, far from being a foundational event, COVID-19 reaffirms a much deeper psychological climate of paranoia, and fear of strangers originally inaugurated by 9/11 and the War on Terror (Korstanje, 2017). Beyond the high productivity and publications revolving around COVID-19, as Francesc Romagosa puts it, the current crisis suggests two important aspects of the pandemic. On one hand, experts have no accuracy or valid information about the evolution of COVID-19 worldwide. On another, this wakes up long-dormant anxieties unless regulated may lead to interclass or ethnic conflicts. In this respect, Mostafanezhad et al. (2020) caution on the anxieties generated by the COVID-19 over recent years. The hostility against Asian tourists in Europe generally and in Italy, in particular, has taken a commonplace in the social imaginary. The circulation of sociopolitical imaginaries historically based on cultural and racial differences with Asian has brought to the foreground. Chinese and other Asian tourists are often associated as potential carriers of the virus, so to speak promoters of the disaster. Unless these stereotypes and prejudice are not controlled the dialogue between West and East grinds into a halt. To the economic effects of COVID-19, scholars should think in the sociocultural problems of racism and xenophobia against Chinese tourists, a covert but always present issue in the Western social imaginary. As Florian Kock et al. (2020) demonstrated, coronavirus disease is often associated with landscapes of high infectability. Their research shows how these types of (feared) spaces are interlinked to beliefs based on ethnocentrism, fear of strangers (xenophobia), and crowdedness. Last but not least, in earlier works, Maximiliano Korstanje (2020) coins the term "wicked-gaze" to denote the significant cultural changes the western imaginary experiences after the COVID-19. Per his thesis, the tourist gaze—which is a concept originally imagined by Urry (2002) to explain the fascination of persons to explore new sensations and landscapes in a hypermobile world—sets the pace to a new (wicked) gaze where the tourist is considered as a hidden enemy who looms to attack the society anytime. The question of whether the tourist gaze resulted from the rise of a globalized economy seems to be irrelevant after the COVID-19 outbreak. A broader process of feudalization started in 2001 is now prompted western economies to an unparalleled economic downturn. As a product of 9/11 and the War on Terror, two events that undermined the social trust, human relations are subdued into a solipsist climate of individualism where the "Non-Western Other" is neglected. Korstanje goes beyond, arguing that "foreign tourists" are now "undesired guests." The borders between selfhood and its alterity were radically changed, withered away in a feudal world. With this in mind, Korstanje adds, hospitality is dying. If the terrorist was a potential enemy living within the society, now any one of us was a potential terrorist, a potential carrier of an invisible enemy. Terrorists look and behave like us, and probably he would be a friend, a relative, or a neighbor. This happens simply because the doctrine of living with the enemy inside affirms the logic of 9/11 and counterterrorism protocols. This raises a more than a pungent question: Is tourism research possible in a post-COVID-19 world?

8.3 THE FUTURE OF TOURISM RESEARCH IN A WORLD WITHOUT TOURISTS

Doubtless, applied research and theory are inextricably intertwined. From its inception, tourism debated with other subdisciplines different conceptual paradigms as well as methodologies and epistemologies. Multidisciplinary research engages an active dialogue among disciplines to resolve the practical problems and dangers the industry faces day-to-day (Coles et al.,

2006). With the benefits of hindsight, the discipline developed a fascination to measure the economic impact of tourism in communities. Two books, which have widely influenced in the epistemology of tourism research, were *Tourism: passport to development?* (Emanuel de Kadt, 1979) and *Tourism, dependency and development* (authored by Stephen Britton). While De Kadt (1979) emphasized on the importance to measure the social maladies proper of the acculturation process, Britton focuses on the economic dependency between the developed and the underdeveloped economies. Tourism, far from revitalizing the asymmetries of local economies, affirms the financial dependency of the poor respecting rich countries. The same tendency coincides with Jafar Jafari's contributions. He kicks off a new tradition that alternates the urgency to protect the environment and the scientification of tourism (Jafari, 2005; Xiao, 2013). As he notably observes, the consolidation of the discipline depends on the possibilities to deal and efficiently resolve the problems of the industry creating a bridge between science and management (Jafari, 2005). In recent decades, tourism research has advanced with leaps and bounds with notable publications, which reflected in the number of journals, doctorate theses, books, and conferences. To some extent, some experts acknowledge that tourism research reached a stage of maturation (Sheldon, 1991; Xiao and Smith, 2006; Beckendorff and Zherer, 2013; Botterill and Platenkamp, 2012), but all in all the economic-based paradigm remains present. Centered on the logic of measurement—instead of understanding—the managerial view monopolized a set of quantitative-related methods in detriment of other methodologies as ethnography or content analysis. While the figure of the tourist is situated as the most important agent of the system, its fear, expectancies, and experiences became in the priority of applied field working. What tourists feel, believe, or think is the basis toward a scientific status, the economic-centered theory aspired (Riley and Love, 2000). As Gale Jennings (2007) eloquently remarks, a closer look on the history of tourism research reveals not only a postpositivist tradition oriented to understand the scientific evidence for what persons overtly say but in a managerial viewpoint which validates a much deeper western-led epistemology. As a result of this, other voices and knowledge which would play a leading role in the configuration of the discipline were systematically overlooked or silenced. Critical voices on the economic-based theory are rarely found in the Anglo-phone world. Rather, they come from other cultures and languages. To wit, Korstanje has exerted a caustic critique on economic-based paradigm because of the following two reasons. At the first glimpse, there is a cognitive dissociation between what interviewees say and finally do. Under some conditions, lay people lie to protect their interests or are unfamiliar with their inner world. What would respond a gangster if I ask for his profession? He would say I am businessman sir! In what the following question surfaces, is he really a businessman? The economic-centered theory appears to be hand-tied to resolve this point. Secondly and most importantly, the managerial viewpoint creates "pseudo-correlation" or what epistemologists known as "ecological fallacies" which means a formal correlation between two variables that lack explanatory capacity. To set a clear but hypothetical example, a survey shows that women are more prone to risk perception than men. The final outcome reveals women are more sensitive to risks than men at the time of planning holidays. But is this real? A new insight—alternating with qualitative methods—explains that men are socialized to repress their feelings, transforming fears into a rage, while women are educated to express more genuinely their fears. Scientifically speaking, risk perception does not correlate directly with the genre but marginally. In a nutshell, the economic-based paradigm and the metrics culture only access to the peak of the iceberg, a singled version of reality (Korstanje, 2018). As discussed, if we live in a world without tourism, as one might infer, the tourism research traverses a state of crisis.

In the next lines, we shall discuss new theoretical and methodological approaches which are fertile ground toward the adoption of new methods in the fields of tourism in a post-COVID-19 world. In a seminal book entitled *Tourism Research Frontiers: Beyond the Boundaries of Knowledge*, Chambers and Rakic (2015) advocate for the creation of new episteme and methods instead of undermining the tents of the discipline. Per their impression, at the time some issues are illustrated through the step of scientific research, a dark-side remains unchecked. The term *research frontier* speaks us of those liminoid borders of disciplines which gained further legitimacy than others. The history and the sociocultural background not only alter these frontiers but the dominant discourses forged by the academic elite. It is safe to say that the crisis opens new opportunities for epistemology to articulate new methodologies while adopting new fresh approaches.

8.4 THE FUTURE OF TOURISM RESEARCH

The crisis generated by the pandemic evinced some major limitations of applied research and the economic-centered theory to make more resilient destinations. In fact, tourism research was going a crisis of sense years earlier than COVID-19 outbreak in China. The following lines of investigation are fertile grounds toward the discipline maturation. This section invariably seeks to respond to what is the future of tourism beyond the problems revolving around tourist centricity.

8.5 THE USE OF DIGITAL TECHNOLOGIES AND VIRTUAL TOURISM

Studies framing the promising function of digital technologies in shaping new virtual landscape interrogate readers on the nature of tourism. In former decades, the notion of physical displacement was of importance to determine the nature of tourism. Nowadays, digital technologies have packed and passed out this notion to the dust of oblivion. Gretzel et al. (2020) allude to the application of electronic datasets and digital technologies to transform the epistemologies of tourism research. Beyond the devastating effects of COVID-19 in the economy, digital technology serves as a catalyst to resolve the emerging pressing problems that affect tourism today. E-tourism, doubtless, has evolved toward more maturated forms paving the ways for the rise of a homogenous body of knowledge. Per their stance, e-tourism seems to be centered on six founding pillars: the continuity of history to understand the past, familiarity with the factors which coadjuvate for knowledge production, transparency in the obtained outcomes, plurality to the diversity of topics and issues to be investigated, creativity to lays the foundations of new approaches, and sensitivity to differential opportunities to take the lead in e-tourism research. In this way, Seytoglu and Ivanov (2020) call the attention to employ robots and AI to assist tourists in a new world where social distance is the rule, not the exception. Last but not least, virtual or e-tourism is not a direct result of the COVID-19. Interesting publications on virtuality and dark tourism can be found in Kaelber (2007), McDaniel (2018), Korstanje and George (2017), or Handayani and Korstanje (2020). But no less true is that COVID-19 and the imposed lockdowns inaugurated a new climate of relations where social distances not only separated peoples but created more impersonal liaisons. One of the most terrifying aspects of the fear COVID-19 awakened up can be explained because we are struggling a war against an invisible enemy. What remains covert generates higher levels of anxiety for an ocular-centric society. The hidden and unsaid represent sources of greater fear and panic. The gap created by social distancing is filled with digital technologies.

Expatriates and tourist stranded abroad: As everybody knows, COVID-19 outbreak suddenly surprised to Europe and the USA disseminating fast through the globe. Thousands of tourists stranded in the most important airports, starting a crisis in geopolitical relations. It is unfortunate that studies focusing on expatriates and stranded tourists do not abound in tourism research. However, this field is a fertile ground for the next approaches in the years to come. COVID-19 showed two important issues. On one hand, travels which were taken for granted were canceled, and borders which marked a dialogue between neighbor countries were closed, in the same ways that the foreigner becomes an object of mistrust and fear. On another, the life stories of expatriates give fresh insight to the applied research. Expatriates are considered compatriots who living abroad were forced to return to home. Another interesting point to discuss is the forced return of stranded tourists. Expatriates' encounters generate interesting rivalries and conflicts which need further attention.

8.6 XENOPHOBIC EXPRESSIONS AND RACISM

The question of racism, chauvinism, or xenophobia is not new, and it is not necessarily related to the pandemic. Some interesting works recently emphasized the effects of terrorism in the articulation of countless travel-ban campaign or fear against the foreigner. Some manifestations as Windrush scandal, the construction of a wall to regulate the USA–Mexico border, as well as the tourist-phobia manifestations are a clear sign that marks hospitality is dying (Korstanje, 2017). Some recently published works alert on the anti-Chinese discourse in western societies. Chinese tourists are often viewed as potential carriers of the COVID-19 and real threats to the public health system (Zheng et al., 2020; Roberto et al., 2020). Tourism, travels, and mobilities are considered potential venues to the dissemination of the pandemic. The restrictions around travels not only activated some long-dormant prejudices but also affirmed an anti-Eastern sentiment finely enrooted in the Western lifestyle.

8.7 CONCLUSIONS

Nobody can travel to a place where he is not welcomed. For that, hospitality is a key factor for the host–guest encounter. Tourism, which is one of the most growing industries of the world, needs from hospitality to survive. To some extent, the COVID-19 outbreaks not only evinced the asymmetries of health systems in the West but also the philosophical dilemma revolving around the figure of the non-Western other. To put the same in bluntly, to the economic problems generated by the pandemic, specialists should add new sociological issues related to the hostility for the "other." In this present chapter, we have enumerated a line of new topics which help in expanding the constellations of applied research in tourism fields. Or what is more important, how can researchers investigate tourism in a world without tourists. This present chapter analyzed the advantages and disadvantages of the economic-based theory and what Adrian Franklin dubbed as "tourist centricity!" Though in its infancy, applied research in COVID-19 is divided into two clear waves. While some scholars claim on the urgency to adopt more resilient and sustainable forms of consumption, which mitigate the impact of climate change, others suggest the pandemic exhibits a great opportunity to overcome the discipline from the epistemological crisis it lies. This chapter provides with a fresh list of potential themes to research which ranges from xenophobia and racism to the use of digital technologies to mediate between hosts and guest, without mentioning the importance of understanding the phenomenological world of expatriates or stranded tourists who were forced to return to their homes. All these are fascinating themes to understand the essence of tourism. As a social construe, tourism should not be taken for granted as a universal right but as a condition imposed by the nation-state. The pandemic resulted invariably into new normality where all given rights were suspended and unilaterally canceled.

KEYWORDS

- **resiliency**
- **tourism**
- **COVID-19**
- **emergency**
- **crisis**

REFERENCES

Becken, S.; Hughey, K. F. Linking Tourism into Emergency Management Structures to Enhance Disaster Risk Reduction. *Tourism Management* **2013,** *36*, 77–85.

Benckendorff, P.; Zehrer, A. A Network Analysis of Tourism Research. *Ann. Tour. Res.* **2013,** *43*, 121–149.

Bianchi, R. Tourism and the Globalisation of Fear: Analysing the Politics of Risk and (in) Security in Global Travel. *Tour. Hospital. Res.* **2006,** *7* (1), 64–74.

Botterill, D.; Platenkamp, V. *Key Concepts in Tourism Research*; Sage: London, 2012.

Britton, S. G. *Tourism, Dependency and Development: A Mode of Analysis* (No. 23), 1981.

Carr, A. COVID-19, Indigenous Peoples and Tourism: A View from New Zealand. *Tour. Geog.* **2020,** 1–12.

Chambers, D.; Rakic, T. (Eds.). *Tourism Research Frontiers: Beyond the Boundaries of Knowledge*; Emerald Group Publishing: Wagon Lane, 2015.

Coles, T.; Hall, C. M.; Duval, D. T. Tourism and Post-Disciplinary Enquiry. *Curr. Iss. Tour.* **2006,** *9* (4–5), 293–319.

De Kadt, E. Tourism: Passport to Development. *Perspectives on the Social and Cultural Effects of Tourism in Developing Countries*; World Bank Press: New York, 1979.

Franklin, A. The problem with tourism theory. In *The Critical Turn in Tourism Studies*; Altejevic, I., Pritchard, A., Morgan, N., Eds.; Routledge: Abingdon, 2007; pp. 153–170.

Gössling, S.; Scott, D.; Hall, C. M. Pandemics, Tourism and Global Change: A Rapid Assessment of COVID-19. *J. Sustain. Tour.* **2020**. DOI: https://doi.org/10.1080/09669582.2020.1758708

Gretzel, U.; Fuchs, M.; Baggio, R.; Hoepken, W.; Law, R.; Neidhardt, J. et al. e-Tourism Beyond COVID-19: A Call for Transformative Research. *Inf. Technol. Tour.* **2020,** 1.

Hall, M. Reflexivity and Tourism Research. In *Qualitative Research in Tourism: Ontologies, Epistemologies and Methodologies*; Philimore, J., Goodson, L., Eds.; Routledge: Abingdon, 2004; pp 137–155.

Handayani, B.; Korstanje, M. E. Virtual Dark Tourism: The Role of Sound Branding and Augmented Reality for Death Sites. In *Destination Management and Marketing: Breakthroughs in Research and Practice*. IGI Global, Hershey, 2020; pp 183–203.

Harris, C.; Wilson, E.; Altejevic, I. Structural Entanglements and the Strategy of Audiencing as a Reflexive Technique. In *The Critical Turn in Tourism Studies*; Altejevic, I., Pritchard, A., Morgan, N., Eds.; Routledge: Abingdon, 2007; pp 41–56.

Higgins-Desbiolles, F. Socialising Tourism for Social and Ecological Justice After COVID-19. *Tour. Geog.* **2020,** 1–14.

Ioannides, D.; Gyimóthy, S. The COVID-19 Crisis as an Opportunity for Escaping the Unsustainable Global Tourism Path. *Tour. Geog.* **2020.** DOI: 10.1080/14616688.2020.1763445

Jafari, J. Bridging Out, Nesting Affield: Powering a New Platform [Opinion Piece]. *J. Tour. Studies* **2005,** *16* (2), 1–15.

Jennings, G. R. Advances in Tourism Research: Theoretical Paradigms and Accountability. In *Advances in Modern Tourism Research*; Physica-Verlag HD: Berlin, 2007; pp 9–35.

Kaelber, L. A Memorial as Virtual Traumascape: Darkest Tourism in 3D and cyber-Space to the Gas Chambers of Auschwitz. *Ertr, eRev. Tour. Res.* **2007,** *5* (2), 24–33.

Kock, F.; Nørfelt, A.; Josiassen, A.; Assaf, A. G.; Tsionas, M. Understanding the COVID-19 Tourist Psyche: The Evolutionary Tourism Paradigm. *Ann. Tour. Res.* **2020,** 103053.

Korstanje, M. E. *Critical Essays in Tourism Research*; Nova Science, Publishing: Hauppauge, 2018.

Korstanje, M. E. Passage from the Tourist Gaze to the Wicked Gaze: A Case Study on COVID19 with Special Reference to Argentina. In *International Case Studies in the Management of Disasters: Natural–Manmade Calamities and Pandemics*; George, B., Mahar, U. D. Q., Eds.; Emerald Group Publishing: Wagon Lane, 2020; pp 9–35.

Korstanje, M. E. *Terrorism, Tourism and the End of Hospitality in the West*; Palgrave Macmillan: Basingstoke, 2017.

Korstanje, M.; George, B., Eds. *Virtual Traumascapes and Exploring the Roots of Dark Tourism*; IGI Global: Hershey, 2017.

Kozak, M.; Crotts, J. C.; Law, R. The Impact of the Perception of Risk on International Travellers. *Int. J. Tour. Res.* **2007,** *9* (4), 233–242.

Laws, E.; Prideaux, B.; Chon, K. S., Eds. *Crisis Management in Tourism*; CABI: Wallingford, 2007.

McDaniel, K. N., Ed. *Virtual Dark Tourism: Ghost Roads*; Springer: New York, 2018.

McKercher, B.; Chon, K. The Over-Reaction to SARS and the Collapse of Asian Tourism. *Ann. Tour. Res.* **2004,** *31* (3), 716.

Mostafanezhad, M.; Cheer, J. M.; Sin, H. L. Geopolitical Anxieties of Tourism: (Im) Mobilities of the COVID-19 Pandemic. *Dialogues in Human Geography* **2020,** *10* (2), 182–186.

Prideaux, B.; Thompson, M.; Pabel, A. Lessons from COVID-19 Can Prepare Global Tourism for the Economic Transformation Needed to Combat Climate Change. *Tour. Geog.* **2020,** 1–12.

Pritchard, A.; Morgan, N. De-Centering Tourism's Intellectual Universe, or Traversing the Dialogue Between Change and Tradition. In *The Critical Turn in Tourism Studies: Innovative Research Methodologies*; Elsevier: Oxford, 2007; pp 11–28.

Rack, J.; Wichmann, O.; Kamara, B.; Günther, M.; Cramer, J.; Schönfeld, C. et al. Risk and Spectrum of Diseases in Travelers to Popular Tourist Destinations. *J. Travel Med.* **2005,** *12* (5), 248–253.

Riley, R. W.; Love, L. L. The State of Qualitative Tourism Research. *Ann. Tour. Res.* **2000,** *27* (1), 164–187.

Roberto, K. J.; Johnson, A. F.; Rauhaus, B. M. Stigmatization and Prejudice During the COVID-19 Pandemic. *Admin. Theor. Praxis* **2020,** *42* (3), 364–378.

Romagosa, F. The COVID-19 Crisis: Opportunities for Sustainable and Proximity Tourism. *Tour. Geog.* **2020,** 1–5.

Saha, S.; Yap, G. The Moderation Effects of Political Instability and Terrorism on Tourism Development: A Cross-Country Panel Analysis. *J. Travel Res.* **2014,** *53* (4), 509–521.

Seyitoğlu, F.; Ivanov, S. Service Robots as a Tool for Physical Distancing in Tourism. *Curr. Iss. Tour.* **2020,** 1–4.

Sheldon, P. J. An Authorship Analysis of Tourism Research. *Ann. Tour. Res.* **1991,** *18* (3), 473–484.

Sigala, M. Tourism and COVID-19: Impacts and Implications for Advancing and Resetting Industry and Research. *J. Busi. Res.* **2020,** *117*, 312–321

Tribe, J. The Indiscipline of Tourism. *Ann. Tour. Res.* **1997,** *24* (3), 638–657.

Tribe, J. Tribes, Territories and Networks in the Tourism Academy. *Ann. Tour. Res.* **2010,** *37* (1), 7–33.

Tribe, J. Critical Tourism: Rules and Resistance. In *The Critical Turn in Tourism Studies*; Altejevic, I., Pritchard, A., Morgan, N., Eds.; Abingdon: Routledge, 2007; pp 29–39.

Urry, J. *The Tourist Gaze*; Sage: London, 2002.

Wen, J.; Wang, W.; Kozak, M.; Lui, X.; Hou, H. Many Brains Are Better Than One: The Importance of Interdisciplinary Studies on COVID-19 in and Beyond Tourism. *Tour. Recreat. Res.* **2020**. DOI: https://doi.org/10.1080/02508281.2020.1761120

Williams, A. M.; Baláž, V. Tourism Risk and Uncertainty: Theoretical Reflections. *J. Travel Res.* **2015,** *54* (3), 27–35.

Xiao, H. Jafar Jafari: The Platform Builder. *Anatolia* **2013,** *24* (2), 288–296.

Xiao, H.; Smith, S. L. The Maturation of Tourism Research: Evidence from a Content Analysis. *Tour. Analy.* **2006,** *10* (4), 335–348.

Yeh, S. S. Tourism Recovery Strategy Against COVID-19 Pandemic. *Tour. Recreat. Res.* **2020,** 1–7.

Zheng, Y.; Goh, E.; Wen, J. The Effects of Misleading Media Reports About COVID-19 on Chinese Tourists' Mental Health: A Perspective Article. *Anatolia* **2020,** *31* (2), 337–340.

CHAPTER 9

HOW TO COPE WITH THE DEVASTATING IMPACTS OF COVID-19 IN TOURISM DESTINATION GOVERNANCE

NAGIHAN CAKMAKOGLU ARICI

Republic of Turkey, Turkish Consulate General in Stuttgart, Stuttgart, Germany

ABSTRACT

Drawing on the approach of crisis management and on the case of Turkey tourism destination governance, this chapter extends understanding on the main impacts, coping strategies, and potential outcomes carried out by destination practitioners during and after the COVID-19 outbreak. Using hermeneutic approach, by reviewing the extended literature of crisis management, pandemic, tourism destination governance, and transformational leadership, this chapter provides recommendations about long-term strategies to tackle with the COVID-19 recently and similar pandemic crisis in the future. Moreover, in light of the previous coping strategies adopted toward past crisis including severe acute respiratory syndrome and Ebola, this chapter suggests strategies for effective destination and crisis management in the new normal. More specifically, it reveals existing measures and implementations done by the Turkey during the pandemic. To the best of the author's knowledge, this chapter is among the first studies focusing on the destination governance during and after COVID-19 outbreak. Future studies can expand our knowledge by considering other tourism destinations and their coping strategies against the COVID-19 outbreak.

Post-COVID Tourism and Hospitality Dynamics: Recovery, Revival, and Re-Start.
Umendra Narayan Shukla & Sharad Kumar Kulshreshtha (Eds.)

9.1 INTRODUCTION

The coronavirus (COVID-19) pandemic, which emerged first in Wuhan City of China and was defined as a pandemic by the World Health Organization (WHO) on March 11, 2020, is one of the most stunning incidents in the 21st century with devastating impacts on tourism. At the very beginning, due to lack of vaccine and rapid spread of the pandemic, WHO has strongly recommended utilizing nonpharmaceutical interventions including quarantine, social distancing, and self-isolation. Hence, many countries with high infection rate have implemented national and international restrictions in travel and tourism actions. However, it has soon been understood that it is not possible to sustain the lockdown making millions of people self-isolated at home, so the authorities and the policymakers have uttered that global tourism be normalized but in a new way. This new normal necessitates adopting the social distancing in the long run in our lives. In this regard, instead of turning back to our previous "normal" after the pandemic, the point is to see this event as an opportunity to transform the understanding of tourism and destination governance. Accordingly, many governments have stepped into a controlled normalization process in order to minimize the devastating impacts of the pandemic. Within this process, governments have restarted travel and tourism activities in a careful and safe way, entailing specific adaptations and reorganizations in the facilities and services.

As for the academic reflections, there can be seen many studies relating the epidemiological and clinical results of this pandemic (e.g., Cortegiani et al., 2020), with a few investigations about the impacts on tourism (e.g., Zenker and Kock, 2020), but there are few studies revealing the impact on tourism destination governance (e.g., Çakar and Uzut, 2020). With the intention of filling this gap, this chapter has aimed to explore how destination governments would transform their tourism policies into this new normal to minimize the devastating results of the coronavirus pandemic.

The COVID-19 pandemic forced governments and health authorities to take an array of precautions from nonpharmaceutical interventions including quarantine, social distancing, and self-isolation to vaccine trials due to its nature of spreading very fast through mobility of the people. The suspense of international travel, tourism, and leisure activities has led the highly vulnerable industry of travel and hospitality to struggle with numerous economic risks even more than previous crises and outbreaks like terrorism, earthquakes, Ebola, severe acute respiratory syndrome (SARS), and Zika. The effects of COVID-19 upon international tourism have been specifically so devastating that the social mobility of the people has been mostly stopped.

On the other hand, the main actors of tourism sector, such as airline companies, hotels, and restaurants, have been unprecedentedly influenced, with an estimation of up to 400 bn Euro (Ioannides and Gyimóthy, 2020; Nicolas, 2020). More specifically, this global crisis has huge impact on a worldwide depression with estimation of a sharp decrease in international tourist arrivals about 78%, which means losing US$ 1.2 trillion tourism revenues. Most of the businesses globally, the revenues of which are directly or indirectly bound to mobility of the traveler and tourists, have had to suspend their service, and for some of them, it is so difficult to operate again. Relatedly, the employees of these businesses have been affected harshly that a great deal has lost their jobs (UNWTO, 2020). It seems that tourism industry has been one of the most affected sectors and will also be one of the last to recover from the COVID-19-related crisis. Within this context, tourism industry as an important recruiter and GDP contributor is in the heart of discussion for numerous destination governors, policymakers, organization leaders, scholars, and researchers. Furthermore, the interventions of the governments to the tourism and related service operations and also financial supports like stimulus packages and tax reduction have been other characteristics of this crisis. Therefore, seeing the unique trait of this pandemic, there is an urgent need for strategic destination governance.

Based on the discussion, the study has adopted hermeneutic approach in understanding the transformation of tourism destination management in the case of Turkey, which is a favorable and highly preferred tourism destination. The current research on COVID-19 and its effects on several areas have been reviewed including the legal regulations to interpret the meaning and understanding of the phenomenon. By examining the effects of the pandemic on destination management procedures from the perspective of Turkey with a hermeneutic approach, this chapter discusses and demonstrates novel practical and legislative implications which might inspire the other destination marketers, contributing to our knowledge on virus outbreaks and tourism management.

9.2 DESTINATION GOVERNANCE

The phenomenon of governance refers to "the interactions among structures, processes and traditions that determine how power and responsibilities are exercised, how decisions are taken, and how citizens or other stakeholders have their say" (Graham et al., 2003: 2–3). On the other hand, destination governance can be defined as a system through which all the internal and external powers, processes, and mechanisms constructed for the sake of protecting all the stakeholders (Beritelle et al., 2007). Related organizations and individuals of a tourist destination are defined as stakeholders that participate in developing and implementing strategies through including local people in the process (Friedman and Miles, 2006). Key stakeholders of tourism industry can be governorships, municipalities, local administrations, NGOs, chambers, several industry organizations, customers, management and shareholders of organizations, experts, media, and environmental groups. These key actors are to be in cooperation and collaborate with each other in decision-making and implementation of the strategies so as to attain the desired goals for good tourism governance (Hall, 2011). Key features of effective tourism destination governance can be uttered as clearly set actor roles, desired atmosphere, efficient communication and strong network between the stakeholders, high engagement, trust, fairness and equity, promotion of learning and development, and effective leadership (Çakar and Uzut, 2018). The notion of governance is related with the phenomenon of power, through which each stakeholder according to their positions is authorized, which can lead to conflict among the stakeholders (Hall, 2007). In such a challenging crisis, conflicting stakeholders are the last thing we want, so efficient governance and leadership are to be in effect in order to overcome this multidimensional crisis.

9.3 CRISIS MANAGEMENT

The concept of "crisis" refers to an unforeseen unfavorable event which could be self-caused (Alan et al., 2006). The diseases at epidemic or pandemic level like the SARS are evaluated as a kind of tourism crises which are aggravated by naturally caused incidents (Liu et al., 2011). With their fragile resistance to these types of huge crises, tourism destinations urgently need strategic crisis management policies and contingency programs. Various crisis management patterns have been proposed by several scholars in tourism literature (e.g., Clarke and Varma, 2004). According to Murphy and Bayley's (1989) disaster planning model, the crisis management process includes four phases: assessment, warning, impact, and recovery. In the phase of assessment, possible risks are defined and assessed. The warning phase involves the warning of the local authorities regarding the threats and the menaces and advices on coping strategies with these dangers and their potential effects. The phase of impact includes the application of contingency programs for the crisis-caused dangers to survive. The final phase encompasses the restoration of economic, residential, and social stability, which can last years or decades with sometimes help of international associations (Foster, 1980). Similar to this one, there are other models classifying the stages of crisis management, namely, Moe and Pathranarakul's (2006) proactive and reactive approach; Kash and Darling's (1998) four stages, which are prodromal, acute, chronic, and resolution; and Glaesser's (2003) classification of potential, latent, and acute stages. However, it can be easier to identify this process in three main stages of precrisis, crisis, and postcrisis. In such natural or self-inflicted crises, it is quite hard to forecast it and the prestage is already gone when a disaster hits. But the point here is to conduct activities to prevent the disaster. But COVID-19 is so extensive and rapid spreading that it has been almost impossible to prevent it. The words are now going through the crisis stage in the very middle of the it, while struggling with new infected individuals, the casualties, demolishing health services and facilities, vaccine trials, and at the same time with the protests of nonbelievers of the coronavirus pandemic. It can be of value for the time being to talk about the recovery programs instead of prevention, which can be returned back after the short-term recovery. Thus, in the following section, you can find discussions regarding the management of the pandemics in the context of tourism, a kind of leadership called transformative leadership, which can aid in these challenging times and for the future and previous experiences of pandemics in tourism context with a specific example of Turkey's implementations and lastly some implications for the future.

9.4 MANAGING THE PANDEMIC CRISIS

The globalization has resulted in a fast growth of tourism industry in an international arena with a large market share and effectiveness. This expansion, on the other hand, has brought about a global risk which industry could face more frequent than before, since globalization is usually considered complicated and turbulent (Jessop, 1999). Due to the fast development in the globalization of tourism industry, small-scale crisis in a destination of the world could influence other destinations of the world. The outbreak of a disease in a region of the world could importantly decrease people's travel intentions and tendencies to other regions of the world as experienced by the COVID-19 pandemic. Hence, tourism is substantially vulnerable to global determinants and events in the broader marketing environment. One of the most important abilities of tourism practitioners is, therefore, to manage such crises in an effective way (Ritchie, 2004).

Furthermore, tourism has significant economic contributions for many destinations that are relied upon tourism sector for their development and continuity. This leads to enhancing stress and tension for destination governors and practitioners, associated with tourism regard to the effects of global crises on the sector and foster actions and approaches to cope with the influences to preserve tourism organizations and destinations in general. There is a need to comprehend the global events and seek approaches which could be adopted to control and mitigate their influences on an increasing and substantial sector. Crisis management needs to become a fundamental ability for tourism destination managers and practitioners. Therefore, the first step is to comprehend the nature and meaning of a crisis in conceiving how to execute and mitigate the devastating effects of the pandemic crisis.

Although there has not been commonly accepted definition of crisis in the tourism literature due to several definitions of "tourism crisis" (Paraskevas and Altinay, 2013; Ritchie, 2004), Sönmez et al.'s (1994) definition of tourism crisis could be considered as the more extensive definition. They defined tourism crisis as

> any occurrence which can threaten the normal operation and conduct of tourism related businesses; damage a tourist destination's overall reputation for safety, attractiveness and comfort by negatively affecting visitors' perceptions of that destination; and, in turn, cause a downturn in the local travel and tourism economy and interrupt the continuity of business operations for the local travel and tourism industry by the reduction in tourist arrivals and expenditures. (Sönmez et al., 1994: 22)

In the crisis management in tourism literature, there are four broadly accepted study trends (Paraskevas and Altinay, 2013). The first research trend concentrates on the effects of crises on tourism industry. The second trend concentrates on the improvement of crisis management through reconstructing the image of tourism destination by adopting applicable crisis management and marketing activities. The third trend sheds light to the significance of precrisis step and claims that tourism destinations and businesses should consider the antecedents and outcomes of past crises so as to develop an applicable strategy for the future ones. The last trend of crisis management literature integrates the expand knowledge of crisis management and suggests concise strategies, approaches, schemes, and structures for a "holistic" crisis management in the industry. Drawing on insights of such accepted trends in crisis management literature, this chapter first focuses on clarifying avenues by which tourism territories could rebuild visitor safety and trust (Armstrong and Ritchie, 2008) and by conducting specific recovery strategies in the destination governance structure. In this sense, this chapter clarifies the significance of the proactive crisis management models and adopting new-generation technologies in destination governance for the recuperation of customers from the devastating influences of the crisis at the postcrisis period.

During crisis times, the reaction of tourism destinations could vary from chaos to a controlled, oriented, and well-managed crisis solution, relying on the features of management (Bowers et al., 2017). The reaction of the managements to the crisis will designate the way and method of recovery as well as future of the business. During such difficult times as the pandemic circumstances, people trust managements to execute the event successfully and move individuals out of harm's way (Demiroz and Kapucu, 2012).

Scholars suggested that crisis and leadership are interlaced as both approaches possess a nature to complete one another (Arici et al., 2020; Nomm et al., 2020). It is the leader's obligation to react to the global events and chaos resulting from a crisis like COVID-19. Transforming everything to the normal is the main challenge of leaders. The current pandemic has the potential to become an effective instrument for encouraging the transformation essential in tourism industry. It could aid present a services marketing structure which addresses change and innovation. This will necessitate reconsidering roles of current actors (e.g., stakeholders) and potential new actors (e.g., new-generation technology) concerning coordinated responses. In sum, despite the devastating impacts of the crisis, it is vital to consider the fact that crises create possibilities to transform organizational structures and long-lasting procedures. The crises have endorsed that the current structure does not function during crisis period, so fundamental changes should be accomplished in the sector and company. Such changes should lead to institutional transformation. Particularly, transformational leaders are expected to take action in these critical times by sharing authority and triggering collaboration.

Efficient leaders can weaken the effects of these concerns and can overcome the fundamental problems of development and coordination. Reconsidering service structure will necessitate an innovative approach which goes beyond current situation. Transforming is necessary. Therefore, transformation will necessitate transformational leadership if it is to develop a new service structure and technology-oriented avenues for the crisis response. The characteristics of transformational leadership, like considering the need for revolution, generating a novel vision, and institutionalizing change, are fundamental elements for developing planned organizational reactions to the pandemic. Such characteristics could be considered as steps which are essential for a successful transformation of service structure (Van Wart, 2005). Bass and Riggio (2006) transformational leaders are the ones who evaluate goals as more noteworthy than rules. Instead, they do not avoid from changing the rules when necessary. Without the transformational leadership to foster this transformation in tourism, the existing structure will continue, with small-scale revisions and precautions made in some countries to ensure sanitary and hygienic conditions in tourism companies. This might function in smaller scale crises or epidemics. However, when a global pandemic like the COVID-19 takes place again—many scholars have argued that we will experience such pandemics in the future—the existing structure will not work, and so we will seek solutions to cope with such difficulties as we are doing this exceptional period. It can be useful to have a look some examples from past endemic or pandemic diseased and some coping strategies of the governments and the health authorities with regard to the tourism aspect with this global crisis.

Previous experiences posed that pandemics could have unfavorable financial effects on tourism. For example, SARS starting in China in December 2002 impacted China. Tourism activities of China had to almost stop between March and July, which caused the revenue of domestic tourism to decline (Wen et al., 2005). During the epidemic of SARS, about 8096 people were infected and 774 of these, or 9.6%, died (WHO, 2004).

In order to combat with SARS, some methods were applied in the entries of travelers, such as personal declaration, quarantine of suspects, flight checks, and thermal imaging (Kuo et al., 2008). But these were all done to minimize the short-term effects of the crisis. Nonetheless, SARS has influenced daily life and travel behavior of people to a great extent in the long term. With high sensitivity of the travelers, the traveling activities almost stopped, which caused fluctuation in the industry. One point learnt from this crisis was to develop crisis projection systems to diminish the loss and the negative effects. Better informing of the public could reduce the fear for crisis. For example, as Grunig (1993) suggested not to conceal information in crisis times, Taiwanese government, one of the affected areas from SARS, utilized a two-way symmetrical public relations model during this crisis, by which all travelers were informed and the government worked with public for optimum outcomes, with a control of travelers on the decision of the government about the air travel. But the press-agentry model that China used made the public be unaware of the real numbers and the situation, because the government did not release information for the public (Liu et al., 2011).

Furthermore, the travel tendencies of the tourists also turned into more isolated like outdoor and eco-tourism. It was also learned that hygiene and national standards for tourist attractions, transportation, and accommodation facilities were inevitable parts of tourism. The Tourism Bureau of Taiwan put a tourism recuperation strategy into action to recover the country's tourism industry. This plan included particular measures for developing safe travel with safety inspection tours with the help of tour operators, promoting Taiwan's key touristic products, and offering tours with premium safety with the cooperation of tour operators (Tourism Bureau, 2003). However, as Chen et al. (2007) suggested developing prevention programs for recurrent similar crisis is the key in combating with the negative effects of pandemics. As for another disease, in the Ebola virus disease, which first emerged near the Ebola River in Congo in 1976 and in South Sudan in 1979, virus is transmitted by direct contact with the blood, body fluids, and tissues of infected animals or individuals (Feldmann, 2004; Ftika and Maltezou, 2013). The average fatality rate is 50% and until now around 11,000 people died from this virus (WHO, 2020). Within the combating strategies, some countries, which were not affected, prepared several travel advisories or suggestions for the six countries affected by Ebola (Hao-Jun et al., 2015). But it is true to say that Ebola cannot be compared with the magnitude of COVID-19.

As for Turkey, a favorable and highly preferred tourism destination, in order to better understand what Turkey has done to cope with COVID-19, the research has been conducted under three main titles, which are physical resources, human resources, and financial resources. First, physical resources are mostly related with hospitality organizations of a destination. As physical resources, service structure, hygiene, cleaning, and health precautions have been discussed. Namely, regarding service structure of the tourism organizations, necessary transformations because of the pandemic included the first social distancing, which turned out as a number one of people's lives after March 2020. Moreover, especially in hotels, restaurants, and cafes, department-specific implementations, such as one-way corridors, direction indicators, no open-buffet service, and instead personal plate service, have been practiced. Turkish hoteliers had to reserve some of the hotel rooms as COVID-19 rooms in case of emergencies and rooms could be checked in after 2-h ventilation. For premium hygiene, cleaning, and health precautions, organizations in Turkey like other countries had to follow hygiene rules of the Ministry of Health including ventilation, disinfection, wearing mask, social distancing, measuring body temperature, monitoring with thermal cameras, and contactless thermometer. Another good example applied in Turkey is the Safe Tourism Certification Program, which identifies and recommends a number of measures regarding the transportation and accommodation of Turkish and international visitors during their holiday in Turkey. Accredited facilities have to follow several international standards and criteria, which have been developed by the Ministry of Health, Ministry of Internal Affairs, Ministry of Foreign Affairs, and related stakeholders, during their operations of accommodation, food and beverage service, culture, art, and congress activities (TGA, 2020). Regarding the human resources of the tourism organizations, the practices, such as e-trainings about the pandemic and other operations for developing employee awareness, virtual platform using, teleconferencing for the managerial meetings, and working from home for the white collars have been promoted. In the aspect of financial resources, Turkish Government applied short-time work and salary compensation for the employees who were suspended during the pandemic as well as tax delay, insurance premium delay bank loans, and insolvency aids for the tourism organizations. However, all of these above are interim applications and precautions to tackle with the short-term results of the coronavirus pandemic. The key prospect should be for the long-term revival of the post-COVID tourism and hospitality sector.

Pandemic-induced crises as in the COVID-19 example trigger other socio-political and economic crises because of their global magnitude. Thus, their effects and results become huge in the same or greater extend. In such massive crises, the stakeholders become the countries not smaller entities of a sector. Nevertheless, this pandemic has showed everybody that the world is on the edge of a big change globally evolving into a more digital, socially distant, and in need of sustainability more than ever. Therefore, the strategies and policies developed by the destination governance organs are to absolutely include new generation technologies, such as artificial intelligence, smart devices, and service robots. The labor-intensive tourism industry can transform into techno-labor-intensive sector with necessary adaptations for a full-fledged digital transformation. A total digital tourism sector seems not possible, seeing that the costumers of this sector seek for authenticity and originality. Thus, the best practice can be an efficient integration of technology and human labor. As another recommendation for tourism organizations and policymakers, new market search with comprehensive research and development can have a positive effect on the revival of pandemic. For instance, domestic travelers can be profitable target audience for a fast start-up. More specifically, with the experience of this pandemic, tourists can have a tendency of not traveling as they used to. The point here is to reconstruct their perception safety in tourism act. The examples like Safe Tourism Certificate of Turkey can diminish the fear of travelers against tourism participation. For effective destination governance, a proactive crisis management approach and a techno-labor integrated model could function as an icebreaker between tourist and tourism organizations to adapt in this new normal.

KEYWORDS

- **COVID-19**
- **global crisis**
- **transformation**
- **tourism destination governance**
- **crisis management**

REFERENCES

Alan, C. B.; So, S.; Sin, L. Crisis Management and Recovery: How Restaurants in Hong Kong Responded to SARS. *Int. J. Hospital. Manage.* **2006,** *25* (1), 3–11.

Arici, H. E.; Arasli, H.; Cakmakoglu Arici, N. The Effect of Nepotism on Tolerance to Workplace Incivility: Mediating Role of Psychological Contract Violation and Moderating Role of Authentic Leadership. *Leadership Org. Dev. J.* **2020,** *41* (4), 597–613.

Armstrong, E. K.; Ritchie, B. W. The Heart Recovery Marketing Campaign: Destination Recovery After a Major Bushfire in Australia's National Capital. *J. Travel Tour. Market.* **2008,** *23* (2/4), 175–189.

Bass, B. M.; Riggio, R. E. *Transformational Leadership*, 2nd ed.; Lawrence Erlbaum Associates: London, 2006.

Bowers, M. R.; Hall, J. R.; Srinivasan, M. M. Organizational Culture and Leadership Style: The Missing Combination for Selecting the Right Leader for Effective Crisis Management. *Busi. Horiz.* **2017,** *60* (4), 551–563.

Çakar, K.; Uzut, S. Exploring the Stakeholder's Role in Sustainable Degrowth within the Context of Tourist Destination Governance: The Case of Istanbul, Turkey. *J. Travel Tour. Market.* **2020,** 1–16. https://doi.org/10.1080/10548408.2020.1782307

Chen, M.-H.; Jang, S. C.; Kim, W. G. The Impact of the SARS Outbreak on Taiwanese Hotel Stock Performance: An Event-Study Approach. *Int. J. Hospital. Manage.* **2007,** *26*, 200–212.

Clarke, C. J.; Varma, S. Strategic Risk Management: The New Competitive Edge. *Long Range Plan.* **2004,** *32* (4), 414–424.

Cortegiani, A.; Ingoglia, G.; Ippolito, M.; Giarratano, A.; Einav, S. A Systematic Review on the Efficacy and Safety of Chloroquine for the Treatment of COVID19. *J. Crit. Care* **2020,** *57* (1), 279–283. DOI: 10.1016/j.jcrc.2020.03.005

Demiroz, F.; Kapucu, N. The Role of Leadership in Managing Emergencies and Disasters. *Eur. J. Econ. Polit. Studies* **2012,** *5* (1), 91–101.

Dredge, D. Policy Networks and the Local Organization of Tourism. *Tour. Manage.* **2006,** *27* (2), 269–280. DOI: 10.1016/j.tourman.2004.10.003

Feldmann, H. Ebola—A Growing Threat? *New Engl. J. Med.* **2014,** *371* (15), 1375–1378. DOI: 10.1056/NEJMp1405314.

Foster, H. D. *Disaster Planning: The Preservation of Life and Property*; Springer Verlag: New York, 1980.

Freeman, R. E.; Harrison, J. S.; Wicks, A. C.; Parmar, B. L.; De Colle, S. *Stakeholder Theory: The State of Art*; Cambridge University Press: New York, 2010.

Friedman, A. L.; Miles, S. *Stakeholders: Theory and Practice*; Oxford University Press: Oxford, 2006.

Ftika, L.; Maltezou, H. C. Viral Hemorrhagic Fevers in Healthcare Settings. *J. Hospital Infect.* **2013,** *83* (3), 185–192. 10.1016/j.jhin.2012.10.013

Glaesser, D. *Crisis Management in the Tourism Industry*; Butterworth-Heinemann: Burlington, 2003.

Graham, J.; Amos, B.; Plumptre, T. W. *Governance Principles for Protected Areas in the 21st Century*; Institute on Governance, Governance Principles for Protected Areas: Ottawa, 2003.

Grunig, J. E. Public Relations and International Affairs: Effects, Ethics and Responsibility. *J. Int. Affairs* **1993,** *47* (1), 137–163.

Hall, C. M. Tourism, Governance and the (Mis-)Location of Power. In *Tourism, Power and Space*; Church, A., Coles, T., Eds.; Routledge, 2007; pp 247–268.

Hall, C. M. A Typology of Governance and Its Implications for Tourism Policy Analysis. *J. Sustain. Tour.* **2011,** *19* (4–5), 437–457. DOI: 10.1080/09669582.2011.570346

Hao-Jun, F.; Hong-Wei, G.; Ding, H.; Bi-Ke, Z.; Hou, S. K. The Ebola Threat: China's Response to the West African Epidemic and National Development of Prevention and Control Policies and Infrastructure. *Disaster Med. Public Health Preparedness* **2015,** *9* (1), 64–65.

Ioannides, D.; Gyimóthy, S. The COVID-19 Crisis as an Opportunity for Escaping the Unsustainable Global Tourism Path. *Tour. Geog.* **2020,** *22* (3), 624–632.

Jessop, B. Reflections on Globalisation and Its (il)Logic (s). In *Globalisation and the Asia-Pacific: Contested Territories*; Olds, K., Dicken, P., Kelly, P. F., Kong, L., Yeung, H. W., Eds.; Routledge: London, 1999; pp 19–38.

Kash, T. J.; Darling, J. R. Crisis Management: Prevention, Diagnosis and Intervention. *Leadership Org. Dev. J.* **1998,** *19* (4), 179–186. https://doi.org/10.1108/01437739810217151

Kuo, H. I.; Chen, C. C.; Tseng, W. C.; Ju, L. F.; Huang, B. W. Assessing Impacts of SARS and Avian Flu on International Tourism Demand to Asia. *Tour. Manage.* **2008,** *29* (5), 917–928.

Liu, J.; Moss, S. E.; Zhang, J. The Life Cycle of a Pandemic Crisis: SARS Impact on Air Travel. *J. Int. Busi. Res.* **2011,** *10* (2), 63–78.

Moe, T. L.; Pathranarakul, P. An Integrated Approach to Natural Disaster Management: Public Project Management and Its Critical Success. *Disaster Prevent. Manage. Int. J.* **2006,** *15* (3), 396–413. https://doi.org/10.1108/09653560610669882.

Monterrubio, J. C. Short-Term Economic Impacts of Influenza A (H1N1) and Government Reaction on the Mexican Tourism Industry: An Analysis of the Media. *Int. J. Tour. Policy* **2010,** *3* (1), 1–15.

Murphy, P. E.; Bayley, R. Tourism and Disaster Planning. *Geog. Rev.* **1989,** *79* (1), 36–46. DOI: 10.2307/215681

Nicolas, E. S. EU Pledges Help, as Tourism Faces e400bn hit. *EU Observer*, 2020. https://euobserver.com/coronavirus/148137 on 24/04/2020 (accessed on Oct 30, 2020).

Nomm, A. H.; Albrecht, J. N.; Lovelock, B. Advocacy and Community Leadership as Functions in National and Regional Level Destination Management. *Tour. Manage. Perspect.* **2020,** *35*, 100682.

Paraskevas, A.; Altinay, L. Signal Detection as the First Line of Defence in Tourism Crisis Management. *Tour. Manage.* **2013,** *34*, 158–171

Ritchie, B. W. Chaos, Crises and Disasters: A Strategic Approach to Crisis Management in the Tourism Industry. *Tour. Manage.* **2004,** *25* (6), 669–683.

Sönmez, S. F.; Backman, S. J.; Allen, L. *Managing Tourism Crises: A Guidebook*; Department of Parks, Recreation and Tourism Management, Clemson University, 1994.

Stieb, J. A. Assessing Freeman's Stakeholder Theory. *J. Busi. Ethics* **2009,** *87* (3), 401–414. DOI: 10.1007/s10551–008–9928–4

TGA, Turkey Tourism Promotion and Development Agency. Safe Tourism Certification Programme, 2020. https://www.tga.gov.tr/about-safe-tourism-program/ (accessed on Dec 8, 2020).

Tourism Bureau. *Annual Report in Tourism*; Taiwan, ROC: Tourism Bureau, 2003.

UNWTO. International Tourism Growth Continues to Outplace the Global Economy, 2020. https://unwto.org/international-tourism-growth-continues-to-outpace-the-economy (accessed on Aug 12, 2020).

Van Wart, Montgomery. *Dynamics of Leadership in Public Service: Theory and Practice*; M. E. Sharpe: New York, 2005.

World Health Organisation. *Summary of Probable SARS Cases with Onset of Illness 1 November 2002 to 31 July 2003* 2004. http:// www.who.int/csr/sars/country/table2004_04_21/en/ (accessed on Dec 4, 2020).

World Health Organization Ebola Virus Disease. Latest Numbers and Situation Reports, 2020. https://www.who.int/health-topics/ebola/#tab=tab_1 (accessed on Dec 06, 2020).

Wen, Z.; Huimin, G.; Kavanaugh, R. R. The Impacts of SARS on the Consumer Behaviour of Chinese Domestic Tourists. *Curr. Iss. Tour.* **2005,** *8* (1), 22–38. DOI: 10.1080/13683500508668203

Zenker, S.; Kock, F. The Coronavirus Pandemic—A Critical Discussion of a Tourism Research Agenda. *Tour. Manage.* **2020,** *81*, 104–164. DOI: 10.1016/j.tourman.2020.104164

CHAPTER 10

COVID-19: CHANGES IN TOURIST PREFERENCE IN INDONESIA

EVA RACHMAWATI

Department of Forest Resources Conservation and Ecotourism, Faculty of Forestry and Environment, IPB University, Indonesia

ABSTRACT

The covid-19 pandemic has changed human behaviour and social lives, including tourist behavior and the desire to conduct tourism activities. Since tourism is a crucial development tool for many developing countries, the stakeholders should start thinking about various efforts to bring positive changes to the tourism industry after this pandemic. This study aims to identify changes in community preferences for travel, especially preferences for the location; the area visited, and the activities undertaken. The method used is the online survey on various social media. The number of respondents is 253 respondents. The data was then analyzed by using descriptive quantitative. The study result indicated that Covid-19 had changed community preferences in tourism behavior. The results showed that 70% of respondents said that Covid-19 would change their travel habits, including the tour's location to be visited (72.9%), the tourist facilities that will be used (67.2%), and the activities that will be carried out during the tour (60.4%). Respondents switch locations far from home to those near the home for tourist sites. For facilities, there was a decrease in the use of restaurant and toilet facilities. There is a change from mass tourism to special interest activities for tourism activities. Related to the changes, there is an implication for tourism destination management. Tourism stakeholders should change their approach and management. The tourism industry should adjust its management and be

Post-COVID Tourism and Hospitality Dynamics: Recovery, Revival, and Re-Start.
Umendra Narayan Shukla & Sharad Kumar Kulshreshtha (Eds.)

more prepared to fulfill tourist needs. The tourism industry has to adjust and require the government's role to strengthen it with relevant policies.

10.1 INTRODUCTION

The chapter will discuss the changes in Indonesian community preferences in doing their tourism or leisure activities due to the Covid-19 pandemic based on the writer's quick research. Literature has noted that tourism is one of the activities most affected by this pandemic (Babu, 2020; Romagosa, 2020; Gössling et al., 2020; Wen et al., 2020; World Tourism Organization, 2020b; Zenker and Kock, 2020). Global panic, environmental pressure, and also change in regulations caused by Covid-19, which in turn impact tourism conditions, both related to distribution and packaging (Wen et al., 2020). Even though the full pattern still unclear, however, certainly, the tourism will certainly experience a change from its previous form (Babu, 2020).

Covid-19 pandemic has changed human behavior and social lives (Babu, 2020; Romagosa, 2020). Covid-19 seems to affect human life from sociological, economic, and psychological conditions (Gössling et al., 2020; Wen et al., 2020; Zenker and Kock, 2020). The changes in socioeconomic would have a significant impact on tourism (Romagosa, 2020) including tourist behavior and desires in conducting tourism activities (Wen et al., 2020). As mentioned by Lee and Chen (2011) that as a result of pandemic restrictions, human behavior is influenced by personal economic, disposable income, costs changing, perceived health risks, and changing consumption capacities.

Covid-19 could easily cripple tourists' emotional stability (Wen et al., 2020). Covid-19 also could changes tourist mobility, socialization and consumption patterns, and many other dimensions of people social lives (Romagosa, 2020). Whereas economic conditions have a greater and faster chance of recovery when compared to the recovery of sociological and psychological impacts (Wen et al., 2020).

Tourism has a vital role in the world economy, accountable for millions of occupations and billions of dollars in income (Wen et al., 2020; World Tourism Organization, 2020b). Tourism is also a crucial development tool for many communities, especially in developing countries. Therefore, policymakers, industry practitioners, and travelers should reconsiderate tourist behavior, tourism operators' market, product development, travel industry rules, policies, and regulations to promote ongoing sustainability (Wen et al., 2020). Stakeholders should start thinking about various efforts to bring positive changes to the tourism industry after this pandemic.

The existence of the Covid-19 pandemic creates an urgent need for businesses, including the tourism business, to change and not return to business as usual when the crisis ends. The situation could be an opportunity to reconsider transforming the global tourism system to align with the SDGs. However, any changes to the tourism system will raise various questions and research needs, such as domestic tourism role in recovery and response to demand for tourist behavior in the short and long term (Gössling et al., 2020).

To get empirical evidence from these assumptions and also develop baseline to design an appropriate strategy to develop future tourism, a survey of community preferences related to tourism activities was conducted. This study aims to identify changes in community preferences in travel, especially changes in preferences for the location; the area visited, and the activities are undertaken. Research related to behavior change in tourism is one factor in determining the tourism recovery strategy in a country. Knowing what has changed what funds will not help determine what to plan and do to restore tourism conditions.

10.2 RESEARCH METHODOLOGY

The method used is the online survey on various social media. The limitation of conducting surveys using the online method is that there is bias compared to the direct interview method. The resulting analysis is a description of individuals who have voluntarily participated in the survey and do not represent all Indonesian society's conditions. In the analysis in this publication, weighers' use has not been carried out, although it could be considered to reduce information bias.

Questions asked in the survey are: will Covid-19 change their habits in doing tourism activities after covid ends?; what is about to change?; the location of tourism destination they often visited before Covid-19, and they would be visiting after Covid-19; the tourist activities they do before and after Covid-19, and the facilities they used before and after Covid-19. They also have been asked the reason for their answer.

The number of respondents gets for this study is 253 respondents, which is relatively small when compared to the total Indonesian population. Based on the Population Data for Semester I 2020 from the Ministry of Internal Affairs, Indonesia's total population of 30 June 2020 was 268,583,016 people and spread across 34 provinces. Table 10.1 shows the data regarding the number of respondents and the distribution of their residence.

TABLE 10.1 Total or Respondents and Their Domicile.

Province	Frequency (n = 253)	Percentage
1. DKI Jakarta	29	11.5
2. West Java	108	42.7
3. Banten	22	8.7
4. DI Yogyakarta	4	1.6
5. Central Java	13	5.1
6. East Java	22	8.7
7. Bali	2	.8
8. South Sumatra	5	2.0
9. Riau Islands	2	.8
10. Jambi	11	4.3
11. West Sumatra	2	.8
12. Riau	3	1.2
13. Aceh	1	.4
14. North Sumatra	1	.4
15. Bengkulu	3	1.2
16. South Kalimantan	3	1.2
17. East Kalimantan	3	1.2
18. West Kalimantan	9	3.6
19. Central Kalimantan	3	1.2
20. Central Sulawesi	2	.8
21. South Sulawesi	2	.8
22. West Papua	2	.8
23. North Maluku	1	.4

The researcher lives in the West Java Province, which might be why the percentage of respondents from West Java is very high. This result indicated that the researcher's domicile could affect the survey instrument's distribution in an online survey. Although the number and distribution of respondents obtained cannot represent the distribution of Indonesian society, this result can illustrate whether there has been a change in community preferences regarding tourism activities after the existence of Covid-19 and the reasons for these changes. The data gained then analyze by using descriptive quantitative.

10.3 RESULT AND DISCUSSION

10.3.1 THE CHANGES IN TOURIST PREFERENCES

The study result indicated that Covid-19 has changes community preferences in tourism behavior. The results showed that 70% of respondents said that Covid-19 would change their travel habits, 12.6% said No, and 17.4% chose not sure whether they would change or not. Their reason for choosing yes is for personal reasons such as safety and security reasons, the changes in their income, and the regulation applied by the government. This result reinforces Kourgiantakis et al. (2020) research result, which said strong preference patterns toward safety during holidays because of the pandemic.

Interestingly, there are still a few people who said that their preferences would not changes. In their opinion, the post-Covid-19 means there is no Covid anymore, which means they could do the same activities just like before. While those who said not sure whether they would change or not, they do not know what would happen in the future.

Things that will change include the tour's location to be visited (72.9%), the tourist facilities that will be used (67.2%), and the activities that will be carried out during the tour (60.4%). For tourist sites, respondents switch from locations far from home to those near the home. For facilities, there was a decrease in the use of restaurant and toilet facilities. For tourism activities, there is a change from mass tourism to special interest activities.

10.3.1.1 THE LOCATION OF TOURISM'S DESTINATION

The survey result indicated several changes in tourism destination location choose by respondents. More people choose destinations closer to their home while fewer respondents choose out of their city, overseas, and within their city (but far from their home).

Figure 10.1 shows that the most changes in tourism destination location are that more respondents choose not to visit the destination that located out of their city (decrease from 186 to 116 respondents). More respondents choose to visit the destinations that are close to home (increase from 93 to 120 respondents). This result is similar with the Kourgiantakis et al. (2020) and Rice et al. (2020) finding which shows that almost half of their respondents express positive preferences toward destinations that are relatively close to their place of residence. Helble and Fink (2020) also said that many people still want to travel and visit a farther destination but, at this time, they prefer to stay closer to home to evade public transportation, given the infection risk and general uncertainty.

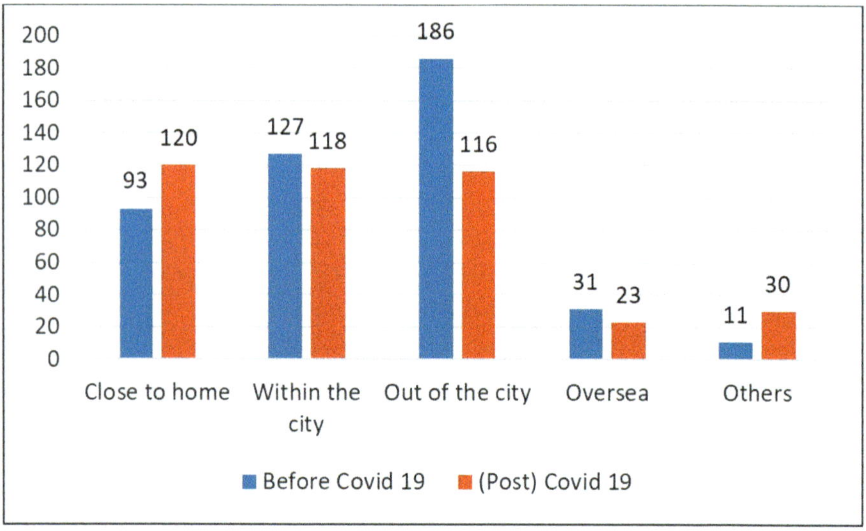

FIGURE 10.1 The changes in respondent preferences in the tourism destination location.

Before Covid, the selection of tourist locations to be visited was based on tourism activities that could be carried out in the area (69.8%), tourism objects in the area (68.3%), and the costs incurred for traveling in that location (59.5%). After Covid happened, a few changes related to the respondent's consideration in choosing the tourist location to visit. Apart from considering the tourism activities that can be done in the area (64.6%), the tourism attractions offered (53.5%), and the costs (51%), the facilities available in the area are also one of the primary considerations in choosing the location to be visited (63.8%). If previously, the respondent did not think about facilities' availability, especially those related to cleaning facilities, then after Covid-19, respondents made this their primary consideration. It can be seen from the respondents' reasons in choosing the area, namely that they wanted an area that provided facilities that support hygiene and health (e.g., a handwashing area in the area), then the facilities provided had also complied with health protocols. This result is reinforced by Assaf and Scuderi (2020), suggesting that cleaning and sanitizing should be emphasized and promoted as part of the supply for most tourism facilities. Apart from that, several respondents expected semiprivate facilities or that many people do not use.

Related to cost, it is still become a primary consideration for choosing tourism destination, before and postpandemic. This result is similar to Yuni (2020) finding, which shows that finances and travel costs are still considered a significant consideration in postpandemic travel.

From Figure 10.2, it can be seen that there are several changes in the respondent's preference for the conditions of the tourist destinations that will be visited after Covid-19. There has been a decrease in the number of respondents interested in visiting tourist destinations that offer game rides (from 99 to 48 people) and areas that are busy or full of visitors (from 89 respondents to 24 people). Despite what the literature has indicated, the surprising result is that respondents who are interested in visiting areas that offer nature-based activities also experience a decrease, although slightly (from 167 respondents to 146 people). This result is opposing to what is suggested in the literature that most people want to carry out activities based on nature. Therefore, further research is needed regarding the preferences of the Indonesian community regarding natural tourism activities. It is also essential to conduct more study on the forms of tourism activities or tourist objects like what the people want after Covid-19.

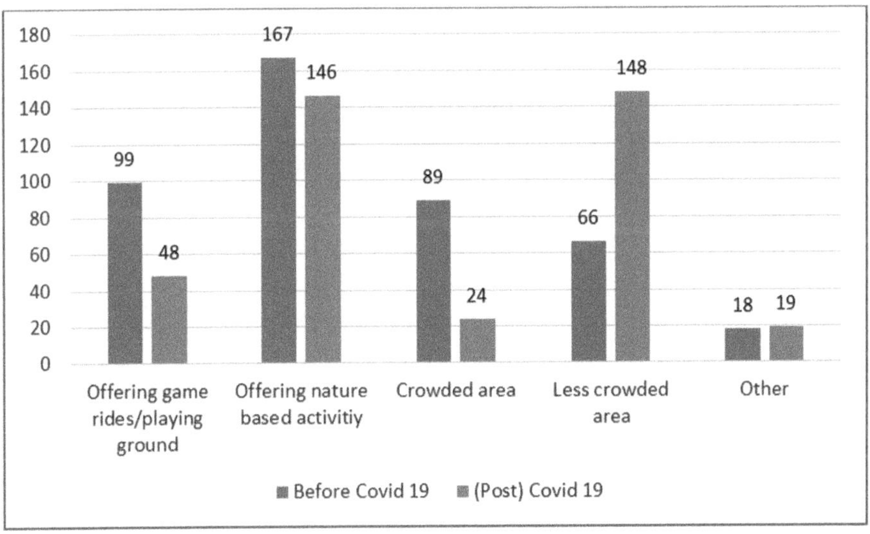

FIGURE 10.2 The changes in respondents' preferences in tourism destination.

Figure 10.2 also shows an increase in visitors' interest to visit tourist areas with low visitors (from 66 people to 148 people). This is following what is conveyed in various literatures.

10.3.1.2 THE TOURIST FACILITIES THAT WILL BE USED

The survey result indicated the decrease of respondent preference in using tourist facilities in a tourism destination. The decrease occurred in almost all facilities such as playing ground, accommodation/hotel, restaurant/café, parking area, and toilet (Figure 10.3).

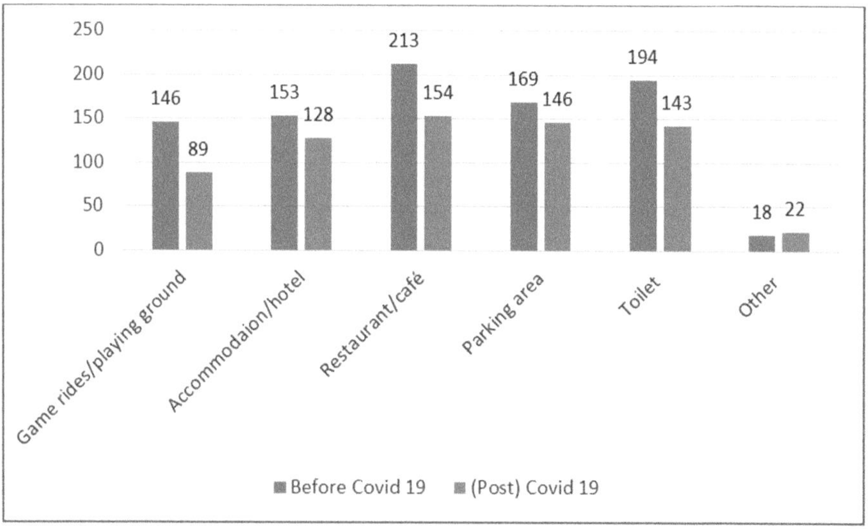

FIGURE 10.3 The changes in respondents' preference for tourist facilities.

The largest decrease occurred in children's games (decreased from 146 people to 89 people), restaurants, cafes (decreased from 213 to 154 people), and toilets (decreased from 194 to 143 people). All of these facilities are facilities that have a higher risk of being a dispersal agent than others. As one of the respondents stated that:

> "Changes in options to minimize encounters with people and reduce the use of places where hygiene is difficult to control" (Respondents_81)

The result is similar with Kourgiantakis et al. (2020) finding. They report that more than a third of their respondents intend to minimize their visits to restaurants or coffee shops, with similar consideration. Previously, the destinations' hygiene and cleanliness, health facilities, and population denseness (including residents and visitors) when creating travel-related decisions considered as the overlooked factors; namely, potential tourists are expected to prompt newfound interest in (Wen et al., 2020).

The result indicated that following the Covid-19 outbreak, the respondents prefer destinations with adequate infrastructure and high-quality medical facilities. This means the tourism destination manager's efforts should include reassuring potential visitors to the availability of essential supplies and publicizing clinic and hospital locations. As noted by Wen et al. (2020) that the medical facilities quantity and quality in tourism destinations could to shape tourist destination image and enhance destinations' power to attract potential tourists after Covid-19.

With decreasing public interest in visiting restaurants, restaurant managers must start thinking about the latest innovations. Wen et al. (2020) suggested that restaurants could offer takeaway and shows their assurance to safety and hygiene procedures. Food is a crucial driver behind tourists' travel and destination choices. The cleanliness and food quality standards of the restaurants are essential to reassuring tourists after Covid-19 to visit their restaurant. Another alternative is to offer packed and sanitized food. The manager should also not fully book seats to reduce human contacts and to maintain sanitation.

10.3.1.3 THE ACTIVITIES THAT WILL BE CARRIED OUT DURING THE TOUR

The survey result shows the changes in respondents' preferences toward tourism activities. The activities that involve a lot of people in the same time such as family gathering experience only slight decrease. Walking activity, culinary, and boating experience quite decrease as well. However, it is quite surprising since the number of decreases of family gathering is smaller than other activities (Figure 10.4).

The Indonesian culture, which has collectivist values, seems to influence this result. In most Indonesian communities, family gathering is a very important event. Moreover, it seems the pandemic could not inhibit their intention to conduct the gathering still. The respondents said that they know their family's health condition; therefore, it would be safe to conduct family gatherings. They also said that family gathering is a safe and fun activity. A significant decrease happened in boating activities. It might because, when boating, they would gather with other people (other than their family) in a small space. In their opinion, it would be dangerous since there is no physical distancing. However, the preference for rafting, which is an almost similar activity with boating, is almost similar. There is a need for a more in-depth investigation related to the reason for this result.

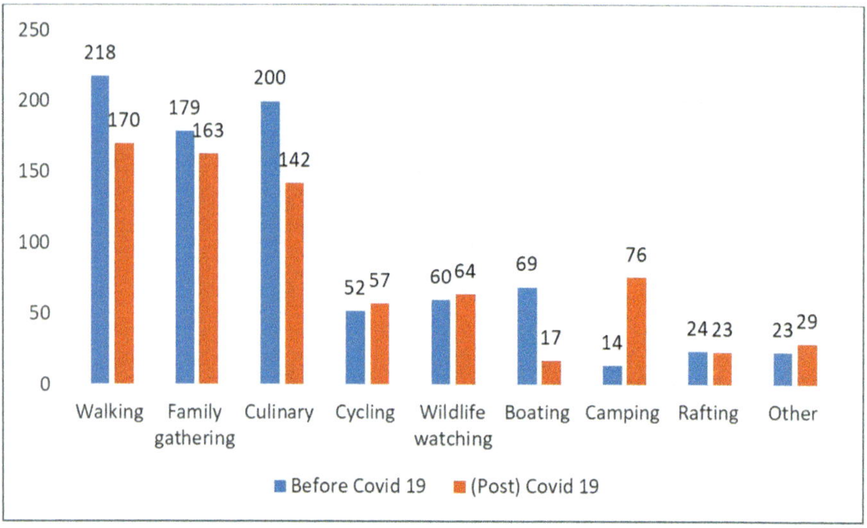

FIGURE 10.4 The changes in tourism activities.

The most chosen tourist activity after the Covid-19 was camping. This is because respondents can still do tourism activities together, but they are still distant from other tourists. Also, cycling and wildlife watching have increased, although the numbers are not too large. In general, the respondents are more likely to choose the tourism activities in the outdoor area. The result is similar to Yuni (2020) research result who noted that in post-Covid pandemic, tourist choose natural tourism such as enjoying the beauty of nature, the sea, mountains, and outdoor activities. However, Rice et al. (2020) research result showed that camping saw the greatest decreases in participation by respondents as a result of the Covid-19 pandemic, while this research result indicated the contrary result. Wildlife watching experiences the slight increase which is similar to Rice et al. (2020) research result..

The result above shows that there is a need for adjustments related to the activities offered in the area to fulfil the community's wishes. As stated by Kourgiantakis et al. (2020), the occurrence of Covid-19 will be changing in the aspects of policy and management of tourist areas. The existence of Covid-19 implies that there must be a quick adjustment, both from a policy and managerial perspective, related to tourism services to serve the desires of the community who may experience changes.

The changes above will further have implications for developing and managing tourist areas, especially tourism products and tourism facilities. The tourism destination should partner with health and safety brands and health providers and generates alliances with health centers and providers to fulfill the need for safety and security (World Tourism Organization, 2020a).

10.3.2 INDONESIAN FUTURE TOURISM DEVELOPMENT

The study finding suggested several things that would change related to tourism destinations. Those suggestion are: (1) domestic tourisms have more opportunity to revive; (2) rural area will be a new favorite of tourism destination; (3) there will be a change in tourism product, which is from mass tourism to special interest (nature-based tourism), (4) tourist will need more space during their activity; and (5) tourist will need more private facilities.

10.3.2.1 POST COVID-19 TOURISM PRODUCT AND ACTIVITY

10.3.2.1.1 Domestic Tourism

The result indicated that respondents are more likely to travel closer to their home, which means domestic tourism could be a choice for future tourism development. This result is reinforces to the World Tourism Organization (2020c) notes that most countries prioritize domestic tourism and thus prepare and adopt actions to encourage travel within their countries. Domestic tourism represents an opportunity for both developed and developing countries to recover from the Covid-19 pandemic impacts on social and economic conditions (World Tourism Organization, 2020d).

Domestic tourism is likely to recover earlier and faster because the travelers consider shorter journeys and destinations closer to homebased in the early phases of travel normalization (World Tourism Organization, 2020c). With the imposition of strict restrictions for international mobility and flights, domestic tourism is the best alternative for most tourist destinations, at least temporarily (Kourgiantakis et al., 2020).

Domestic tourism could be defined as resident-visitor activities within their country of reference (World Tourism Organization, 2020c). Redirecting tourists from overseas destinations to domestic tourism is expected to fill the gap in around half of the economies (Helble and Fink, 2020), which is one of the reasons why many countries are considering domestic tourism to help accelerate economic recovery (Helble and Fink, 2020; World Tourism Organization, 2020c).

The study also indicated that there are strong domestic tourism markets in the Indonesian community. The markets size of domestic tourism is generally the result of numerous factors, including demographics, income, transport, and hospitality infrastructures (World Tourism Organization, 2020c). However, promoting domestic tourism is not straightforward. Due to the pandemic, many people lost their job. They experienced a decrease in income, making them have a smaller amount of disposable income for vacation activities. As shown in the study, the cost is still one primary factor in considering tourism activities (Helble and Fink, 2020). To overcome this issue, there is a need for appropriate financing aspects in running domestic tourism.

Another factor that becomes a critical consideration in developing domestic tourism is social distancing, and other control measures may cause it difficult or unappealing. The tourism destination manager should design an innovative approach to implement the condition. Also, the manager needs more effort and time to reorient their management were mostly geared toward foreign markets since, in most cases, domestic tourist expenditure is different from the foreign tourists (Helble and Fink, 2020).

Related to domestic tourism, UNWTO launch global guidelines for restart tourism. The organization recommends several steps such as: (1) Encourage proximity and domestic tourism in the short term to enhance the local value chain; (2) Implement marketing campaigns to communicate corporate values and support to local providers and workers; (3) Partner with local providers, local destination management organizations, and other players in the sector (World Tourism Organization, 2020a).

The recent crisis presents an opportunity for many countries to re-evaluate their domestic tourism and apply policies that boost domestic travel (World Tourism Organization, 2020c). Data related to domestic tourism in Indonesia is still lacking in analyzing the opportunities and challenges of this tourism development more deeply. Therefore, a more in-depth research is needed in this regard, such as the demand characteristics of domestic tourism, visitor preferences, etc.

10.3.2.1.2 Rural Tourism

The result of this study indicated the preferences for rural tourism. The survey showed that some of the respondents like to do outdoor activities, mostly offered by rural tourism destinations. The result reinforces what the UNWTO noted in their publications. Many countries start developing specific products, mainly related to nature and activities in the open area, far from congested areas. The new tourism product includes wellness tourism, natural protected areas based tourism, adventure, ecotourism, cultural, and rural tourism (World Tourism Organization, 2020c).

World Tourism Organization (2020e) noted that in the Covid-19 pandemic context, rural tourism's role is more relevant than ever. Tourism in rural areas offers significant recovery opportunities since tourists prefer less populated destinations and open-air experiences and activities. Simultaneously, rural communities are much less prepared to deal with the direct and indirect impacts of the crisis. Which means rural tourism development is essential to support the community to face the economic and social implications of the pandemic.

World Tourism Organization (2020e) recommends several steps for rural tourism development such as: (1) positioning tourism as a strategic pillar in rural development policies; (2) tourism for the rural communities benefit and wellbeing: building an inclusive, fair, resilient and sustainable sector; (3) new opportunities for rural development and tourism: technology, innovation, and digitalization; (4) product development and value-chain integration for a sustainable and improved travel experience; and (5) encouraging sustainable strategies and practices in rural destinations.

Related to product development, UNWTO in global guidelines for restart tourism, recommends to add new services or product such as food delivery or takeaway service, co-working spaces, family experiences, drive and stay packages; develop segmented and sustainable products focused on rural areas, nature, and culture: eco-tourism, individual sports or small group, history, traditional routes, bird-watching tourism; introduce storytelling to create new experiences in tourism; create personalized and small group trips and packages (World Tourism Organization, 2020a).

World Tourism Organization (2020e) recommends several aspects in developing rural tourism product:

- The development of tourism products in rural areas should be based on assessing rural destinations' strengths and opportunities to define their potential. They could then create a proper enabling environment (i.e., infrastructure, resources management, investments, human capital) and a long-term development plan and strategy that can be sustainable and economically feasible. To this end, it is essential to implement market intelligence initiatives that provide relevant insights on trends and rural travelers' needs to adapt products and marketing according to the targeted market while maintaining local identity.

- The development of tourism products and experiences in countryside areas should follow an all-inclusive and fully integrated perception engaging all other sectors, stakeholders, and socio-economic activities in the rural and tourism value chain. Rural communities should be intrinsically involved in the co-creation of social and environmentally sustainable, and economically viable tourism products. Collaboration within the community is essential, and the stimulation or creation of networks of producers and cooperatives into working together to design and develop tourism products, agricultural produce, and artisanship are highly recommended. The rural community must be proud of what they are, what they do, and their way of life. They should feel ownership of their territory, their resources, and lifestyle, and therefore also be empowered to welcome visitors and create valuable experiences.
- National, regional, and local governments and the private sector should encourage thematic routes, road trips, island connections, and national routes that link different rural destinations. Collaboration between various rural destinations to develop joint tourism experiences and connectivity is encouraged for a more practical tourism development approach and added value for visitors.
- National, regional and local governments, private sector, and communities should promote new and authentic experiences unique to rural areas, facilitating contact with nature and culture in uncrowded environments, in line with current consumer trends. The full variety of rural resources should be considered. This includes tourism activities related to culture, gastronomy and beverage, health and wellness, adventure and sports, agrotourism and ecotourism, mountain, coastal, marine, and natural protected areas post-visit follow-ups and access to rural produce. Focus on product development, emphasizing an interpretation that provides visitors' experiences, in terms of cultural learning and environmental conservation in rural areas, should be given priority.

10.3.2.2 POST COVID-19 TOURIST FACILITY

If it was before the Covid-19 pandemic, the needs of tourism consumers, is based on meeting physiological (recreational) needs, so during this pandemic, the need for security and physical survival becomes a priority. Restrictions in preventing the spread of this virus make security needs dominate as a necessity (Suprihatin, 2020). Cleaning and sanitizing should be emphasized and opportunely promoted as part of the supply. Cleaning protocols will likely change and personal protective equipment will be required for most tourism facilities (Assaf and Scuderi, 2020).

10.4 RESEARCH IMPLICATION

The study result indicated the changes on community preferences in tourism location, activities, and facilities. Related to the changes, there is an implication for tourism destination management. Tourism stakeholders should change their approach and management. For example, as mentioned by Wen et al. (2020) that practitioners, including tourism industry should devise tailored strategies to restore visitors' confidence in traveling domestically and overseas. The tourism industry should improve their booking mechanism to more innovative online booking, prepare private facilities, conduct interpretation programs to enhance tourist experience, and design destination zoning.

The tourism industry should adjust their management and more prepare to fulfill the tourist needs. The tourism industry has to adjust and require the government's role to strengthen it with relevant policies. Producers must adapt and activate tourist consumer motivation by adjusting tourist needs, changing what tourists think, and understanding changes in consumer behavior. The producer's role, which is to activate consumer motivation, changes what is felt so that it becomes a need that must end with a purchase decision (Suprihatin, 2020). Several things that need to be done by World Tourism Organization (2020a) for the tourism industry include: (1) implement global and harmonized health, safety, and hygiene protocols; (2) train the personnel in safety and security protocols and safe-service delivery; (3) communicate protocols and responsibilities to the customers, keep them informed onsite and through digital and social media; (4) clean all surfaces more frequently, use products and disinfectants that meet requirements with particular attention to high-touch surfaces; (5) set up a plan in the event a guest or employee falls ill on site; (6) create an internal Covid-19 coordination committee to follow-up on the implementation and execution of protocols and service delivery; (7) redesign processes toward more digital touchless operations in all possible touch-points; (8) Keep a record of statistical information for tourism planning and management in full coordination with local and national tourism and statistical authorities; (9) promote social dialogue and work to ensure the health and wellbeing of workers and the working environment (support employees in monitoring their health and adapt Human Resources policies); and (10) engage in public/ private sector dialogue and recovery mechanism.

The tourism industry should also improve its online booking mechanism. UNWTO regulates several things related to bookings during this pandemic. Some things that must be done regarding this booking include: (1) information on health requirements and procedures from the origin country, travel providers and destination; (2) information on cancellation and policies if sick while traveling or at the destination; (3) and sharing data (national tracing apps) (World Tourism Organization, 2020a).

Simultaneously, the Government has to prepare a policy to support domestic and rural tourism to become more sustainable tourism. Several regulations should be designed, such as regulations related to tourist quota, carrying capacity, and health protocol—procedure, reward, and sanction. For a country whose economy is very dependent on the tourism industry, it is natural for the government to give great support to tourism companies to rebuild and reform the industry. Most governments have designed and established programs to stimulate the economy to mitigate Covid-19. These programs are expected to help the tourism industry recover after the crisis. Programs geared specifically to tourism include establishing health and sanitation protocols, certification, marketing campaigns, and special incentives to increase demand. Governments also need to invest and promote investments that help ensure the long-term sustainability of their tourism industries and seize opportunities for recovery (Helble and Fink, 2020).

UNWTO recommends several things that could be done by the government to promote domestic travel and re-establish confidence in the tourism sector. Those steps are: (1) financial incentives; (2) marketing and promotion; (3) product development; (4) partnerships; (5) market intelligence; and (6) capacity building and training (World Tourism Organization, 2020c). The Indonesian government implemented several policies to accelerate or attract 736 thousand tourists from market focus countries with high category average spending per arrival (ASPA), namely Australia at USD 1800, Middle East at USD 2200, and other markets such as America, Europe, and Asia; generate foreign exchange of IDR 13 trillion. It is expected that the government's incentive scheme will be the best solution for the tourism sector to impact the number of foreign tourists visiting and provide more equitable welfare for the community through tourism activities (Sugihamretha, 2020).

Another policies that have been taken by the Indonesian government include an additional allocation of the Ministry of Tourism and Creative Economy's State Revenue and Expenditure Budget of IDR 298.5 billion;

Incentives for airlines, travel agents; incentives in joint promotion schemes, tourism promotion activities, familiarization trip (fam-trip), and influencer; Encourage incentives under the association's suggestion that the rates for hotel and restaurant taxes in 10 tourist destinations should be zero. The government will subsidize or provide grants to local governments affected by tax rates for hotels and restaurants in the regions by IDR 3.3 trillion (Sugihamretha, 2020).

The government also offer a domestic ticket discount that can be used by domestic tourists on average are 51.44% for 25% of the seating capacity in one flight, including 30% discount for 25% quota of seats on every flight to 10 tourist destinations. This is valid for three months, namely March, April, and May 2020. There is an additional discount of 15.8% Avtur from Pertamina and a 5.64% discount on flight fares (PJP2u/PSC and NAV) from AP and Airnav. This policy will last for three months. This discount applies to domestic airlines with destinations for Batam, Belitung, Bintan, Denpasar, Manado, Labuan Bajo, Lombok, Lake Toba, Malang, and Yogyakarta (Sugihamretha, 2020). The government also encouraging domestic tourists' movement by continuing to carry out domestic tourism promotion campaigns, encouraging meetings of government agencies and corporations in domestic tourist destinations; promote existing events (music, sports) in the country. On the other hand, efforts were also made to attract the potential of MICE and international events to be held in Indonesia (Sugihamretha, 2020).

10.5 CONCLUSIONS

This research shows that some Indonesians have experienced several changes in tourism activities after the Covid-19 occurrence. The survey results indicate that several preferences are not the same as those expressed in various literatures, including the public's interest in nature-based activities. Besides, even with social distancing, some people still like activities that involve many people, such as family gatherings. This shows that the context, habits, or culture of the community influence changing people's preferences regarding tourist habits.

This study also indicates several things that need to be done further or require further research. The research that needs to be done includes deepening the understanding of the forms of tourism activities that the community wants after the Covid-19. Apart from that, stakeholders also need to make more efforts based on the available guidelines, both at national and international scales.

KEYWORDS

- **tourist preference**
- **tourist destination**
- **tourist facilities**
- **tourism activity**
- **online survey**

REFERENCES

Assaf, A.; Scuderi, R. Covid-19 and the Recovery of the Tourism Industry; SAGE Publications Sage: London, 2020.

Babu, S. Six Ways How Covid-19 Will Change Tourism Industry, 2020. https://www.policycircle.org/economy/industry/six-ways-how-covid-19-will-change-tourism-industry/

Gössling, S.; Scott, D.; Hall, C. M. Pandemics, Tourism and Global Change: A Rapid Assessment of Covid-19. *J. Sustain. Tour*. **2020**.

Helble, M.; Fink, A. Reviving Tourism Amid the Covid-19 Pandemic. *ADB BRIEFS NO. 150*, 2020.

Kourgiantakis, M.; Apostolakis, A.; Dimou, I. Covid-19 and Holiday Intentions: The Case of Crete, Greece. *Anatolia* **2002,** 1–4. DOI: 10.1080/13032917.2020.1781221

Lee, C.-C.; Chen, C.-J. The Reaction of Elderly Asian Tourists to Avian Influenza and SARS. *Tour. Manage.* **2011,** *32* (6), 1421–1422. https://doi.org/10.1016/j.tourman.2010.12.009

Rice, W. L.; Meyer, C.; Lawhon, B.; Taff, B. D.; Mateer, T.; Reigner, N.; Newman, P. The Covid-19 Pandemic Is Changing the Way People Recreate Outdoors: Preliminary Report on a National Survey of Outdoor Enthusiasts Amid the Covid-19 Pandemic, 2020.

Romagosa, F. The Covid-19 Crisis: Opportunities for Sustainable and Proximity Tourism. *Tour. Geog.* **2020,** 1–5.

Sugihamretha, I. D. G. Respon Kebijakan: Mitigasi Dampak Wabah Covid-19 Pada Sektor Pariwisata. *Indonesian J. Dev. Plan.* **2020,** *4* (2), 191–206.

Suprihatin, W. Analisis Perilaku Konsumen Wisatawan Era Pandemi Covid-19 (Studi Kasus Pariwisata di Nusa Tenggara Barat). *BESTARI* **2020,** *1* (1), 56–66.

Wen, J.; Kozak, M.; Yang, S.; Liu, F. Covid-19: Potential Effects on Chinese Citizens' Lifestyle and Travel. *Tourism Review*, 2020.

World Tourism Organization. *Global Guidelines to Restart Tourism*; World Tourism Organization (UNWTO): Madrid, Spain, 2020a.

World Tourism Organization. *UNWTO Briefing Note—Tourism and Covid-19, Issue 1. How Are Countries Supporting Tourism Recovery?* World Tourism Organization (UNWTO): Madrid, Spain, 2020b.

World Tourism Organization. *UNWTO Briefing Note—Tourism and Covid-19, Issue 3. Understanding Domestic Tourism and Seizing Its Opportunities*; World Tourism Organization (UNWTO): Madrid, Spain, 2020c.

World Tourism Organization. UNWTO Highlights Potential of Domestic Tourism to Help Drive Economic Recovery in Destinations Worldwide; World Tourism Organization (UNWTO): Madrid, Spain, 2020d.

World Tourism Organization. *UNWTO Recommendations on Tourism and Rural Development: A Guide to Making Tourism an Effective Tool for Rural Development*; World Tourism Organization (UNWTO): Madrid, Spain, 2020e.

Yuni, L. H. K. Analysis of Domestic Tourist Travel Preferences Post-Covid-19 Pandemic. *J. Appl. Sci. Travel Hospital.* **2020,** *3* (2), 80–88.

Zenker, S.; Kock, F. The Coronavirus Pandemic—A Critical Discussion of a Tourism Research Agenda. *Tour. Manage.* **2020,** *81*, 104164–104164. DOI: 10.1016/j.tourman.2020.104164

CHAPTER 11

EXAMINING REVIEWS ON TRIPADVISOR TO UNDERSTAND THE IMPORTANCE OF MINI-CLUBS TO THE CONTEMPORARY HOSPITALITY SECTOR, AND POTENTIAL IMPACTS OF COVID-19

HUGUES SERAPHIN[1] and MAXIMILIANO E. KORSTANJE[2]

[1]*Senior Lecturer in tourism and Event Management Studies, The University of Winchester, UK*

[2]*Lecturer at Departments of Economic, University of Palermo, Buenos Aires, Argentina*

ABSTRACT

The study reveals three important things regarding resort Kids' clubs or mini-clubs: First, families write significant number of reviews regarding resort mini-clubs on TripAdvisor; second, the experience of children at resort mini-clubs contributes to the overall experience of families; and finally, reviews on resort mini-clubs contribute to the overall performance (customer satisfaction) of resorts. Overall, resort mini-clubs can be said to be services part of the performance of resorts or service apart from the performance of resorts. However, the breakout of COVID-19 impediment the ability of resort mini-clubs to contribute fully to customer satisfaction. COVID-19 could be said to have disabled the potential of resort mini-clubs. Radical innovation is therefore needed for resort mini-clubs to play again its central role in customer satisfaction.

Post-COVID Tourism and Hospitality Dynamics: Recovery, Revival, and Re-Start.
Umendra Narayan Shukla & Sharad Kumar Kulshreshtha (Eds.)

11.1 INTRODUCTION

The hospitality sector which is "the world's fastest growing industry, and also the one of the leading earners of foreign currency" (Boella and Goss-Turner, 2020: 5), is also playing an important societal role, as meeting basic human needs (namely sleeping and eating) is made of multinational and locally operated small- and medium-sized enterprises (Clarke and Chen, 2007). This sector which is a major provider of employment worldwide is also part of a bigger sector often referred to as hotel, catering and tourism sector, or HCTS (Nickson, 2013). The accommodation (sleeping) aspect of the hospitality sector, which is the key and only focus of this study, is segmented into five main types of hotels, known as budget hotels; airport hotels; business hotels; resort hotels; and boutique hotels (Clarke and Chen, 2007).

Resort hotels which are at the heart of this study have grown as a result of the increase of leisure time; annual weekend breaks; and interest in sport, health, etc. (Clarke and Chen, 2007). Those resort hotels can mainly be found in ski and beaches resorts, but also in theme parks; and are mainly accommodating leisure customers (Clarke and Chen, 2007). Among those customers are children, a group (among others) considered as passive and powerlessness (Hutton, 2016), and undermined by academic research in hospitality and cognate sectors despite their importance (Canosa et al., 2016, 2018a, 2018b; Lugosi et al., 2016). Alongside this group undermined by academic research, is also a service provided specifically to them, which is also totally undermined by research, namely resort mini-clubs, also referred as Kids' club (Séraphin and Yallop, 2019a, 2019b).

The purpose of this research is first to show that resort mini-clubs contribute to both customer satisfaction and the competitive advantage of resorts (Séraphin and Vo Thanh, 2020; Séraphin and Yallop, 2019a, 2019b); second, to investigate the impacts of COVID-19 on resort mini-clubs, with a focus on customer satisfaction.

11.2 CONTEXTUAL FRAMEWORK

11.2.1 ONLINE REVIEWS AND USER-GENERATED CONTENT

Online reviews posted by users of a particular product or service that could be viewed as a form of word-of-mouth is important source of information for other potential customers. They are more trusted than information provided by the organization itself (Hudson and Hudson, 2017; Martin-Fuentes et al.,

2020; Yu et al., 2017). Beyond the fact that online user reviews influence others decision, it also appears that the number of reviews also contributes to the visibility; popularity; the image; the number of bookings that an hotel gets (Martin-Fuentes et al., 2020). It is also worth highlighting the fact that there is a correlation between the number of reviews and the rating of organizations on platforms such as TripAdvisor, Booking.com, etc. (Martin-Fuentes et al., 2020). This obviously works positively if the reviews are good. It has the opposite impacts if reviews are negative (Martin-Fuentes et al., 2020). The preceded information is all the more the case for the hospitality sector; hence, the reasons the reviews are closely monitored by organizations (Martin-Fuentes et al., 2020), as it is important for them to understand what customers liked or disliked (Yu et al., 2017), due to the fact that customer satisfaction is central for the survival of organizations in the service industry (Yu et al., 2017).

Ladsawut and Nunkoo (2018) are arguing that social media are also quite important for the hospitality sector. Customers use this media to share their experiences by providing reviews on hotels/resorts they stayed in. As for hoteliers, these reviews are perceived as marketing tools as they are evidence of customers interests with the services provided, and are also a way to collect qualitative and quantitative feedback to evaluate performance in terms of customer satisfaction. Beyond the marketing aspect, social media could be considered as a media which has contributed to operate a shift of power in the hospitality sector, as now the power is shared between marketers and customers (Ladsawut and Nunkoo, 2018). Equally important, because hotel managers and marketers in the hospitality sector have no control over the frequency and accuracy of content that appears online, they need to monitor and reply to online reviews daily (Ladsawut and Nunkoo, 2018).

11.2.2 COVID-19 AND THE TOURISM INDUSTRY (AND RELATED SECTORS)

With the breakout of COVID-19, the tourism industry finds itself forced to navigate and perform into a negative context, where travel is associated with disease, and sometimes death (Jamal and Budke, 2020). All the sectors of the industry are impacted, some more than others (Gossling et al., 2020). The social media sometimes contributes to make the situation even worst, by spreading fear, stress, and distress (Yu et al., 2020). Some ethnic groups are sometimes victimized (Zheng et al., 2020). That said, for some, the breakout of the virus is offering the industry an opportunity to reinvent itself for the best (Rivera, 2020). How does that translate in the case of resort mini-clubs? Two things are going to be a problem: first, the fact that the outbreak of the virus has highlighted how important health is (Cooper and Alderman, 2020), as a result, individual deeply rooted evolutionary protection mechanisms are activated (Kock et al., 2020); second, the enforcement of social distances everywhere, and for all types of activities (gov.uk[Online]; WHO [Online]).

11.3 METHODOLOGY

11.3.1 OVERALL APPROACH

This study is based on a deductive approach, which aims to test hypothesis, theory, or knowledge already gained (Brunt et al., 2017) in a top-down fashion (Hammond and Wellington, 2013) using gathered data (Fox et al., 2014). This approach is also associated with exploratory analyses (Banasiewicz, 2013). Data for this study is collected from TripAdvisor. Collecting data from online platform where people share their experiences seem to be a trend, as well as a methodological approach in academic research (Mkono and Hughes, 2020). The first step of this study follows Mkono and Hughes (2020) method in studying online confessions expressing guilt and shame using Google keyword searches. Instead of the search engine Google, this study is using TripAdvisor as a data source.

Step 1: The keyword *Kids club* was entered in the search box. The searches generated hundreds of resorts over 34 web pages. These were then narrowed to the first 120 resorts that came up from the search. This convenient sample presents some limitations, the first being that it is based on a nonprobability sampling, and subsequently, the second being that it is not representative of an entire group (Fox et al., 2014; Quinlan, 2011).

Step 2: For each resort (120 in total), the total number of reviews as well as the number of reviews with mention to Kids club are recorded in a table (Appendix 11.1) for correlation analysis.

Step 3: One review with mention to Kids club per resort (120 reviews altogether) is copied and pasted on a word document for content and cluster analysis.

From step 2 (quantitative data) and 3 (qualitative data), it appears that this study is based on mixed methods, which is "a combination of typically, quantitative and qualitative methods to provide complementary and perhaps contrasting perspective of a phenomenon" (Hammond and Wellington, 2013: 171).

11.3.2 DATA ANALYSIS (QUANTITATIVE DATA)

Once the data collected (step 1) and recorded (step 2) in the table (Appendix 11.1), the focus was first placed on identifying the importance (in percentage) of reviews specific to Kids' club with regards to overall reviews. For this, Excel was used to calculate:

a) the overall ratio (for the 120 resorts): total number of reviews specific to Kids' club/total overall reviews;
b) ratio per resort: total number of reviews specific to Kids' club for the resort/total number of reviews (regardless Kids' club) for the resort;
c) SPSS (Pearson Correlations) was then used to establish some correlation between the total number of reviews and reviews specific to Kids' club. Indeed, "the Pearson Product Moment Correlation is used to measure the degree to which two variables are correlated or associated with each other" (Silver et al., 2013: 216).

11.3.3 DATA ANALYSIS (QUANTITATIVE DATA)

Once the data collected (step 3), they were then analyzed using NVIVO (Figure 11.1). The software supports qualitative data analysis, as well as thematic analysis (Brunt et al., 2017), so doing it allows to code data efficiently and built theory effectively (Brunt et al., 2017). The software was applied to code the reviews. Following the approach used by researchers such as Fan et al. (2017), during coding, meaningful units in reviewers' comments were captured and utilized to formulate key themes regarding customers (more specifically parents), views on Kids' club.

11.4 RESULTS

11.4.1 QUANTITATIVE DATA ANALYSIS

Reviews related to Kids' club represent a relatively significant part of the total number of reviews received by resorts. Indeed, they represent on average 11% of the overall number of reviews. In some cases, they represent between 44% (Paradise Resort Gold Coast) and 38% (Club Med Ixtapa Pacific) of reviews. Other resorts within this bracket include to mention a few: Club Med Bintan Island; Bali Dynasty Resort Hotel; Movenpick Resort and Spa Jimbarab Bali; etc. (Appendix 2).

FIGURE 11.1 Template showing number of overall reviews and review specific to kids' club.

Source: TripAdvisor.

The Pearson Product Moment Correlation reveals a weak correlation between the total number of reviews received by resorts and the number of reviews specific to Kids' club (Figure 11.4). A correlation is considered weak when between 0.21 and 0.40 (Silver et al., 2013). In the case of the study, the correlation is 0.268. (Figure 11.2).

Figure 11.3 enables a visualization of the situation.

From the results of the quantitative data it appears that Kids' club are part of the services that contribute to the experience of customers staying in resorts, as there are a significant number of reviews on this service. Resort Kids' club and subsequently should be considered as important as any other service delivered by resorts, and not considered as by-services as it is the case now (Séraphin and Yallop, 2019a, 2019b). Additionally, those reviews are expression of lived experiences (Mkono and Hughes, 2020).

11.4.2 QUALITATIVE DATA ANALYSIS

In order to discuss the results of the qualitative analysis done using NVIVO, a message attribute-framing framework has been used. An attribute-framing message analyses how the attribute of a product or service can satisfy the needs and wants of customers (Zhang et al., 2018). The attribute-framing messages could be classified into two categories. First, the gain-framed messages which focus on the benefits of utilizing a specific product or service. The emphasis is on the positive outcomes. Second, the loss-framed messages, which focus on how negative outcomes could be avoided utilizing a specific product or service (Zhang et al., 2018). As for the effectiveness of both approaches, there is no consensus in terms of which one is the most effective in terms of customers' evaluation and perception of a product or

Nodes

Name	Files	References
Activities	1	38
Age range	1	8
Convenient to parents	1	8
Cost	1	6
Equipment	1	18
Events	1	3
Flexibility	1	1
Friends	1	1
Other ancilary services	1	3
Repeat customers	1	2
Safety	1	2
Schedule	1	8
Staff	1	72

FIGURE 11.2 Data coding (nodes).

Source: NVIVO (the author).

Correlations

		Total Numb. reviews	Numb reviews on mini-clubs
Total Numb. reviews	Pearson Correlation	1	.268**
	Sig. (2-tailed)		.003
	N	120	120
Numb reviews on mini-clubs	Pearson Correlation	.268**	1
	Sig. (2-tailed)	.003	
	N	120	120

**. Correlation is significant at the 0.01 level (2-tailed).

FIGURE 11.3 Weak correlation. Number of overall reviews and reviews specific to kids' club.

Source: SPSS (the author).

service (Zhang et al., 2018). Alongside this framework could be used the frame contents (aspects of perceived reality of a product/service with a focus on their salience) and frame implication (recommendations) framework developed by Entman (1993).

Thus, the NVIVO analysis (visual representation of the nodes) of reviews reveals that "Staff" and "Activities" are the factors that impact the most on the experience of children and parents.

From the four feedback below (reference 1/2/8/23/52/71), it appears that resort-mini clubs are a must experience for children on holiday with their family. This quality experience relies essentially on the professionalism and personality of the staff (frame contents/gain-framed). Because the parents know that the children are having fun with capable staff, they can as a result have fun and enjoy their holidays (frame implications/gain-framed).

11.5 DISCUSSION AND CONCLUSIONS

11.5.1 MINI-CLUBS AND CUSTOMERS' SATISFACTION

Studying resort mini-clubs is about studying social network. This type of research involves examining social system, environment, and context (Robin, 2015). Indeed, mini-clubs are part of a network of interconnected products and services offered by resorts to their customers. There are two main types of networks, namely, the centralized networks where all the nodes are highly dependent of a main node, meaning that this node influences how the entire organization work; and peripheral networks, with isolated nodes interacting between themselves while still interacting with a main node (Robin, 2015). Mini-clubs are nodes within a peripheral network. Mini-clubs are connecting services and departments such as catering; room services; entertainment/animation; but each of these services/departments are forming its own cluster or subgroup.

As any service, the role of mini-club within the network of services is to satisfy customers wants and needs (Drewery and McCarville, 2017). The latter is the result of positive emotions, which subsequently lead to repeat re-patronage behaviors and positive (e)word of mouth (Drewery and McCarville, 2017). These positive emotions arise as an evaluation of customers regarding whether or not (a) the products and services promised were provided (b) interaction with others, staff included (c) the environment in which the products and services were provided (Drewery and McCarville, 2017). Equally important, Drewery and McCarville (2017: 178), highlighted the fact that: "leisure services may be particularly emotional in nature"; "leisure settings and the nature of the activities themselves all build emotional buy-in on the part of participants." In the same line of thoughts,

Drewery and McCarville (2017: 178), added that positive emotions from customers' reviews are indicated by the type of words they use, such as: "Happy"; "hopeful"; "delighted"; "pleased"; etc., and as a result of the importance of positive emotions' satisfaction, they suggested that service providers should design services that facilitate those positive emotions. This view is further supported by Séraphin et al. (2018) who argued that effective tools in terms of positively influence the perception of customers are the one combining elements of the "ethos" (credibility and trustworthiness); "pathos" (emotional appeal of the message); and finally, "logos" (the capacity to inform).

Resort mini-clubs appear as suitable products and services that can contribute to the satisfaction of customers as they are:

- Generating positive emotions and therefore positive e-word-of-mouth on TripAdvisor. The messages are framed around ethos, logos, and pathos, and as a result, leads to the intention of parents to stay again in the same resort
- Promoting interaction between customers (children–children; parent–parents; children–parents), but also interaction with staff (child reps-children; child reps-parents), in a safe and pleasant environment
- Considered as leisure activities (Séraphin and Yallop, 2019a, 2019b; Séraphin and Vo Thanh, 2020), as a result are emotional in nature (Drewery and McCarville, 2017).

Ma and Scott (2017) highlighted a certain number of gaps in literature regarding customer delight and satisfaction with the hospitality sector. This study is addressing quite a few of them (either totally of partially):

1. Lack of empirical test to support the conceptual correlations between the emotions and needs.
2. Lack of examination of the nature of affective responses to delightful experience.
3. Lack of empirical test of delight and affective commitment to loyalty.
4. Personal meaning of delight.

Based on the strong potential of resort mini-clubs to satisfy the needs and wants of customers (family segment), this study is suggesting in the following section that resort mini-clubs could be used more effectively by managers. Indeed, because (a) parents and children have developed an affective connection with mini-clubs and staff; because (b) emotions promote learning; (c) because learning guide future actions; and because (d) positive emotions create lasting ties between clients and the organization (Drewery and McCarville, 2017), what could be the role of resort mini-clubs in the environmental, economic, social sustainability of resorts?

11.5.2 MINI-CLUBS AND CUSTOMERS' SATISFACTION IN A COVID-19 CONTEXT

As activities and staff (Figure 11.4) are playing a major role in children (and parents) satisfaction with mini-clubs, with social distancing (gov.uk[Online]; WHO [Online]); fear for children to be sick (Cooper and Alderman, 2020), mini-clubs and their staff are likely to activate a protection mechanism (Kock et al., 2020), which is likely to impact on activities, attitudes, relationships, etc. and therefore customer satisfaction. Table 11.1 describes how COVDI-19 negatively impact every single aspect of the factors that used to contribute to children and parents' satisfaction with mini-clubs.

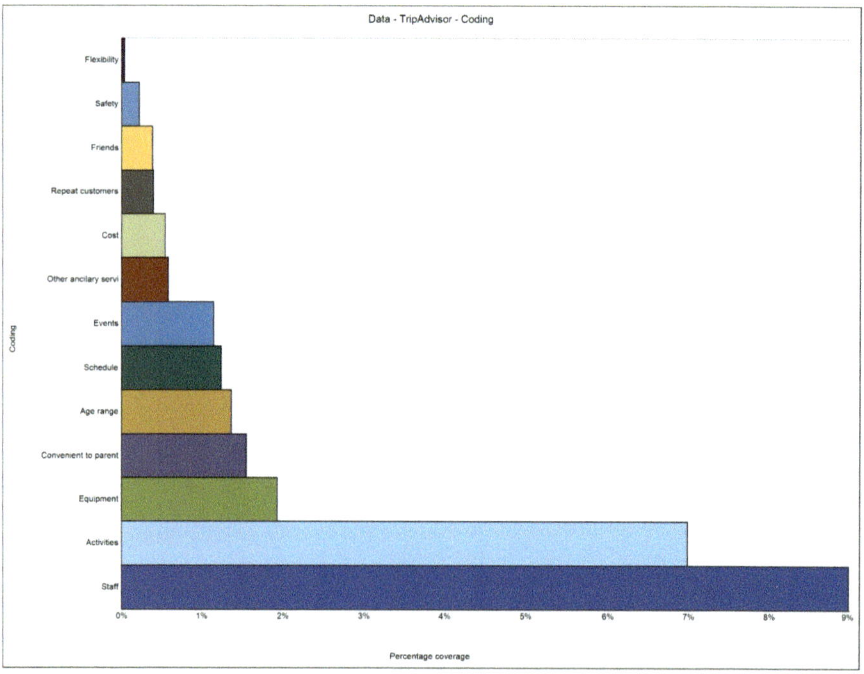

FIGURE 11.4 Factors impacting the satisfaction of children and parents.

Source: NVIVO (the author).

TABLE 11.1 Impacts of COVID-19 on Factors Contributing to Children (and Parents) Satisfaction with Mini-Clubs.

Factors contributing to children (and parents) satisfaction with mini-clubs	Impacts of COVID-19
Staff	Staff will have to enforce social distancing, and will be wearing masks. It will therefore be more difficult for them to demonstrate their friendliness, and emotions will not be conveyed the same way as facial expression will be hidden by a mask.
Activities	Some activities will be delivered differently due to the limited number of children mini clubs will be allowed to accommodate. Additionally, children won't be able to have fun the same way. Social capital is likely to be developed in a different way.
Equipment	Health and safety equipment are going to be more present, and children might not be allowed to access all the equipment they used to.
Convenient to parents	Parents might not be willing to send their children to mini-clubs
Age range	Some age group will probably be banned (and the number of children reduced to a maximum)
Schedule	Mini-clubs might start working on rota to accommodate all children if there is a demand. Working hours might remain the same.
Events	Events are likely to be cancelled, as it is currently the case for large events.
Other ancillary services	Depending of the nature of these services, they might be discontinued.
Cost	To encourage more parents to use mini-clubs, management team might decide to reduce the cost of mini-clubs.
Repeat customers	Staff in mini-clubs will have to be particularly imaginative to keep up the level of satisfaction.
Friends	See "activities"
Safety	See "equipment"
Flexibility	See "staff" and "repeat customer"

Source: Hugues Séraphin

11.5.3 MINI-CLUBS AND THE FUTURE OF THE HOSPITALITY AND TOURISM INDUSTRY

As any other service, mini-clubs need to renew their process to remain competitive, but also to contribute to the stability and profitability of resorts (Osman, 2018; Séraphin and Yallop 2019a, 2019b). In this line of thoughts, Séraphin and Vo Thanh (2020) suggested that mini-clubs should be used as a sustainability educational tool to turn current children on holiday into future responsible tourists. Séraphin and Vo Thanh (2020) also suggested that mini-club should deliver activities that are related to the destination where the resort is, so that children can learn about the visited destination and have a unique and authentic experience. This need for authenticity is all the more important as for the younger generation, as every day has to be exceptional. Young consumers differentiate themselves from others through their experiences, such as holidays abroad; cultural events attended; cutting edge leisure activities; etc. (Yeoman, 2013). All the preceded elements combined should help resorts to retain children (and parents/guardians) as customers, and equally important, maintain the link with the family so that children, when adult become customers. This is quite possible as positive emotions create lasting ties between clients and the organization (Drewery and McCarville, 2017), this is all the more the case in times of crisis, such as COVID-19 (Hang et al., 2020). Additionally, nostalgia is also considered as a powerful tool to attract some customers keen to step back in time (Yeoman, 2013). In that instance, and in relation to theories related to childhood experiences, which argues that there is a relationship between childhood experiences or circumstances, and adult behaviour, attitude, achievements; etc. (Cox and Cooper, 1989; Drennan et al., 2005; Simoes et al., 2015), childhood experience in mini-clubs could (at some point of adulthood) play the role of *Madeline de Proust*, in reference to Proust's experience of recalling his past after tasting a madeleine dipped in tea (Bray, 2013).

The study reveals three important things regarding resort Kids' clubs or mini-clubs: First, families write significant number of reviews regarding resort mini-clubs on TripAdvisor; second, the experience of children at resort mini-clubs contributes to the overall experience of families; and finally, reviews on resort mini-clubs contribute to the overall performance (customer satisfaction) of resorts. Overall, resort mini-clubs can be said to be services part of the performance of resorts or service apart from the performance of resorts. However, the breakout of COVID-19 impediments the ability of resort mini-clubs to contribute fully to customer satisfaction. COVID-19 could be said to have disabled the potential of resort mini-clubs. Radical innovation is therefore needed for resort mini-clubs to play again its central role in customer satisfaction.

KEYWORDS

- **COVID-19**
- **crisis**
- **TripAdvisor**
- **customer satisfaction**
- **mini-club resorts**

REFERENCES

Banasiewicz, A. D. *Marketing Database Analytics: Transforming Data for Competitive Advantage*; Routledge, Abingdon, 2013.

Bray, P. Forgetting the Madeleine: Proust and the Neurosciences. *Progress Brain Res.* **2013,** 205, 41–53

Brunt, P.; Horner, S.; Semley, N. *Research Methods in Tourism, Hospitality and Events Management*; Sage: London, 2017.

Canosa, A.; Graham, A. Ethical Tourism Research Involving Children. *Ann. Tour. Res.* **2016,** *61* (15), 1–6.

Canosa, A.; Graham, A.; Wilson, E. Growing up in a Tourist Destination: Developing an Environmental Sensitivity. *Environ. Educ. Res.* **2020**. doi.org/10.1080/13504622.2020.1768224

Canosa, A.; Graham, A.; Wilson, E. Progressing a Child-Centred Research Agenda in Tourism Studies. *Tour. Analy.* **2019,** *24* (1), 95–100.

Canosa, A.; Graham, A.; Wilson, E. Reflexivity and Ethical Mindfulness in Participatory Research with Children: What Does It Really Look Like? *Childhood* **2018a,** *25* (3), 400–415.

Canosa, A.; Graham, A.; Wilson, E. Child-Centred Approaches in Tourism and Hospitality Research: Methodological Opportunities and Ethical Challenges. In *Handbook of Research Methods for Tourism and Hospitality Management*; Nunkoo, R., Ed.; Edward Elgar: Cheltenham, 2018b; pp 519–529.

Canosa, A.; Moyle, B. D.; Wray, M. Can Anybody Hear Me? A Critical Analysis of Young Residents' Voices in Tourism Studies. *Tour. Analy.* **2016,** *21* (2–3), 325–337.

Clarke, A.; Chen, W. *International Hospitality Management: Concepts and Cases*,1st ed.; Butterworth-Heinemann: Oxford, 2007.

Cooper, J. A.; Alterman, D. H. Cancelling March Madness Exposes Opportunities for a More Sustainable Sports Tourism Economy. *Tour. Geog.* **2020**. doi.org/10.1080/14616688.2020.1759135

Cox, C. J.; Copper, C. L. The Making of the British CEO: Childhood, Work Experience, Personality, and Management Style. *Acad. Manage. Perspect.* **1989,** *3* (3), 241–245. DOI: 10.5465/ame.1989.4274744

Drennan, J.; Kennedy, J.; Renfrow, P. Impact of Childhood Experiences on the Development of Entrepreneurial Intentions. *Int. J. Entrepren. Innov.* **2005,** *6* (4), 231–238. DOI: 10.5367/000000005775179801

Drewery, D.; McCarville, R. Service, Emotional Satisfaction and Behaviour Patterns. In *The Routledge Handbook of Consumer Behaviour in Hospitality and Tourism*, Dixit, S. K., Ed.; Routledge: Abingdon, 2017; pp 177–184.

Entman, R. Framing: Toward Clarification of a Fractured Paradigm. *J. Commun.* **1993,** *43* (4), 51–58.

Fan, D. X. F.; Zhang, H. Q.; Jenkins, C. L.; Tavitiyaman, P. Tourist Typology in Social Contact: An Addition to Existing Theories. *Tour. Manage.* **2017,** *60*, 357–366.

Fox, D.; Gouthro, M. B.; Morakabati, Y.; Brackstone, J. *Doing Events Research: From Theory to Practice*; Routledge: Abingdon, 2014.

Gossling, S.; Scott, D.; Hall, C. M. Pandemics, Tourism and Global Change: A Rapid Assessment of COVID-19. *J. Sustain. Tour.* **2020**. doi.org/10.1080/09669582.2020.1758708

Gov.uk. Closing Certain Businesses and Venues, n.d. https://www.gov.uk/government/publications/further-businesses-and-premises-to-close/further-businesses-and-premises-to-close-guidance

Gov.uk. Government to Offer Economic Assistance to the Events Industry During COVID-19. https://petition.parliament.uk/petitions/301186

Hang, H.; Aroean, L.; Chen, Z. Building Emotional Attachment During COVID-19. *Ann. Tour. Res.* 2020. DOI: 10.1016/j.annals.2020.103006

Hudson, S.; Hudson, L. *Marketing for Tourism, Hospitality and Events*; London: Sage, 2017.

Hutton, M. Neither Passive Nor Powerless: Reframing Economic Vulnerability via Resilient Pathways. *J. Market. Manage.* **2017,** *32* (3–4): 252–274.

Jamal, T.; Budke, C. Tourism in a World with Pandemics: Local-Global Responsibility and Action. *J. Tour. Futures* **2020**. https://doi.org/10.1108/JTF-02-2020-0014

Kock, F.; Norfelt, A.; Josiassen, A.; Assaf, A. G.; Tsionas, M. G. Understanding the COVID-19 Tourist Psyche: The Evolutionary Tourism Paradigm. *Ann. Tour. Res.* **2020**. DOI: 10.1016/j.annals.2020.103053

Lee, H.; Jai, T. M.; Li, X. Guests' Perception of Green Hotel Practices and Management Responses on TripAdvisor. *J. Hospital. Tour. Technol.* **2016,** 7 (2), 182–199.

Ladsawut, J.; Nunkoo, R. Consumer Empowerment in the Hospitality Industry. In *The Routledge Handbook of Destination Marketing*; Routledge: Abingdon, 2018; pp 480–487.

Lugosi, P.; Robinson, R. N. S.; Golubovskaya, M.; Foley, L. The Hospitality Consumption Experiences of Parents and Careers with Children: A Qualitative Study of Foodservice Settings. *Int. J. Hospital. Manage.* **2016,** *54*, 84–94.

Ma, J.; Scott, N. Customer Delight from Hospitality and Tourism Experience. In *The Routledge Handbook of Consumer Behaviour in Hospitality and Tourism*; Dixit, S. K., Ed.; Routledge: Abingdon, 2017; pp 218–230.

Martin-Fuentes, E.; Mateu, C.; Fernandez, C. The More the Merrier? Number of Reviews Versus Score on TripAdvisor and Booking.com. *Int. J. Hospital. Tour. Admin.* **2020,** *21* (1), 1–14.

Minto-Coy, I. Diasporas and Development: An Assessment of the Irish Experience for the Caribbean, Caribbean Papers 7. Centre for International Governance Innovation, London, 2009.

Minto-Coy, I. Diaspora Engagement for Development in the Caribbean. In *Diasporas, Development and Governance*; Chikanda, A., Crush, J., Walton-Roberts, M., Eds.; Springer: New York, 2016; pp 121–139.

Mkono, M.; Hughes, k. Eco-guilt and Eco-Shame in Tourism Consumption Contexts: Understanding the Triggers and Responses. *J. Sustain. Tour.* **2020**. DOI: 10.1080/09669582.2020.1730388

Nickson, D. *Human Resources Management for the Hospitality and Tourism Industries*; Routledge: Abingdon, 2013.

Osman, O. A. E. Innovation in Product/Service Development and Delivery. In *The Routledge Handbook of Destination Marketing*; Gursoy, D., Chi, C. G., Eds.; **2018,** pp 429–438.

Page, S. J. *Tourism Management: An Introduction*; Routledge, New York, 2013.

Quinlan, C. *Business Research Methods*. Cengage: London, 2011.

Rivera, M. A. Hitting the Reset Button for Hospitality Research in Times of Crisis: Covid19 and Beyond. *International Journal Hospitality Management.* **2020**. DOI: 10.1016/j.ijhm.2020.102528

Robins, G. *Doing Social Network Research*; Sage: London, 2015.

Séraphin, H.; Yallop, A. An Analysis of Children' Play in Resort Mini-Clubs: Strategic Implications for the Hospitality and Tourism Industry, *World Leisure J.* **2019a**. DOI: 10.1080/16078055.2019.1669216

Séraphin, H.; Yallop, A. Proposed Framework for the Management of Resorts Mini Clubs: An Ambidextrous Approach. *Leisure Studies* **2019b,** *38* (4), 535–547.

Séraphin, H., Green, S. The Significance of the Contribution of Children to Conceptualising and Branding the Smart Destination of the Future. *Int. J. Tour. Cities* **2019,** *5* (4), 544–559.

Séraphin, H. & Dosquet, F.; Ambaye, M., Capatina, A. DRA Model and VOLM in Tourism. *Int. J. Hospital. Tour. Syst.* **2017,** *11*, (1), 1–13.

Séraphin, H. & Dosquet, F. Travel Agencies and Tour Operators at *Kidzania*: A Proactive and Innovative Approach. In *Handbook of Research on International Travel Agency and Tour Operation Management*; Dhiman, M. C., Chauhan, V., Eds.; Hershey: IGI Global, 2019.

Séraphin, H.; Vo-Than, T. Investigating the Application of the Principles for Responsible Management Education to Resort Mini-Clubs. *Int. J. Manage. Educ.* **2020.** DOI: org/10.1016/j.ijme.2020.100377

Silver, L.; Stevens, R.; Wrenn, B.; Loudon, D. *The Essentials of Marketing Research*; Routledge: Abingdon, 2013.

Simoes, N.; Crespo, N.; Moreira, S. B. Individual Determinants of Self-Employment Entry: What Do We Really Know? *J. Econ. Surv.* **2016,** *30* (4), 783–806. doi:10.1111/joes.12111

UNICEF. Learning Through Play, 2018. https://www.unicef.org/sites/default/files/2018–12/UNICEF-Lego-Foundation-Learning-through-Play.pdf

WHO. COVID-19 Technical Guidance for Schools, Workplaces and Institutions, 2019. https://www.who.int/emergencies/diseases/novel-coronavirus-2019/technical-guidance/guidance-for-schools-workplaces-institutions.

Yeoman, I. A Futurist's Thoughts on Consumer Trends Shaping Future Festivals and Events. *Int. J. Events Fest. Manage.* **2013,** *4* (3), 249–260.

Yi, S.; LI, X.; Jai, T. M. Hotel Guests' Perception of Best Green Practices: A Content Analysis of Online Reviews. *Tour. Hospital. Res.***2018,** *18* (2), 191–202.

Yu, M.; Li, Z.; He, J.; Zhou, J. Communication Related Health Crisis on Social Media: A Case of COVID-19 Outbreak. *Curr. Iss. Tour.* **2020**. DOI: 10.1080/13683500.2020.1752632

Yu, X.; Li, X.; Jai, T. M. The Impact of Green Experience on Customer Satisfaction: Evidence from TripAdvisor. *Int. J. Contemp. Hospital. Manage.* **2017,** *29* (5), 1340–1361.

Zhang, M.; Zhang, G. Y.; Gursoy, D.; Fu, X. R. Message Framing and Regulatory Focus Effects on Destination Image Formation. *Tour. Manage.* **2018,** *69*, 397–407.

Zheng, Y.; Goh, E.; Wen, J. The Effects of Misleading Media Report About COVID-19 on Chinese Tourists' Mental Health: A Perspective Article, *Anatolia* **2020**. DOI: 10.1080/13032917.2020.1747208

APPENDIX 11.1

Total Number of Reviews and Reviews That Are Kids Club Focus.

Order of appearance	Name of resort	Destination	Total number of reviews	Number of reviews on kids club
1	Club Med Cancun	Mexico	10,475	1852
2	Melia Bali	Indonesia	8320	455
3	Sofitel Bali Nusa Dua Beach Resort	Indonesia	6085	716
4	Barcelo Maya Colonial	Mexico	9143	375
5	Club Med Bintan Island	Indonesia	4911	1332
6	Panama Jack Resorts Cancun	Mexico	10,843	595
7	Sofitel Fiji Resort & Spa	Fiji	4630	508
8	Address Dubai Mall	UAE	4238	68
9	Bali Dynasty Resort Hotel	Indonesia	3561	880
10	Rixos The Palm Dubai Hotel & Suites	UAE	6038	328
11	InterContinental Danang Sun Peninsula Resort	Vietnam	4652	205
12	Golden Sands Resort by Shangri-La	Malaysia	3936	301
13	Sandos Caracol Eco Resort	Mexico	19,813	1942
14	Seadust Cancun Family Resort	Mexico	3993	160
15	Club Med Bali	Indonesia	6301	1222
16	Holiday Inn Resort Phuket	Thailand	4488	521
17	The Westin Resort Nusa Dua, Bali	Indonesia	3787	546
18	Club Med Ixtapa Pacific	Mexico	4414	1660
19	Baia Lara Hotel	Turkey	3659	283
20	Vogue Hotel Supreme Bodrum	Turkey	43,030	115
21	Rixos Bab Al Bahr	UEA	6101	408

APPENDIX 11.1 *(Continued)*

Order of appearance	Name of resort	Destination	Total number of reviews	Number of reviews on kids club
22	Phuket Graceland Resort & Spa	Thailand	3516	207
23	Wyndham Resort Denarau Island	Fiji	1614	224
24	Movenpick Resort & Spa Jimbarab Bali	Indonesia	1401	324
25	Marival Emotions Resort & Suites	Mexico	6279	628
26	The Westin Snowmass Resort	Colorado	1135	67
27	Yeats Country Hotel Spa and Leisure Centre	Ireland	1312	233
28	Don Carlos Leisure Resort & Spa	Spain	1918	76
29	West Bay Club	Turks & Caicos	1265	101
30	Voayge Torba	Turkey	6380	114
31	Paradise Resort Gold Coast	Australia	1680	746
32	JW Marriottt Phuket Resort & Spa	Australia	5447	563
33	Liberty Hotel Lykia	Turkey	3251	415
34	Sol Katmandu Park & Resort	Spain	2692	397
35	Swiss-Garden Beach Resort Damai Laut	Malaysia	1102	83
36	Club Med Punta Cana	Dominican Republic	12,940	1698
37	Iberostar Selection Playa Pilar	Cuba	2469	155
38	TUI Blue Alcudia Pins	Spain	2138	392
39	Rixos Premium Saadiyat Island	UEA	1580	138
40	The Royal Cancun All Suites Resorts	Mexico	3348	146
41	Aliathon Resort	Cyprus	3266	923
42	Hard Rock Hotel Tenerife	Spain	4836	353
43	Buganvillas Vacation Club	Mexico	984	20

APPENDIX 11.1 *(Continued)*

Order of appearance	Name of resort	Destination	Total number of reviews	Number of reviews on kids club
44	Club Med Sandpiper Bay	Florida	6031	2067
45	Holiday Village	Spain	3286	1419
46	Grand Oasis Cancun	Mexico	13,420	165
47	Club Alize	Turkey	2453	720
48	Iberostar Selection Playa Mita	Mexico	4451	347
49	Grand Memories Splash	Dominican Republic	10,385	907
50	Novotel Bali Nusa DuAa Hotel & Residences	Bali	3578	723
51	Club Tuana Fethiye	Turkey	3851	494
52	Sandos Playacar	Mexico	19,655	1040
53	Adventure Zone	Malyasia	313	13
54	Salalah Rotana Resort	Oman	3536	47
55	Centara Karon Resort Phuket	Thailand	5659	353
56	Finest Playa Mujeres	Mexico	7459	1459
57	Club Med Cherating Malaysia	Malaysia	3976	902
58	Crown Paradise Club Camcum	Mexico	12,719	2266
59	Melia Puerto Vallarta	Mexico	5197	397
60	Barcelo Puerto Vallarta	Mexico	7047	230
61	Royal Solaris Los Cabos	Mexico	8349	639
62	Club Mac Alcudia	Spain	8034	2328
63	Suncity Hotel 7 Beach Club	Turkey	3164	219
64	Habtoor Grand Resort Autograph Collection	UAE	5378	198
65	Padma Resort Legian	Indonesia	8164	828

APPENDIX 11.1 *(Continued)*

Order of appearance	Name of resort	Destination	Total number of reviews	Number of reviews on kids club
66	Iberostar Selection Paraiso Maya	Mexico	6618	494
67	The Westin Kuala Lumpur	Malyasia	3499	108
68	Occidental Nuevo Valarta	Mexico	6099	193
69	Centara Grand Beach Resort Phuket	Thailand	4399	572
70	Hotel the Volcan Lanzarote	Spain	4912	151
71	Club Regina Pueto Vallarta	Mexico	1210	49
72	Crown Paradise Puerto Vallarta	Mexico	3467	432
73	Grand Hyatt Bali	Indonesia	5140	308
74	Atlantis The Palm	UAE	27,455	1860
75	Dremas Sands Cancun Resort & Spa	Mexico	8209	341
76	Majestic Colonial Punta Cana	Dominican Republic	21,479	1093
77	Paradisus Playa Del Carmen La Esmeralda	Mexico	11,315	1199
78	Club Med Kani Maldives	Maldives	6510	303
79	Holiday Inn Resort Krabi Ao Nang Beach	Thailand	2879	301
80	Grand Palladium Kantenah Resort & Spa`	Mexico	7842	312
81	The Ritz-Carlton Dubai	UAE	4214	187
82	The Ritz Carlton Bali	Indonesia	3430	261
83	Grand Solmar at Rancho SanLucas	Mexico	555	66
84	Occidental at Xcaret Destination	Mexico	18,868	1108
85	Hard Rock Hotel Bali	Indonesia	7980	927
86	Grand Mirage Resort & Thalsasso Spa Bali	Indionesia	5279	572
87	St James's Club Morgan Bay	St Lucia	7382	496

APPENDIX 11.1 *(Continued)*

Order of appearance	Name of resort	Destination	Total number of reviews	Number of reviews on kids club
88	Belle Vue Club	Spain	9598	623
89	Club Med Valmorel French Alp	France	2986	128
90	Hervey Bay RSL	Australia	343	28
91	Bahia Principe Grand Coba	Mexico	16,205	711
92	Hyatt Ziva Los Cabos	Mexico	8212	385
93	Kenzi Club Agdal Medina	Morocco	7217	90
94	Le Meridien Mina Seyahi Beach Resort & Marina	UAE	3245	195
95	Sherwoos Exclusive Lara	Turkey	4611	250
96	Royalton Negril	Jamaica	3876	343
97	Sofitel Dubai Downtown	UEA	5112	52
98	Royalton Bavaro Resort & Spa	Dominican Republic	4479	571
99	Shangri-La's Rosa Sentosa Resort & Spa	Singapore	4450	649
100	Landmar Hotel Costa Los Gigantes	Spain	8759	1137
101	Royalton Punta Cana Resort & Casino	Dominican Republic	13,188	915
102	Iberostar Club Cala Barca	Spain	3230	341
103	Hard Rock Hotel Penang	Malaysia	7685	532
104	Dreams Los Cabos Suite Golf Resort & Spa	Mexico	3188	271
105	Bulgari Resort Dubai	UAE	354	16
106	Los Zocos Club Resort	Spain	3386	320
107	Centara Grand Mirage Beach Resort Pattaya	Thailand	5777	356
108	Marival Distinct Luxury Residences	Mexico	3183	209
109	Novotel Bogor Resort & Covention Center	Indonesia	2365	63

APPENDIX 11.1 *(Continued)*

Order of appearance	Name of resort	Destination	Total number of reviews	Number of reviews on kids club
110	Ocean El Faro	Dominican Republic	2434	104
111	Kustur Club Holiday Village	Turkey	2392	93
112	The Grand at Moon Palace	Mexico	9423	357
113	Radisson Blu Resort Fiji Denarau Island	Fiji	6807	1226
114	Royalton Riviera Cancun Resort 7 Spa	Mexico	14,109	15,052
115	Grand Hyatt Dubai	UEA	6063	225
116	Hotle Flamingo Oasis	Spain	5640	644
117	Insotel Cala Mandia Resort & Spa	Spain	3042	430
118	Magic Aqua Rock Gardens	Spain	3864	155
119	Sunwing Kamala Beach	Thailand	1058	131
120	TUI Family Life Orquidea	Spain	2390	546

Source: The author

Data collected in February 2020

APPENDIX 2

Significance of reviews on Kids' club

Name of resort	Destination	Total number of reviews	Number of reviews on mini-clubs	Ratio reviews mini-clubs/total reviews
Club Med Cancun	Mexico	10,475	1852	18%
Melia Bali	Indonesia	8320	455	5%
Sofitel Bali Nusa Dua Beach Resort	Indonesia	6085	716	12%
Barcelo Maya Colonial	Mexico	9143	375	4%
Club Med Bintan Island	Indonesia	4911	1332	27%
Panama Jack Resorts Cancun	Mexico	10,843	595	5%
Sofitel Fiji Resort & Spa	Fiji	4630	508	11%
Address Dubai Mall	UAE	4238	68	2%
Bali Dynasty Resort Hotel	Indonesia	3561	880	25%
Rixos The Palm Dubai Hotel & Suites	UAE	6038	328	5%
InterContinental Danang Sun Peninsula Resort	Vietnam	4652	205	4%
Golden Sands Resort by Shangri-La	Malaysia	3936	301	8%
Sandos Caracol Eco Resort	Mexico	19,813	1942	10%
Seadust Cancun Family Resort	Mexico	3993	160	4%
Club Med Bali	Indonesia	6301	1222	19%
Holiday Inn Resort Phuket	Thailand	4488	521	12%
The Westin Resort Nusa Dua, Bali	Indonesia	3787	546	14%
Club Med Ixtapa Pacific	Mexico	4414	1660	38%
Baia Lara Hotel	Turkey	3659	283	8%
Vogue Hotel Supreme Bodrum	Turkey	43,030	115	0%
Rixos Bab Al Bahr	UEA	6101	408	7%

APPENDIX 11.2 *(Continued)*

Name of resort	Destination	Total number of reviews	Number of reviews on mini-clubs	Ratio reviews mini-clubs/total reviews
Phuket Graceland Resort & Spa	Thailand	3516	207	6%
Wyndham Resort Denarau Island	Fiji	1614	224	14%
Movenpick Resort & Spa Jimbarab Bali	Indonesia	1401	324	23%
Marival Emotions Resort & Suites	Mexico	6279	628	10%
The Westin Snowmass Resort	Colorado	1135	67	6%
Yeats Country Hotel Spa and Leisure Centre	Ireland	1312	233	18%
Don Carlos Leisure Resort & Spa	Spain	1918	76	4%
West Bay Club	Turks & Caicos	1265	101	8%
Voayge Torba	Turkey	6380	114	2%
Paradise Resort Gold Coast	Australia	1680	746	44%
JW Marriottt Phuket Resort & Spa	Australia	5447	563	10%
Liberty Hotel Lykia	Turkey	3251	415	13%
Sol Katmandu Park & Resort	Spain	2692	397	15%
Swiss-Garden Beach Resort Damai Laut	Malaysia	1102	83	8%
Club Med Punta Cana	Dominican Republic	12,940	1698	13%
Iberostar Selection Playa Pilar	Cuba	2469	155	6%
TUI Blue Alcudia Pins	Spain	2138	392	18%
Rixos Premium Saadiyat Island	UEA	1580	138	9%
The Royal Cancun All Suites Resorts	Mexico	3348	146	4%
Aliathon Resort	Cyprus	3266	923	28%
Hard Rock Hotel Tenerife	Spain	4836	353	7%
Buganvillas Vacation Club	Mexico	984	20	2%

APPENDIX 11.2 *(Continued)*

Name of resort	Destination	Total number of reviews	Number of reviews on mini-clubs	Ratio reviews mini-clubs/total reviews
Club Med Sandpiper Bay	Florida	6031	2067	34%
Holiday Village	Spain	3286	1419	43%
Grand Oasis Cancun	Mexico	13,420	165	1%
Club Alize	Turkey	2453	720	29%
Iberostar Selection Playa Mita	Mexico	4451	347	8%
Grand Memories Splash	Dominican Republic	10,385	907	9%
Novotel Bali Nusa DuAa Hotel & Residences	Bali	3578	723	20%
Club Tuana Fethiye	Turkey	3851	494	13%
Sandos Playacar	Mexico	19,655	1040	5%
Adventure Zone	Malyasia	313	13	4%
Salalah Rotana Resort	Oman	3536	47	1%
Centara Karon Resort Phuket	Thailand	5659	353	6%
Finest Playa Mujeres	Mexico	7459	1459	20%
Club Med Cherating Malaysia	Malaysia	3976	902	23%
Crown Paradise Club Camcum	Mexico	12,719	2266	18%
Melia Puerto Vallarta	Mexico	5197	397	8%
Barcelo Puerto Vallarta	Mexico	7047	230	3%
Royal Solaris Los Cabos	Mexico	8349	639	8%
Club Mac Alcudia	Spain	8034	2328	29%
Suncity Hotel 7 Beach Club	Turkey	3164	219	7%
Habtoor Grand Resort Autograph Collection	UAE	5378	198	4%

APPENDIX 11.2 *(Continued)*

Name of resort	Destination	Total number of reviews	Number of reviews on mini-clubs	Ratio reviews mini-clubs/total reviews
Padma Resort Legian	Indonesia	8164	828	10%
Iberostar Selection Paraiso Maya	Mexico	6618	494	7%
The Westin Kuala Lumpur	Malyasia	3499	108	3%
Occidental Nuevo Valarta	Mexico	6099	193	3%
Centara Grand Beach Resort Phuket	Thailand	4399	572	13%
Hotel the Volcan Lanzarote	Spain	4912	151	3%
Club Regina Pueto Vallarta	Mexico	1210	49	4%
Crown Paradise Puerto Vallarta	Mexico	3467	432	12%
Grand Hyatt Bali	Indonesia	5140	308	6%
Atlantis The Palm	UAE	27,455	1860	7%
Dremas Sands Cancun Resort & Spa	Mexico	8209	341	4%
Majestic Colonial Punta Cana	Dominican Republic	21,479	1093	5%
Paradisus Playa Del Carmen La Esmeralda	Mexico	11,315	1199	11%
Club Med Kani Maldives	Maldives	6510	303	5%
Holiday Inn Resort Krabi Ao Nang Beach	Thailand	2879	301	10%
Grand Palladium Kantenah Resort & Spa`	Mexico	7842	312	4%
The Ritz-Carlton Dubai	UAE	4214	187	4%
The Ritz Carlton Bali	Indonesia	3430	261	8%
Grand Solmar at Rancho SanLucas	Mexico	555	66	12%
Occidental at Xcaret Destination	Mexico	18,868	1108	6%
Hard Rock Hotel Bali	Indonesia	7980	927	12%
Grand Mirage Resort & Thalsasso Spa Bali	Indionesia	5279	572	11%

APPENDIX 11.2 *(Continued)*

Name of resort	Destination	Total number of reviews	Number of reviews on mini-clubs	Ratio reviews mini-clubs/total reviews
St James's Club Morgan Bay	St Lucia	7382	496	7%
Belle Vue Club	Spain	9598	623	6%
Club Med Valmorel French Alp	France	2986	128	4%
Hervey Bay RSL	Australia	343	28	8%
Bahia Principe Grand Coba	Mexico	16,205	711	4%
Hyatt Ziva Los Cabos	Mexico	8212	385	5%
Kenzi Club Agdal Medina	Morocco	7217	90	1%
Le Meridien Mina Seyahi Beach Resort & Marina	UAE	3245	195	6%
Sherwoos Exclusive Lara	Turkey	4611	250	5%
Royalton Negril	Jamaica	3876	343	9%
Sofitel Dubai Downtown	UEA	5112	52	1%
Royalton Bavaro Resort & Spa	Dominican Republic	4479	571	13%
Shangri-La's Rosa Sentosa Resort & Spa	Singapore	4450	649	15%
Landmar Hotel Costa Los Gigantes	Spain	8759	1137	13%
Royalton Punta Cana Resort & Casino	Dominican Republic	13,188	915	7%
Iberostar Club Cala Barca	Spain	3230	341	11%
Hard Rock Hotel Penang	Malaysia	7685	532	7%
Dreams Los Cabos Suite Golf Resort & Spa	Mexico	3188	271	9%
Bulgari Resort Dubai	UAE	354	16	5%
Los Zocos Club Resort	Spain	3386	320	9%
Centara Grand Mirage Beach Resort Pattaya	Thailand	5777	356	6%
Marival Distinct Luxury Residences	Mexico	3183	209	7%

APPENDIX 11.2 *(Continued)*

Name of resort	**Destination**	**Total number of reviews**	**Number of reviews on mini-clubs**	**Ratio reviews mini-clubs/total reviews**
Novotel Bogor Resort & Covention Center	Indonesia	2365	63	3%
Ocean El Faro	Dominican Republic	2434	104	4%
Kustur Club Holiday Village	Turkey	2392	93	4%
The Grand at Moon Palace	Mexico	9423	357	4%
Radisson Blu Resort Fiji Denarau Island	Fiji	6807	1226	18%
Royalton Riviera Cancun Resort 7 Spa	Mexico	14,109	15,052	107%
Grand Hyatt Dubai	UEA	6063	225	4%
Hotle Flamingo Oasis	Spain	5640	644	11%
Insotel Cala Mandia Resort & Spa	Spain	3042	430	14%
Magic Aqua Rock Gardens	Spain	3864	155	4%
Sunwing Kamala Beach	Thailand	1058	131	12%
TUI Family Life Orquidea	Spain	2390	546	23%
	Ratio	**6137.4**	**655.975**	**11%**

Source: The author

Data collected in February 2020

CHAPTER 12

REBOOTING AND TRANSFORMING TOURISM IN POST-COVID-19 PANDEMIC: A PERSPECTIVE FOR INDIA

MANISHA SEAL[1] and SHWETASAIBAL SAMANTA SAHOO[2]

[1]*Tourism & Travel Management, Jyoti Nivas College Autonomous, Koramangala, Bengaluru, Karnataka, India*

[2]*Faculty of Management Studies, Sri Sri University, Sri Sri Bihar, Cuttack, Odisha*

ABSTRACT

Purpose—The COVID-19 pandemic has an adverse influence in the global tourism industry worldwide. The need of the hour is to take early and appropriate measures to overcome the slowdown in tourism industry by analyzing its long-term impacts. In this context, the purpose of this study is to identify the impact of a pandemic on the Indian tourism and travel industry and to formulate a strategic framework to reinstate and rejuvenate the tourism industry postpandemic by emphasizing on global travel regulations and health prerequisites for domestic and international travelers.

Design/Methodology/Approach—The study is exploratory in nature and based upon a qualitative research approach. To address the issue, the sources of data collection embraced are both primary and secondary data. Opinion of and discussion with tourism professionals and extensive review of literature was done in order to identify present and future scenario of tourism industry in new normal of COVID-19.

Post-COVID Tourism and Hospitality Dynamics: Recovery, Revival, and Re-Start.
Umendra Narayan Shukla & Sharad Kumar Kulshreshtha (Eds.)

Findings—The investigation exposes that tourism which may experience numerous complications in resurrection but this does not mean that it has been collapsed completely. As the work discusses and emphasizes on rebooting and rebuilding of the tourism sector in the post-COVID-19 pandemic, it advocates on emphasizing on promotion of alternative tourism in COVID-19 and encouraging "domestic tourism" with being the present focus. Tourism companies need to redesign their products and other facilities to align with the post-corona world. The study elucidates strategic measures to promote tourism post-COVID-19 pandemic with regulated travel procedures, framework, international travel policy, and medical kiosk along with health prerequisites for safe and secure international traveling postpandemic global scenario.

Originality/Value—Being health and safety the prime concern, the researcher has developed and suggested a 3S (SSS) model emphasizing screening, sanitation, and social distancing that will provide valuable insight on guidelines for the central government, state governments, and other stakeholders in developing required operational strategy so to minimize the evils of pandemic on Indian tourism. The relevance of discontinuity theory of immunity also has been discussed to encounter a pandemic.

12.1 INTRODUCTION

COVID-19 outbreak, being originated in Wuhan city of China, is now the major problem of the world. It was accepted as pandemic by World Health Organization (WHO) on March 11, 2020. Keeping in mind the latest information, it can be righty said that the world has never encountered such a pandemic in the last century. Tourism is considered as the most sensitive sector to crises such as natural disasters, terrorist attacks, wars, and other kinds of unanticipated phenomena. The global tourism industry is dynamic in nature. The rapid change in various occurrences in global environment brings a huge impact on country's tourism and travel industry. The enduring COVID-19 pandemic has triggered a severe crisis not only for tourism industry but also caused heavy injury to the global economy. The tourism and travel industry has become one of the worst sufferers of the crisis, as countries across the globe imposed travel restrictions on tourist's movement by closing their borders and suspending their international air carriers. During this global pandemic, the world is forced to survive under complete lockdown or partial lockdown by restricting the human movement.

The world has completely grounded the tourism industry which was in the midst of a dream run. There is a sudden fall in tourism demand that can be viewed in the backdrop of the robustness experienced by the industry in the last two decades due to this ongoing pandemic (UNWTO). In the present scenario, the global tourism industry undergoes the loss of 850 million to 1.1 billion international tourists' inflow worldwide. It also records the degradation of 910 billion to 1.2 trillion US dollars in its export revenue from tourist arrivals. The world's biggest labors intensive industry may experience the loss of 100–120 million jobs pertaining to tourism industry. It is also clear that the impact COVID-19 pandemic is going to be more devastating than damage caused by severe acute respiratory syndrome (SARS) in 2002–2003. India's tourism industry is undergoing a vulnerable phase as its major sources of foreign tourist- generating countries such as the United States of America, the United kingdom, China, Germany, and Russia are the worst hit by the COVID- 19 pandemic. Numerous micros to small- and middle-level tourism stakeholders, entrepreneurs, local vendors, and freelance tourism service caterers are in the verge of losing their livelihood. The world Travel and Tourism Council estimated 12–14% fall in employment status stands true for India as total loss of employment in this sector is projected as 5–6 million in year 2020 (Babu, 2020). The tourist arrival data from government sources suggest that foreign tourist arrivals (FTA) have been dropped 9.3% on a monthly basis. The tourism industry of India receives most of the domestic and international tourists during the summer season. With the cascading effect of coronavirus, there could be seen a deep negative impact on tourism industry that has indeed started to suffer huge losses. Since tourism products are not the physical products, the demand for the same cannot be postponed. So, some strategic approach is required to deal with such pandemic's impact keeping in mind the present and future survival of tourism industry. The recovery is expected to start from domestic tourism and gradually the international tourism. The industry needs to be extra cautious and makes special arrangement for international travel with regulated norms. Since the global pandemic risk still continues, sanitation, screening, and social distancing will be primary concern in product development and delivery. The world is going to recover from this COVID-19 pandemic as it is been previously recovered from any of such crises. This is the first time it is seen that this global pandemic COVID-19 of such unparalleled scale broadcasted in real time. Most of the people of the world would never contract the coronavirus at all and those people will feel fine and recover from the virus quickly. They will be the first virus-free travelers of the world that would aid to breathe new life into tourism industry and the related service sectors. Closing down the borders and lockdowns are not the viable solution of fighting the virus in long run. The governments, stakeholders, and the travelers themselves must be alert and well prepared and take advantages of globalization and technological revolution to pave the path for survival of tourism industry. Against the backdrop of heightened global uncertainty both for tourists and tourism sector, emphasizing on resilience and recovery of the Indian tourism sector is the prerequisite for sustainable and responsible growth of travel and tourism business. In order to revive and restrengthen India's tourism industry an innovative strategic action plan to be identified with new perspective. During the emergence of COVID-19 pandemic worldwide, when the global community is undergoing huge economic setbacks and social instability, the global medical fraternity across the world working constantly in order to identify solution to this global issue. Therefore, in the midst of such crucial timeframe, it is an urgent need for the tourism researchers to identify the innovative ways to deal with the situation and to recover the tourism industry. In this context, the study aims at addressing the issues in revival and rejuvenation of the Indian tourism industry impacted by COVID-19. The main objectives of the study are

- to identify the adverse impact of COVID-19 on tourism industry of India.
- to develop and recommend strategic perceptive on reinstate and rejuvenate tourism industry in India during and post-COVID-19 pandemic.

12.2 REVIEW OF LITERATURE

The human history is perforated with many infectious diseases that left thousands of millions of people dead. Some of the well-known examples are the Great Plague and the Spanish flu. The first pandemic of the 21st century is not the COVID-19. The first one was the SARS of 2002 that infected 8000 people causing 774 deaths in 26 countries of the world (Wilder-Smith, 2006). According to WHO, COVID-19, which is also known as coronavirus, appeared in December 2019. A 61-year-old person in China was diagnosed to have similar symptoms with the SARS virus on January 9, 2020 and on January 11, 2020; the person was confirmed as the first case, announced by the Wuhan Municipal Health Commission ("China reports first dead," 2020). Tourism is the short-term movement of people to destinations outside their normal places of domicile and work. It also includes various activities undertaken by people during their stay at the destination, and the created facilities and services provided to them to meet their needs (Collins-Kreiner and Wall, 2015). Tourism is defined as people traveling to and staying in abroad for a period of over 24 h (Vijaya, 2016). It also includes the motivations, expectations, and experiences of the tourists and acceptance of the host residents and the roles played by the numerous institutions and agencies to facilitate the travel and stay of travelers (Collins-Kreiner and Wall, 2015). Tourism has emerged as a major industry globally with an annual average growth rate of 4–5%. It contributes toward 8% of the global GDP and 10% of employment (WTO, 2020). Global tourism is affected by many types of intrusive events, such as terrorist attacks like 9/11, epidemic outbreaks like SARS-CoV-2, MERS-CoV, Ebola, Swine flu, and so on in the past (Wen et al., 2020). However, the outbreak (COVID-19) originated from China has severely impacted almost every industry, including tourism worldwide (Yeh, 2020). The virus started spreading to all continents through air transportation causing infection exponentially (Nicolaides et al., 2020). Consequently, to stop spreading of such virus, many countries partially/completely shut down their boarder and cancelled all flights and events such as sports, pilgrimage, entertainment, and MICE activities. During the last years, outbreak in diseases has decimated the escalating growth of tourism development and has brought down its contribution to GDP. This is owing to the fact that one of the predefined characteristics of tourism products is "unstable demand" due to unexpected occurrence of natural calamities or outbreak in diseases (Chiranjib, 2020). As compared to 1918 epidemic, the recent COVID-19 pandemic has a particularity. The travel industry in 2020 can never be compared with what the industry was in 1918. People never traveled so much for pleasure, business, or education purposes; not much players were involved to facilitate their travel. So, the pandemic has badly affected the industry by suspending the supply and the demand of tourism services (Papanikos, 2020). The industry is still vulnerable to different types of crises such as economic crises, political disturbances, terrorism, natural disaster, and epidemics (Yozcu and Cetin, 2020). This is because of perishable nature of tourism products. According to UNWTO, the COVID-19 pandemic has had a significant impact on tourism industry, and to respond to pandemic, 100% of the world's tourist destinations have introduced travel restrictions. The first travel restriction due to COVID-19 was imposed by North Korea on January 22. The North Korean administration stated the restriction for tourists who wanted to travel the country entering via China ("North Korea Bans Foreign Tourists," 2020). To prevent the spread of virus, China started to apply travel restrictions on January 24, 2020, in its 13 cities. Thus, after North Korea, it was China and soon after other countries of the world imposed restrictions on tourists entering the country ("China Expands Virus Lockdown," 2020). With the restrictions of travel opportunities, Chinese tourists who spend billions annually were drastically reduced affecting the tourism industry adversely. The Chinese government also ordered closure down of major factories which ultimately reduced the flow of goods through global supply chains (Ozili and Arun, 2020). The virus virtually affected almost all parts of the hospitality and tourism value chain across the countries. Cancellation of flights and events, closing down the attractions, and shutting down booking of the accommodations are immediately felt due to the impact of COVID-19. Restaurants had to close as well, though in some countries, a switch to take-away/delivery sales allowed some to continue operations. Bankruptcies and lay-offs also were reported, with British airline FlyBe succumbing first to market pressure, declaring bankruptcy on March 5, 2020 (Business Insider, 2020). Among the service sectors in India, the tourism and hospitality industry is one of the essential drivers of development. Indian tourism has striking potentials seeing the rich cultural and historical heritage, variety in ecology, terrains, and places of natural beauties spread across the country. Besides generating the foreign exchange earnings, it is also a great source of employment for the people of the country (Patel et al., 2020). As stated by WTTC (World Travel & Tourism Council), India had been ranked third position among 185 nations in terms of its travel and tourism's total contribution to GDP in the year 2018. According to the Travel and Tourism Competitiveness Report 2019 published by the World Economic Forum, India was ranked 34th in travel and tourism (WTTC, 2020). Due to the ban on domestic and as well as international air travel, Indian Tourism in particular has plunged into the darkness during the summer. The COVID-19 pandemic has required and now used as a revolutionary catalyst to bring reformation in various industries including tourism (Mair, 2020). Anytime soon, the travel agencies and tour operators are going to face the financial crisis as tourist arrivals to India will be declined. The way the cancellations have taken place for different types of tourism products, all segments of tourism such as domestic, inbound, outbound, and MICE activities will be affected due to coronavirus. The decrease in FTA to India results in decreasing in the most favorable source for foreign exchange earnings; at the same time, revenue generated by hotel industry also will be affected adversely (Vineet, 2020). The Indian tourism industry is projected to incur a revenue loss of Rs. 1.25 trillion in calendar 2020 as a fall out of the shutdown of hotels and suspension in flight operations after the onset and spread of the coronavirus (COVID-19) pandemic (Das, 2020). This indicates that the COVID-19 pandemic has severely impacted the tourism industry necessitating a strategic approach to deal with the situation.

12.3 METHODOLOGY

The study is exploratory in nature and based upon qualitative research approach. The researchers have adopted the qualitative approach as the quantitative approach will not be suitable in the present situation when the future is not clear and also researchers never had similar experiences before. The source of data collection embraced is both primary and secondary data. Primary data are collected by taking the opinion of various tourism professionals in Facebook chat, WhatsApp chat, and LinkedIn chat and through discussion in webinar. Extensive review of literature from online sources was done to carry out the study. To address noted issues created due to pandemic, the researcher developed 3-S Model (SSS) and discussed discontinuity theory of immunity.

12.4 IMPACT OF COVID-19 ON TOURISM INDUSTRY IN INDIA

Early in 2020, the tourism industry of India was on the brink of accelerating its pace toward development and promotion of tourism in the country. The country was not aware of the impact of the biological disaster coronavirus officially called COVID-19, is a threat originated from Hubei province in Wuhan city of Central China in the month of December 2019, and continues to adversely affect more than 213 countries (WHO, 2020). On January 30, 2020, the first confirmed case of COVID-19 was reported in Kerala and is one of the popular tourism destinations among the foreign tourists, also known as "God's own country." Since January 2020, this virus has spread in almost all the states and union territories of India.

India boasts itself with a rich and diverse envelope of variety of niche tourism such as cultural, religious, historical, rural, adventure, medical, and MICE activities. The honorable Prime Minister of India announced 21 days lockdown on March 24, 2020, and banned the international and domestic arrival and departure points through all mode of transportation. Hotel industry was strictly warned not to entertain any kind of new bookings and check-ins. This lockdown gradually affected all segments of tourism industry adversely and became an inevitable scenario and the industry is expecting to have a job loss of 38 million (Mallapur, 2020). According to the ministry of tourism, with the outbreak of pandemic, the FTA found to be downcast by about 67% yearly in the January–March quarter, while the domestic tourist arrivals were lowered by about 40%. As the situation got worsen, India announced suspension of all tourist visas till April 15 in a bid to prevent the spread of the virus. Outbreak of the pandemic also impacted the domestic tourism badly. As summer travel is the peak travel period in India, most domestic travelers book their flight tickets 2–3 weeks before the travel date; however, the travel companies found more than 30% drop in domestic travel during this summer as compared to the last year. Most of the summer holiday tourist destinations such as Kerala, Rajasthan, Goa, and Himachal Pradesh have been severely affected due to abrupt cancellation of travel reservation which has directly impacted domestic tourism and inbound tourism. The outbound travelers are restricted to travel till uncertain period. This has accounted a loss of business for Indian travel agencies and tour operators. There was a decline in the number of domestic travelers as people with emergency and for business purpose were only traveling. So far, the hotel industry was concerned in the third week of March, the industry experienced more than 65% decline in room occupancy as compared to same period of 2019. Almost all the restaurants were closed, except few were allowed for home delivery. With little relaxation in later period, the restaurants were expected to earn more than 50% revenue. Closure of the borders and effect of lockdown was badly felt on the airlines market, resulting in bankruptcy filing process by many firms (like Virgin Australia and Air Mauritius) for an industry already strained. Many of their fleets were grounded. Some of the airlines like Indigo and Go Air have taken cost-cutting steps including furloughs. Due to travel restrictions enforced by the Indian government as well as governments across the globe, various conferences and leisure travel reservations to foreign destinations have already been canceled or need to be rescheduled to uncertain dates.

The compiled data relating to the present scenario of COVID-19 cases in the country are presented in Figure 12.1 as on December 10, 2020:

The figure depicted that in India over 9.7 million confirmed cases of the coronavirus (COVID-19) and around 9.2 million patients had recovered. This indicates that the recovery rate is better as compared to the fatality rate and people in India have a better immunity to fight against pandemic infection. This will build confidence among the people to undertake travel again.

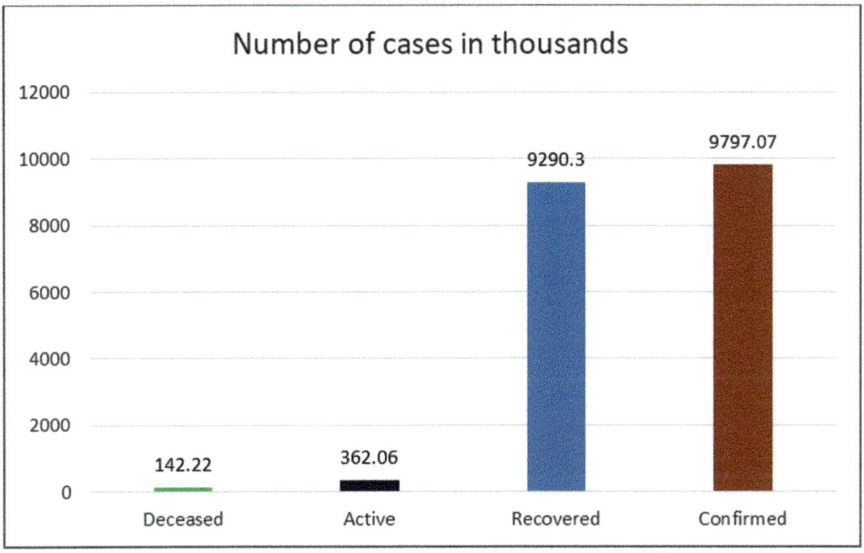

FIGURE 12.1 Diagram showing coronavirus cases in India (as of December 10, 2020).
Source: Compiled by Authors.

12.5 TOURISM SURVIVAL STRATEGY IN POSTPANDEMIC

Nobody knows how long the coronavirus will last. Once the situation gets normal, the industry will pick up. People will be ready to explore the beauty of India again. However, the travel behavior and consumption pattern will not be same as before. People will travel with much precautions and expectations. The country needs to get prepared to manage tourism during this pandemic and postpandemic in such a way that travel would be safer, environment-friendly, responsible, accessible, and affordable for all tourists. Alterative tourism products like rural tourism, wine tourism, adventure tourism, ecotourism, experiential tourism, and homestay tourism, especially lesser known tourism destinations, need to be promoted to serve the tourists and enhance their positive travel experience (Hotelier India, 2020).

- *Increased Emphasis on Domestic Tourism*

With the unlock 1.0 as announced by the Indian government, they had permitted the intrastate and interstate transportation and allowed hotels and restaurants to reopen with certain restrictions. This gradually facilitated the domestic travel. At present, domestic air travel is permitted and many destinations in India are opened up to welcome the tourists. Due to the second wave of coronavirus and restrictions of international air travel, the emphasis must be given on promotion of domestic tourism among various states within the boundaries of the country. The wound inflicted by the pandemic is so deep and hence resurgence of international travel will take some more time during postpandemic. During postpandemic, domestic travel should be encouraged to stimulate the tourism industry. Depending upon the tourist's behavior and travel pattern, pilgrimage or religious travel to various pilgrimage centers is one of the prominent travel patterns that can be viewed among Indians in various parts of the country due to their inherited religious faith. Therefore, pilgrimage tourism is one of the significant forms of domestic tourism in which tourists from every segment of the society would like to participate which enhances opportunities for domestic tourism during the crises as well. Ethnic tourism and VFR (visiting friends and relatives) tourism are also other forms of domestic tourisms which create great motivation for the travelers even during lockdown and ongoing pandemic. Being a social animal, an individual cannot live alone for a longer period of time as an urge for love and belongingness will motivate them to travel to their ethnic places, especially during various important occasions or events.

- *3-S (SSS: screening, sanitization, and social distancing) model illustrating the tourism during postpandemic*

Travel will look different in a number of ways during postpandemic. People would like to explore the world with safety precautions, face masks, and social distancing. Considering these aspects, the study has developed and recommended a model for its implementation during postpandemic scenario known as 3S (SSS) model to deal with the situation. The theory of 3S advocates on standardized procedure of "screening, sanitization, and social distancing" at various phases of travel so as to ensure healthy and safe travel for the tourists as shown in Figure 12.2:

12.5.1 PHASE 1: SCREENING, SANITIZATION, AND SOCIAL DISTANCING (3-S) AT DEPARTURE POINT

It begins with the point of origin, where the traveler must carry and provided with all the safety measures like face mask/shields, hand gloves, and sanitizer as a precautionary measure to travel by any mode of transportation such as bus, train, or air carrier. It is necessary to have sanitization of transportation terminal, passenger sanitization, and carrier sanitization by deploying proper technology for a safe travel in the current situation. There must be medical kiosk equipped with basic technical equipment and trained staff to detect and to handle any medical emergency concerning to COVID-19 pandemic. The norms of social distancing or physical distancing among the travelers must be maintained in the carrier and terminal points to maintain a considerable gap among the travelers.

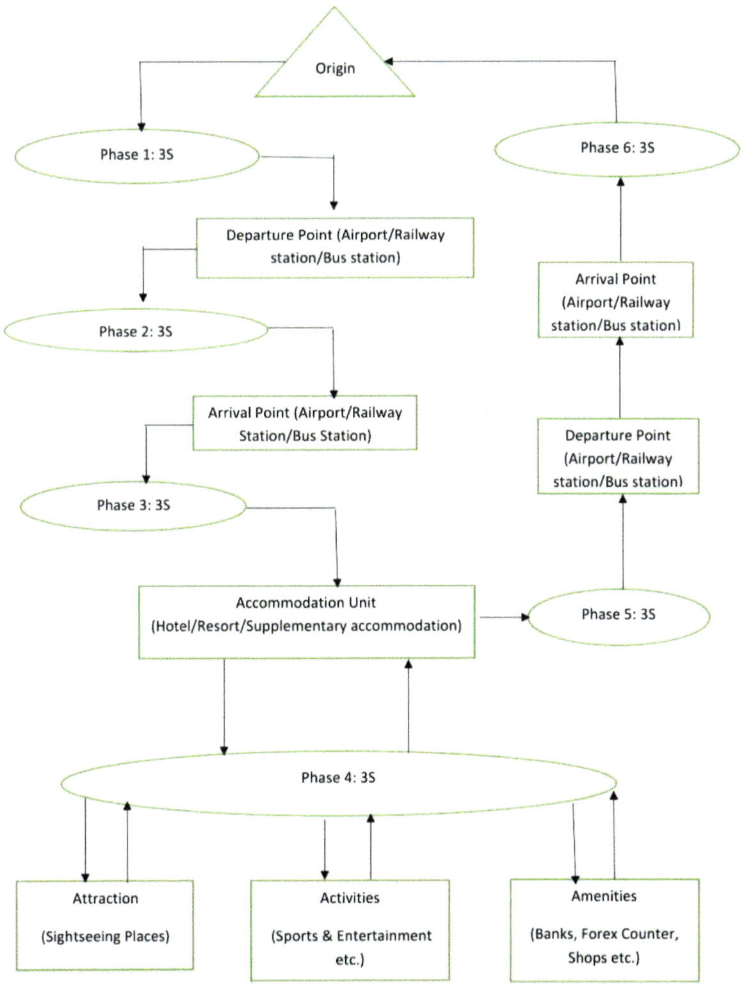

FIGURE 12.2 3-S Model (screening, sanitization, and social distancing).

Source: Own.

12.5.2 PHASE 2: SCREENING, SANITIZATION, AND SOCIAL DISTANCING (3-S) AT ARRIVAL POINT

Like phase 1, the phase 2 sanitization, screening, and social distancing procedure must take place in every arrival point, weather it is a bus station, railway station, or in airport. Simultaneously, social distancing norms must be maintained among the travelers and various service points to safeguard and restrict any spread of virus infection.

12.5.3 PHASE 3: SCREENING, SANITIZATION, AND SOCIAL DISTANCING (3-S) AT ACCOMMODATION UNIT

This phase of 3-S model of sanitization, screening, and social distancing (3-S) procedure is associated with accommodation unit which commences upon arrival and prior to the check-in. The accommodation unit must be equipped with medical kiosk or medical room with various sanitization and screening facilities for undertaking screening and sanitization of each guest before they check-in to the hotel room. At the same time, social distancing as a significant norm must be monitored strictly and followed by the staff at various service delivery points. Touchless services such as digitalized services and robot services must be encouraged wherever possible. Each time the guest goes out for leisure/business activities and comes back to the hotel, again, the sanitization and screening will be required. Guest rooms need to be sanitized thoroughly upon check out/departure of the guests. Public areas and entire hotel premises also required to be sanitized very frequently.

12.5.4 PHASE 4: SCREENING, SANITIZATION, AND SOCIAL DISTANCING (3-S) AT TOURIST ATTRACTIONS, ACTIVITIES, AND TOURIST AMENITIES

Phase 4 comprises 3-S procedure that to be conducted near various tourist attractions, during tourist activities and while using the services of various tourist amenities. The authority/in-charge must develop standardized procedure for screening and social distancing to be followed for each visitor visiting to the tourist destination, while participating in tourism activities and at usage points of amenities. Visitors spending more time at any tourist destination need to be sanitized at a certain interval.

12.5.5 PHASE 5: SANITIZATION, SCREENING, AND SOCIAL DISTANCING (3-S)—DEPARTURE POINT/ARRIVAL POINT

Phase 5 of 3-S model commences once the visitor checks out the accommodation unit and travels back toward the origin which may include all the departure points and transit/stopover points. The 3-S procedure is to be followed as discussed in phases 1 and 2.

12.5.6 PHASE 6: SANITIZATION, SCREENING, AND SOCIAL DISTANCING (3-S)—POSTARRIVAL TOWARD ORIGIN

This phase is the last and final phase of sanitization, screening, and social distancing at the arrival terminals in the origin area before tourist reaches to the place of residence.

12.6 DISCUSSION

The rebooting and transforming tourism in post-COVID-19 pandemic would be reliant upon various assumptions, anticipation, and innovative perspectives based on scientific theories, systemic studies based on past occurrences, and historical references which depict that the world is not unknown to such occurrences, that is, plague, cholera, influenzas, and flu are various forms of pandemics which outbroke in different period of time and impacted vigorously the global human race (McMillen, 2016). These historical phenomena had thrown numerous challenges for the survival of human race in a global platform and recovered miraculously to ensure human survival. At present, the world is undergoing a similar scenario. In this context, the discontinuity theory of immunity comes into picture (Pradeu and Vivier, 2016). The discontinuity theory of immunity reveals about innate and adaptive immunity of human body which throws light on slow adoptability of immune system by indicating long-lasting modifications in human biological system. The theory argues that the chronic exposure to antigens leads to immune tolerance of human body. This theory of immunity is applicable to various chronic diseases such as viral infections, cancers, and allergies which may also be applicable to highly contagious novel coronavirus infection in present scenario. The survivors and the rest of human race in the world who struggled, suffered, and survived this pandemic for 6 months to 1 year of time frame or more than that may be able to develop a natural immunity to fight against COVID-19. This anticipated scenario may bring in normalcy in human life to perform various courses of actions including ways of living, consumer behavior, and their travel pattern for numerous reasons. As per the researcher's understanding about this discontinuity theory of immunity, it advocates only on an appropriate time frame to heal and to bring in natural immunity among the people of the world. Another perspective which advocates on "vaccination" aiming to eradicate COVID-19 pandemic and to recover reinstates as well as rejuvenates human life and its way of living. The newly emerged global warriors such as doctors, scientists, medical researchers, and the whole medical fraternity are constantly investigating and researching to invent an effective, medically, and globally accepted vaccine to combat this pandemic. Therefore, it has been anticipated that the world will soon resilient and recover from the present scenario. However, a traveler is required to obtain and carry a vaccination certificate as a mandatory travel document along with passport, visa, travel insurance, and medical insurance. This will encourage the people and build confidence among them to participate in tourism which will boost up the global tourism industry including India in post-COVID era.

This cannot be denied that the pandemic has impacted the tourism industry badly covering sectors like travel agency, tour operators, hotels, air, water, and surface transportation and many others. Tourism has witnessed a significant decline during 2020. Battered and bruised by the pandemic, many of such sectors are suffering from financial crisis to sustain the tourism industry in post-COVID-19. In such a situation, the Government of India needs to play an important role by easing such financial stress for the survival and growth of industry. The need of the hour is the reduction of GST on various travel trade establishments. Introduction of a tax holiday for such service providers for a specified period may help in revival of the industry. Furthermore, these companies need moratorium on principal amount, working capital, interest payments, loans, and overdrafts for another 1 year (Radhakrishnan, 2020) that will help them rebuild and sustain.

12.7 CONCLUSIONS

The world's history is a testimony to the fact that pandemic cannot refrain travel and tourism phenomena worldwide. It can only create obstacles and hindrance to defer the growth of tourism and travel pattern. The COVID-19 pandemic has caused a life threat and fear among the prospect visitors worldwide which need to be addressed with the recommended strategic measures as discussed in the study to combat this challenge. During and post-COVID-19 pandemic also, the theory of 3-S must be followed by the tourism and travel industry as an important norm globally to prevent any form coronavirus infection as a precautionary measure. Many experts are recommending the development and promotion of domestic tourism in India for the growth and survival. At the same time, the inbound tourism cannot be ignored as it generates lot of economic benefits for the nation. Both domestic and inbound tourism should be promoted in India following the procedure as mentioned in 3-S model in coming days. Further, the travel pattern and behavior of the people are going to be changed. People will more prefer for special interest and alternative forms of tourism and look for the destinations that are safe to travel and less crowded. So, governments and the other stakeholders must design tourism products and develop their policies keeping in mind this new orientation of traveler and promote safe tourism; like democracy tourism is for the people, by the people, and of the people. It is never going to die. The industry is going to restore its lost glory. It is just a matter of time and new ways of looking at tourism with certain restrictions and precautions.

KEYWORDS

- **tourism industry**
- **COVID-19 pandemic**
- **alternative tourism**
- **domestic tourism**
- **3-S model and strategic tourism development**

REFERENCES

Babu, S. Policy Circle, 2020. https://www.policycircle.org/economy/industry/six-ways-how-covid-19-will-change-tourism-industry/

Business Insider. UK Airline Flybe Declares Bankruptcy as Coronavirus Dooms the Already Struggling Carrier, 2020. https://www.businessinsider.de/international/uk-airline-flybe-declares bankruptcyflights-grounded-2020-3/?r=US&IR=T (accessed on Mar 31, 2020).

China Expands Virus Lockdown, Encircling 35 Million. New York Times, Jan 23, 2020. https://www.nytimes.com/2020/01/23/world/asia/china-coronavirus-outbreak.html (accessed on April 15, 2020).

China Reports First Death from Mysterious Outbreak in Wuhan Aljazeera, Jan 11, 2020. https://www.aljazeera.com/news/2020/01/china-reports-death-mysterious-outbreak-wuhan-200111023325546.html (accessed on Mar 27, 2020).

Chiranjib, K. C. Corona (COVID-19) and Tourism: Moving Forward in the Aftermath of the Fall, Articles/Research Note/Preprint, 2020.

Collins-Kreiner, N.; Wall, G. Tourism and Religion: Spiritual Journeys and Their Consequences. In Business Media Dordrecht, 2015. https://doi.org/10.1007/978-94-017-9376-6_34

Das, J. Business Standard, 2020. https://wap.business-standard.com/author/search/keyword/Jayajit-dash (accessed on May 11, 2020).

Hermann, I.; Moscoso, F. V. Voluntourism Best Practices: Promoting Inclusive Community-Based Sustainable Tourism Initiatives, New Zealand Tourism Research Institute, Auckland University of Technology, 2018.

Hotelier India. 2020. https://www.hotelierindia.com/business/12059-india-will-emerge-again-as-a-favourite-tourist-destination-after-covid-19-era-says-prahlad-singh-patel (accessed on Dec 12, 2020).

https://www.who.int/emergencies/diseases/novel-coronavirus-2019

IATA Survey Report, 2020.

Kumar, V. Indian Tourism Industry and COVID-19: Present Scenario. Journal of Tourism and Hospitality Education **2020,** *10*, 179–185. https://doi.org/10.3126/jthe.v10i0.28768

McMillen, C. W. *Pandemics: A Very Short I Introduction*; Oxford University Press, 2016.

Mallapur, C. India's Covid-19 Lockdown May Cause 38 Million Job Losses in Travel and Tourism Industry, 2020. Sroll.in (accessed on May 4, 2020).

Nicolaides, C.; Avraam, D.; Cueto-Felgueroso, L.; González, M. C.; Juanes, R. Hand-Hygiene Mitigation Strategies Against Global Disease Spreading Through the Air Transportation Network. *Risk Analy.* **2020,** *40* (4), 723–740. https://doi. org/10.1111/risa.13438

Mair, S. What Will the World Be Like After Coronavirus? Four Possible Futures. The Conversation Mar 30, 2020. https://theconversation.com/what-will-the-world-be-like-after…d=IwAR2wr9pzssSdBSxjaHaWba9 (accessed on May 2, 2020).

North Korea Bans Foreign Tourists Over China Virus: Tour Agency. Aljazeera, Jan 22, 2020. https://www.aljazeera.com/news/2020/01/north-korea-bans-foreign-tourists-china-virus-tour-agency-200122042416138.html.

Ozili, P.; Arun, T. Spillover of COVID-19: Impact on the Global Economy. *SSRN Electr. J.* **2020**. https://doi.org/10.2139/ssrn.3562570

Patel, P. K.; Shrama, J.; Kharoliwal, S. The Effects of Novel Corona Virus (Covid-19) in the Tourism Industry in India. *IJERT* **2020,** *7* (2).

Pine, R.; McKercher, B. The Impact of SARS on Hong Kong's Tourism Industry. *Int. J. Contemp. Hospital. Manage.* **2004,** *16* (2), 139–143. https://doi.org/10.1108/09596110410520034

Pradeu, T.; Vivier, E. The Discontinuity Theory of Immunity, 2016. https://immunology.sciencemag.org/content/1/1/aag0479

Salvador, M. Volunteer Tourists, Which Are Their Motivations, 2019. http://www.albasud.org/noticia/en/1132/volunteer-tourists-which-are-their-motivations (accessed on June 4, 2020).

Radhakrishnan, A. COVID-19|Post-pandemic, India's Tourism Sector Stares at 70% Job Loss. *The Hindu*, 2020. https://www.thehindu.com/news/national/kerala/coronavirus-tourism-sector-stares-at-70-job-loss/article31310234.ece (accessed on Dec 12, 2020).

The World Economic Forum. 2020. https://www.weforum.org/ agenda/2020/06/4-charts-airline-crisis-covid-way-ahead/

Tourism Economics. Total Travel Industry Impact. Çevrimiçi, 2020. https://www.ustravel.org/ toolkit/covid-19-travel-industry-research, Erişim tarihi: (accessed on April 12, 2020).

UNWTO. World Tourism Barometer: Special Focus on Impact of Covid19, May 2020.

Vijaya, A. G. Tourism Industry in India: Performance Evaluation of Tourism in Andhra Pradesh; A Ph.D Research Design Seminar Submitted to Rayalaseema University Kurnool, Andhra Pradesh, India, 2016.

WHO. Novel Corona Virus Disease (Covid-19) Situation Update Report, 2020.

World Travels and Tourism Council. Impact of Pandemic on Travel and Tourism, 2020. https://www.wttc.org/priorities/crisis-preparedness/pandemics/ (accessed on December 6, 2020).

WTO. International Tourism Highlights. Çevrimiçi, 2020. https://www.e-unwto.org/doi/pdf/10.18111/9789284421152, Erişim tarihi (accessed on April 12, 2020).

Yeh, S. S. Tourism Recovery Strategy Against COVID-19 Pandemic. *Tour. Recreat. Res.* **2020,** 1–7. https://doi. org/10.1080/02508281.2020.1805933

Yozcu, O. K.; Cetin, G. A Strategic Approach to Managing Risk and Crisis at Tourist Destinations. In Tourist Destination Management; 2019; pp 273–287.

CHAPTER 13

POST-COVID TOURISM IN VIETNAM: THE NEW NORM

JUSTIN MATTHEW PANG[1] and ADRIAN WEE[2]

[1]*Royal Melbourne Institute of Technology, Vietnam*

[2]*British University Vietnam, Vietnam*

ABSTRACT

As a country, Vietnam has done well in managing its national COVID-19 infections. Even though, the country was hit by three waves of infections, the Vietnamese government has been quick to respond. They mitigated its spread, contained the virus, and have enjoyed >2 months of being COVID-free. However, despite the prompt and efficient efforts by the government, COVID-19 infections have left an indelible mark on the hospitality and tourism industry in Vietnam resulting in new innovative processes to ensure business survivability. Many hotels have streamlined their operations, reduced their manpower, and leveraged on technology. Restaurants and food providers have reduced their dine-in capacity, stopped buffet offerings, and capitalized on food delivery. Vietnamese airline carriers now take greater measures in their safety and cleaning practices. These efforts have ensured the sustainability of Vietnamese hospitality and tourism businesses during this time of global pandemic. As many economies continue to struggle with the control of the COVID-19 virus, Vietnam, having combated and won the fight against COVID-19, is ready to apply not only what it has learnt from this episode, but also new and innovative processes of work in a post-COVID world, making this the new hospitality business norm.

Post-COVID Tourism and Hospitality Dynamics: Recovery, Revival, and Re-Start.
Umendra Narayan Shukla & Sharad Kumar Kulshreshtha (Eds.)

13.1 INTRODUCTION

13.1.1 BACKGROUND OF THE STUDY

The COVID-19 virus that emerged in December 2019 from China has had a detrimental effect on a global scale to date. The virus caused the death of more than one and a half million people to date and has also devastated the global economy, leaving no industry left unscathed. This pandemic has been attributed by many as the worst global economic crisis in recent decades, shutting down global economies almost overnight (UNWTO, 2020).

While many countries have been left struggling to manage the effects of the pandemic, that is, dealing with the control of the COVID-19 virus, the enforcement of human mobility, the care of ill-stricken citizens, an increased unemployment rate, and a shattered economy, some nations have done well to mitigate the spread of infections. The reasons for their success have been in their decisive application and expedition of pandemic control regulations. One such country is Vietnam, a developing nation located in South East Asia, with a population of 95 million people.

Vietnam, with its northern districts bordering China's Guangxi province was one of the first countries to see cross-border transmissions of COVID-19 in January 2020. These cases were quickly isolated and contained by the authorities with the passing of strict control measures and regulations. However, after a month of zero cases being reported, a resurgence of COVID-19 infections hit Vietnam, with its origin being a local citizen returning to Vietnam from the United Kingdom (VnExpress, 2020). This time, the Vietnamese government imposed stricter measures, putting several major cities on lockdown, and curtailing the movement of its citizens. By May 2020, the lockdown was lifted when no local transmissions were reported. Once again, normalcy returned to the country. However, in late July 2020, due to the illegal human trafficking of Chinese immigrants smuggled into Vietnam caused the city of Danang to go into lockdown with growing clusters of COVID-19 reported within the metropolitan area (Thanh, 2020). Other peripheral provinces also saw new cases of infections appearing in their counties. While the authorities did not want to impose another nationwide lockdown, they expedited immediate measures to contain the spread with harsh penalties for noncompliance. While they were successful in attempting to mitigate the spread, the COVID-19 virus took its toll on Vietnam, resulting in 35 deaths. Nonetheless, this effort was seen as a victory by many countries for their speedy response which made the containment possible within a span of one month.

Despite the continued and occasional recurrence of COVID-19 in Vietnam, the government is constantly able to mitigate and contain its spread, thus preventing an overwhelming situation, where medical staff and resources are unable to cope with the waves of the infected. Through these trying times, Vietnam and her economic industries have come to strengthen their solidarity against possible viral outbreaks and learnt how to better prepare and developed possible innovative contingency efforts and processes for the evitable recurrences of the COVID-19 virus.

13.1.2 SIGNIFICANCE OF THE STUDY

This study highlighted the continued development and evolution of Vietnam's war against the COVID-19 global pandemic virus by both its government and industries. Vietnam that had prior experience in handling a nationwide pandemic when they combated the Severe Acute Respiratory Syndrome (SARS) virus back in 2003, was able to draw from her past experiences. The current situation with COVID-19 has further honed the country's knowledge, and capabilities in managing this new pandemic especially in the current times of social human mobility and open media platforms.

The study also depicted how the Vietnam governments and industries are constantly adapting to the current state-of-affairs in the country and in a COVID-19-ridden world. It looked at new innovative ways to sustain their businesses and to adhere to the new safety and hygienic practices that have been imposed by the state government and international regulatory bodies.

By understanding these new innovative methods taken by Vietnam that had in the past overcome three waves of COVID-19 infections, one would be able to obtain a sample microcosm understanding of how current infected economies will evolve in the future. This will grant readers a possible notion on how industries can adapt to post-COVID world, where new business processes and practices will become the new norm.

13.1.3 OBJECTIVES OF THE STUDY

This study aimed to look at the actions taken by the Vietnamese government, hospitality and tourism industry, and business organizations on how they have evolved in dealing with the viral infections in post-COVID Vietnam. It will look at how these methodologies have enabled the businesses to remain sustainable and agile in their practices during these unprecedented times.

It also looked at how these methodologies could be the new inceptions of possible future processes of the hospitality and tourism industry, especially when Vietnam is hailed as one of the efficient countries in mitigating the spread of COVID-19 on a national level.

13.2 LITERATURE REVIEW

The hospitality and tourism industry has always been one of the first industries to bear the brunt of any global pandemic (Nankervis, 2000). The reason for this is due to the industry's business in connecting travelers to various destinations, the offering of different modes of transportation and accommodations, and the provision of services, food, and entertainment. The industry is also one that is often severely affected by global pandemics. This is because in the event of a viral outbreak, most travel will cease almost immediately due to the fear of the contagion and the imposition of travel bans by most government to mitigate and eradicate the source virus (Pizam, 2003). This phenomenon is not unlike how the Vietnam and its government reacted when faced with the current pandemic.

13.2.1 SARS AND COVID-19 PANDEMICS IN VIETNAM

Vietnam faced its initial pandemic when the SARS occurred in the South East Asian countries in late 2002. Vietnam had never encountered a pandemic prior and was initially unable to manage the pandemic. This resulted in 68 casualties of which five cases were fatal (Roos, 2003). Their economy also took a hit from the pandemic, with the hospitality and tourism having been tremendously affected. Occupancies in hotel were barely 30% for the first two quarters in 2003 (Vietnam Investment Review, 2020).

However, the Vietnamese government was quick to recover from their initial hesitation and instituted measures, formulated by China, that helped to identified potential SARS carriers, monitored their movements and whereabouts, quarantined SARS patients in designated medical facilities, employed proper personal protective equipments (PPEs) for their medical staff, comprehensive identification and isolation of suspected SARS cases and the proper timely and accurate reporting, documentation, and sharing of information within the various agencies and other countries (Anderson et al., 2004; Pang et al., 2003). The success efforts of the Vietnamese government lauded praises from the international community and they were among the first nations to be taken off the World Health Organisation's list of SARS-infected countries (World Health Organisation, 2003). This baptism of fire gave Vietnam the experience and advantage in the handling of future national-level pandemics.

Vietnam's ability to handle pandemics was tested again in late January 2020, when COVID-19 was discovered within the country. Having had prior experiences in dealing with such matters, the Vietnamese government initiated a taskforce, entitled "National Steering Committee on Epidemic Prevention" (Vietnam+, 2020). The taskforce, instituted many directives, including the closure of the national borders, social distancing, and the closure of educational centers, entertainment, and tourism sites. Essential businesses, such as food and beverage outlets, hotels, and the airlines could remain partially opened under strict conditions and regulations. These businesses had to compile with the directives of the government and take initiatives to ensure high sanitation and hygiene levels.

13.2.2 INITIATIVES TAKEN BY HOSPITALITY AND TOURISM BUSINESSES DURING COVID-19

There have been many initiatives taken during this pandemic by hospitality and tourism organizations. These initiatives range from restricting and downsizing of staff (Hao et al., 2020); cost reductions and cash savings (Hotel Business, 2020); and other operational changes (Im et al., 2021). Many of these operational changes, deemed as the most important included, changes in service methodologies (Alonso et al., 2020; Kim and Lee, 2020); new technology adoption (Baum et al., 2020); and compliance to hygiene and safety standards (Sigala, 2020; Gursoy and Chi, 2020).

These operational initiatives help to align guest's perception of safety to the hospitality and tourism establishments' offering, thereby meeting their expectations, and reducing uncertainty to use the service and products offered by the organizations (Shin and Kang, 2020). Among the above-mentioned strategies, the tactics employed by hotels range from the use of technology (Kussmann, 2020). Hotels such as Marriott, Hilton, and Hyatt have installed check-in kiosks, robot cleaning systems, and mobile check-in systems. Gracia (2020) also mentioned that many establishments have adopted advanced cleaning procedures and systems, such as electrostatic sprayers and ultra-violet lights for enhanced sanitation. Hotels have also started to streamline service systems and procedures that reduces contact touchpoints with guest, using policies, practices, processes, and people (Lemon and Verhoef, 2016).

13.3 METHODOLOGY AND DATA COLLECTION

To understand the success of Vietnam and their hospitality and tourism industries in combating the spread of the COVID-19 virus, it was necessary to obtain the different approaches taken by the hospitality and tourism industry as well as the companies within each sector of the said industry. These approaches taken by the different organizations are seen by many as the quintessential best practices in preventing COVID-19 infections due to the national success of the country.

As these measures and approaches are relatively new and currently enforced, the only credible way to collect reliable data would be from news articles featured on notable news websites and from observations conducted by the research team. This form of research noted as "Content Analysis" is arguably the most valid given the current situation where need of information is urgent (Neuman, 2007). Moreover, the informations were validated through a second means of qualitative research where the content obtained from websites will be cross-referenced to actual site visitations, observations noted and interviews conducted with the staff from the respective hospitality and tourism companies where these measures and approaches have been implemented.

The websites that were used are the credible and mainstream news websites from Vietnam. These websites often highlighted the different methodologies and processes taken by the different hospitality and tourism companies, specific to the hotels, restaurants, and aviation sector. These websites are generally reliable given that they are the most common endorsed websites from the Vietnamese people and authorized by both the public and the government bodies. Visitation and site inspections, with interviews from the employees and staff from the hospitality and tourism companies also helped to confirm these findings and allowed greater insight as to why these processes were adopted.

13.3.1 THE EFFECTS OF COVID-19 IN THE HOSPITALITY AND TOURISM SECTORS IN VIETNAM

COVID-19 has had a profound effect on the global economy especially for the hospitality and tourism sector. The spread of the virus has forced governments to take unprecedented action to mitigate the rise in infection, resulting in the grounding of most airline carriers and the closure of international borders. With the hospitality and industry at the forefront and supporting global trade and businesses, it has seen the most challenging times since the 1920s' Great Depression (AHLA, 2020).

However, for Vietnam, it has had the foresight to pre-empt the initial spread of the COVID-19 virus when China first announced about their national epidemic. It went on to successfully deal with three waves of COVID-19 infections, resulting in the country returning to normalcy within a brief period of time. While Vietnam continues to stand guard against the spread of COVID-19 by closing its borders, it is able to depend on the domestic market for tourism, unlike other smaller countries, such as Singapore, which is heavily reliant on an international market. Nonetheless, despite the best efforts by the government, Vietnam's hospitality and tourism encountered resonating challenges in their growth and operations during this COVID-19 pandemic.

13.3.1.1 EFFECTS ON THE HOTEL SECTOR IN VIETNAM

The hotel sector in Vietnam has been growing steadily since 2015 with an increase in properties and rooms. This growth is spurred on by greater demand from international travelers. It also saw an increase in room rates. However, the hotel sector took a severe blow due from the spread of COVID-19 and the subsequent closure of borders. The occupancy and the revenue per available room (RevPAR) of many international luxury hotels plummeted when the first wave of COVID-19 hit Vietnam in February 2020. The occupancy and RevPAR in Vietnam of March 2020 compared to March 2019 were vast differences of −71% and −75%, respectively (STR, 2020a). While the cases in Vietnam were generally as high when compared with other countries such as China and Korea (then), the decrease in occupancy and RevPAR was fuelled more so from fear and speculation.

The decreases in occupancy and RevPAR were further exacerbated when a second wave of COVID-19 spread through Vietnam in April 2020. This time, the Vietnamese government took more drastic measures, with the full closure of borders, implementing a soft lockdown and enforced social distancing among its citizens. The impact of these measures hit the hotels hard, with Vietnam's April 2020 occupancy and RevPAR plummeting by more than 81% and 86%, respectively, compared with the previous year, due to a lack of international visitors traveling to Vietnam (STR, 2020b). To sustain their businesses, hotels were forced to employ lesser staff and close many of their ancillary operations such as restaurants, spas, retail shops to reduce overhead cost.

13.3.1.2 EFFECTS ON THE AVIATION SECTOR IN VIETNAM

The first sector hit by the COVID-19 infection was the aviation sector. The aviation sector in Vietnam was touted to the fastest growing country in terms of airline passengers from 2017 and is speculated to grow at the rate of 7.8% till 2040 (Clark, 2019). This increase in passenger load is due to Vietnam's economic growth as a developing country, with the need for business travel and having more local attractions gaining popularity. With rising income levels, more Vietnamese are able to travel between the two major cities of Hanoi and Ho Chi Minh City and to other provinces.

However, with the spread of COVID-19, Vietnam's national carrier, Vietnam Airlines suffered a loss of 26.5% in revenue which accounts for approximately US$110 million in the first quarter of 2020. The airline expects further losses in 2020, amounting to US$687 million. The national carrier has carried out crisis mitigation measures, such as the cutting of staff salaries and enforcing unpaid leave to minimal the loss of US$193 million. Vietnam Airlines has also appealed to the Vietnamese government to borrow up to US$515 million to prevent the company from going into insolvency (Tu, 2020). Other private Vietnamese carriers, such as Bamboo Airlines and Vietjet have also taken drastic measures with the retrenchment of foreign labor.

13.3.1.3 EFFECTS ON THE FOOD AND BEVERAGE SECTOR IN VIETNAM

The food and beverage sector in Vietnam is a vibrant scene with the country being hailed as a food haven by many food connoisseurs. It is generally dominated many independent small and medium enterprises (SMEs) with more than 540,000 outlets (CNBC, 2018). The projected growth rate of the food and beverage sector in Vietnam is forecasted at about 6.8%, due to the increase in the higher earnings of the Vietnamese and the growing trend of more people partaking in meals from commercial eateries (Ministerie van Landbouw, Natuur en Voedselkwaliteit, 2018; Statista, 2020).

The spread of the COVID-19 virus and the initial lockdown of Vietnam in the first half of 2020, saw the revenues of many food and beverage outlets drop by up to 50%. Even after the lifting of the people movement control order, food and beverage outlets are unable to recover to pre-COVID-19 levels with revenue still trailing at only 80% to 50% of previous earnings (Thong, 2020). The reason for this is due to the lack of international travelers and visitors and the drop in domestic tourism. Moreover, given that most

Vietnamese eateries have the similar characteristics of social dinning, that is, the sharing of common cutleries, communal-style eating, and crammed eating quarters, have put many Vietnamese off from patronizing these small, medium food enterprises due to the encouraged social distancing campaign enforced by the Vietnamese government.

13.3.2 THE CURRENT PRACTICE NORMS OF THE VIETNAMESE HOSPITALITY AND TOURISM INDUSTRIES

COVID-19 having had a detrimental effect on the various sectors in the Vietnamese hospitality and tourism industry, ravaging their desired prospective growth and significantly reducing their earnings. However, businesses within the industry have stayed resilient, and have adhered to the overlining directives of the Vietnamese government to ensure that widespread infections are kept at bay and reduced. The various hospitality and tourism sectors have devised measures to curtail the spread of the infections to the point where currently these approaches are accepted as the norms of Vietnamese society. Below highlights some of these initiatives taken by the various sectors.

13.3.2.1 NEW NORMS AND MEASURES IN VIETNAMESE HOTELS

Having not been as severely affected by COVID-19 as other hotels in COVID-stricken countries, many local and overseas chain-operated properties within Vietnam have still opted to take a strong precautionary and preventive stance against the spread of the virus. According to MovetoVietnam.com (2020), many overseas chained hotels within Vietnam were quick to develop reservations and booking mobile applications that have facilitated the ease of making reservations without having to book through a travel agent or a third party in-person. This personalization of using one's mobile app has reduced the use of communal public computers, thus further reducing face-to-face interaction and the physical touching of public devices.

The use of technology has also been extended to the registration process. Several local hotels within Vietnam have employed the use of self-service kiosks to prevent interactions between staff and guest at check-in (Tran, 2020). These devices require minimal effort on the part of the guest. While the use of technology advocates greater personalization and social distancing, more cleaning and disinfecting are required on the part of the hotels to keep their common operating devices sanitized.

Like many lodging establishments, the Vietnamese hotels have also recognized that interaction between staff and customers would be inevitable especially within the hotel context. As such, staff have been informed to minimal contact with customers. Social distancing practices are strongly encouraged by the hotels. Staff are encouraged to emphasize on showing, rather than telling. Staff are also encouraged to wear PPEs, such as mask, gloves, and face shields. Above and beyond these initiatives, many hotels have erected acrylic glass or transparent rubber sheets at guest interactions "hotspots," such as at the reception and concierge desks (D. Nguyen, personal communication, December 1, 2020).

Hotels have also leveraged on technology with the use of service robots in service operations. These robots have been in array of service functions ranging from cooking to porterage. Service robots have been used on trial in overseas properties, such as in Singapore and in Dubai. In Vietnam, service robots have been used in the room cleaning ad food delivery functions in several chained hotels (Viet Nam News, 2020). These service robots help staff to deliver food as well as clean rooms, thus reducing contact between service staff and guest. This use is particularly useful for hotels that have been designated as quarantine hotels.

From a sectorial standpoint, all Vietnamese hotels have intensified their cleaning routines with additional checks and audits (Savills, 2020). These cleaning cycles have now been increased from two to four times per 8-hour shift for public areas within the hotels. Hand sanitizers and masks are readily available to all visitors. Temperature checks are also a mandatory procedure as advocated by the Vietnamese government.

13.3.2.2 NEW NORMS AND MEASURES IN THE AVIATION SECTOR (VIETNAM AIRLINES)

Vietnam Airlines is the national carrier of Vietnam. This carrier sets the standard by which other Vietnamese airlines, such as Bamboo Airlines and Viet-jet adhere to, as it has government and political links to the country's administration. Given this, the Vietnam Airlines hold an important position as the designated role model in terms of operational and hygiene standards for other airlines to adhere to. Moreover, being the national airline carrier, it carries the country's pride.

In terms of the new norms of air travel in Vietnam, Vietnam Airlines has taken its cleaning and sanitation processes strictly. According to Vietnam

Airlines Heritage Magazine (2020), their aircrafts are subjected to daily deep cleaning where equipment and furnishings are coated with disinfectants overnight so that all microscopic bacteria are exterminated due to the extended exposure to the sanitizing agent. According to Vietnam Airlines (2020a), regulated cleaning is carried out throughout their operation routine by the aircrew. The sanitizer and cleaning agents used meet the approvals of world aviation authority bodies. The cleaning is not only extended to aircrafts, but also service counters and public areas within the airports (Vietnam Airlines, 2020a).

To ensure a hygienic environment in their aircrafts, Vietnam Airlines replaced their onboard air filters to the High-Efficiency Particle Air (HEPA) filters (Vietnam Airlines, 2020a). These HEPA filters compile with European approved standards of removing up to 99% of particles, including dust, pollen, human skill cells, bacteria, and viruses whose diameter is 0.001 micrometer or more. These filters have been reported to minimize the risk of airborne bacteria and viruses on flights. Moreover, Vietnam Airlines in compliance with other airlines have redesigned their air flows such that air enters the cabin from the top. By doing, the flow of air traverses vertically down, reducing the diffusion of airflow across the aircraft cabin.

In terms of service equipment and amenities, headphones are treated to ultraviolet light disinfection before guest usage (Vietnam Airlines, 2020a). All service equipments are also cleaned and disinfected at the end of each flight. During flights, the offering of antibacterial hand wipes to guest are a service norm. This hand wipe contains more than 60% alcohol content, capacity of killing most bacteria. Certain amenities that have a greater common usage have been discontinued. Vietnam Airlines has ceased the provision of newspapers and blankets, to prevent the transmission of germs and bacteria. Travelers are also mandated to wear masks throughout the flights, unless when they are partaking in meals and/or drinks (Vietnam Airlines, 2020a).

As for registration and check-in at the reception desks at the airports, Vietnam Airlines, like many international airlines, have installed plexiglass panels over their countertops to reduce accidentally spread of infection due to the near proximity between staff and guest (Vietnam Airlines, 2020a; Skytrax, 2020). They have also installed more registration self-service kiosks at the airports, minimizing customer-staff contact (JourneyonAir.com, 2020). Customers are also encouraged to download the Vietnam Airlines mobile application on their personal devices (Vietnam Airlines, 2020a). The mobile application can assist consumers to check-in, select seats, and manage their personal accounts and preferences. This use of these applications has also minimized contact time between staff and consumer. Beyond the use of mobile applications, another new norm offered by the airlines would be that of a telephone check-in service, which has recently been revised (Vietnam Airlines, 2020a). This offering also helps to reduce unnecessary face-to-face transactions between staff and customers.

The final norms that are applicable to all travelers (be it traveling inbound, outbound, or domestic) is the checking of body temperatures before boarding and the declaration of one's health status. Travelers also would need to comply the Vietnamese Ministry of Health's mandate of declaring one's health condition before boarding a flight from the embarkation locality (Vietnam Airlines, 2020b). All travelers are to input their health conditions, place of residency, and visit on the ministry's website of tokhaiyte.vn so that the authorities can maintain contact tracing if necessary.

13.3.2.3 NEW NORMS AND MEASURES IN VIETNAMESE RESTAURANTS

As mentioned in the earlier section, Vietnam's food and beverage sector comprises of a myriad of SMEs, totally more than half a million eateries (CNBC, 2018). The service styles of these food outlet are very varied, ranging from buffets to ala-carte plated service. The localities of these eateries are also somewhat different – some might be serving as make-shift hawkers by the road, others might be solely located within a structured permanent premise, such as a coffee shop, while some might be a tenant within a shopping mall. Where these dining outlets are located can influence the hygiene standards, current norms, and the measures taken to prevent the spread of viruses. Street hawkers might not readily have clean water and facilities, reducing their sanitation levels, but in shopping centers, the management of the malls might mandate all food operators to compile with minimal health and hygiene standards. Nonetheless, most of eateries, be they street hawkers or shopping mall eateries, have started to offer alcoholic hand wash to patrons as a norm.

In Vietnam, a very entrenched medium of ordering food has been that of using social applications to order and deliver food (Nhan Dan, 2020). Applications such as GoViet, and Grab Food are readily used by the Vietnamese consumers. However, during the time of Vietnam's lockdown, food deliveries saw an increase in purchasing and ordering. Although, Vietnam is currently COVID-free with no community cases, these online purchases have continued to remain high. This could probably be due to the confidence induced from past purchases during the lockdown, creating a new norm and trend in the country in a post-COVID era.

Another norm is that commonly practiced in Vietnam's restaurants is the high adherence to sanitation. In many mainstream restaurants, especially those located within shopping malls, temperature taking is mandatory for all guest (Vietnam Insider, 2020). Temperatures are taken at the point of entry. On the same note, alcoholic wipes are readily available within the eateries. Service and food preparation staff are also mandated to wear masks at all times during operational hours.

A new norm that has evolved from the recent past and is now actively practiced specifically at buffet restaurants would be the "order-as-you-need" service method (Loh, 2020). Food items are no longer placed in a communal area, where consumers can freely partake. Dishes are now specifically ordered and delivered to the diner. This service style practices social distancing and reduces human transactions and the free mingling of diners with the eateries. This methodology has also been adopted readily as it reduces waste and provides greater savings (T. Nguyen, personal communication, December 1, 2020).

Lastly, another norm that has evolved from the COVID-19 lockdown in Vietnam is the limitation of diners within a restaurant. Many eateries have already leveraged on technology and the familiarity of use of applications to ensure control the number of visitors (Thu, 2017). Many restaurants have recently employed more mobile applications for diners to make reservations (MovetoVietnam.com, 2020). When doing so, guests can choose available time slots for their bookings and are required to indicate their number of diners. This practice not only helps restaurants to mange their food preparations but also allows the control of patronage to their outlets, thereby adhering to social distancing practices.

13.4 DISCUSSIONS ON THE NEW NORMS IN A POST-COVID WORLD

Currently with COVID-19 ravaging the world with multiple waves recurring in many countries, it would be difficult to foresee the immediate short-term norms of a post-COVID world and its impact on the hospitality and tourism industry. Many countries are starting to look at the norms on a long-term basis, having initially thought that it would be over within a few months. They are realizing that the COVID-19 would have a significant impact on not only their economies, but also on the processes that they have been used in the past. They know that the future operating standards and methodologies would have to change and evolved in accordance with these current COVID-19 times and measures.

In the case of Vietnam, the country is fortunate to have had intervals of short-terms infections cum waves, followed by long lulls of noninfections. This recede in COVID-19 cases has been achieved not only due to the pre-emptive and proactive measures taken by the Vietnamese government, but also a compliant and obedient society that trusts the judgment of their leaders. Therefore, with Vietnam having sustained through their COVID-19 infection episodes, and with life returning back to normalcy, the country can be seen as a pre-emptive microcosmic model of what is to come in terms of changes and coping in a society in a post-COVID world. This study of Vietnam can help countries that are currently undermined by COVID-19 infections to look at the Vietnamese measures taken and how the norms of their society will be changed, in order to set yardsticks on their own nations' recovery efforts. Hospitality and tourism companies in other COVID-stricken countries can also look to how the Vietnamese companies have adapted, should they revive from the decrease in COVID-19 cases in their countries.

13.5 CONCLUSIONS

Vietnam having had an extended period of being COVID-19 free has given the world a glimpse of how a country can operate after a pandemic. It is able to demonstrate how hospitality and tourism companies have been able to stay resilient and innovative to address the concerns of the public and to maintain a high level of hygiene and sanitation, leveraging on the openness to learn best practices from other nations and on technology. There is also a huge integration between the Vietnamese government and private companies, displaying intuitive 'public-private-political' initiatives to combat a national viral infection. This combined practice, which seems foreign to many non-Asian countries, is probably the greatest norm for the future, in a post-COVID world, where only with a truly integrated 'public-private-political' approach can bring about changes and new processes especially for the betterment of hospitality and tourism industry on a national and global level.

KEYWORDS

- **Vietnam**
- **COVID-19**
- **measures**
- **normalcy**
- **success**
- **innovation**

REFERENCES

AHLA. *Leisure and Hospitality Industry Proves Hardest Hit by COVID-19*, 2020. AHLA: https://www.ahla.com/COVID-19s-impact-hotel-industry

Alonso, A.; Kok, S.; Bressan, A.; O'Shea, M.; Sakellarios, N.; Koresis, A. et al. COVID-19, Aftermath, Impacts, and Hospitality Firms: An International Perspective. *Int. J. Hospital. Manage.* **2020,** *91*, 102654.

Anderson, R.; Fraser, C.; Ghani, A.; Donnelly, C.; Riley, S.; Ferguson, N. et al. Epidemiology, transmission Dynamics and Control of SARS: The 2002–2003 Epidemic. *Phil. Trans. R. Soc. Lond. B* **2004,** *359* (1447), 1091–1105.

Baum, T.; Mooney, S.; Robinson, R.; Solnet, D. COVID-19's Impact on the Hospitality Workforce-New Crisis or Amplification of the Norm? *Int. J. Contemp. Hospital. Manage.* **2020,** *32 (9)*, 2813–2829.

Clark, J. *Vietnam's Booming Aviation Market—2019 Report* 2019. livinginasia.co: https://livinginasia.co/vietnam-aviation-market/

CNBC. *Why McDonald's Flopped in Vietnam* 2018. CNBC: https://www.youtube.com/watch?v=l9pthhpd7So

Garcia, I. *Hilton, Hyatt and Marriott Will Introduce New Cleaning Protocols*, 2020. House Beautiful: https://www.housebeautiful.com/lifestyle/a32367701/hilton-hyatt-and-marriott-new-cleaning-protocols-coronavirus/

Gursoy, D.; Chi, C. Effects of COVID-19 Pandemic on Hospitality Industry: Review of the Current Situations and a Research Agenda. *J. Hospital. Market. Manage.* **2020,** *29* (5), 527–529.

Hao, F.; Xiao, Q.; Chon, K. COVID-19 and China's Hotel Industry: Impacts, a Disaster Management Framework, and Post-Pandemic Agenda. *Int. J. Hospital. Manage.* **2020,** *90*, 102636.

Hotel Business. *Hotel Companies Implement Changes to Reduce COVID-19 Impact*, 2020. Hotel Business: https://www.hoyelbusiness.com/hotel-companies-implement-changes-to-reduce-COVID-19-impact/

Im, J.; Kim, H.; Miao, L. CEO Letters: Hospitality Corporate Narratives During the COVID-19 Pandemic. *Int. J. Hospital. Manage.* **2021,** *92*, 102701.

Journeyonair.com. *Vietnam Airlines Deploys Self-Service Kiosks To Speed Up Check-in Services*, 2020. Journeyonair.com: https://journeyonair.com/vietnam-travel-news/vietnam-airlines-deploys-self-service-kiosks-speed-check-services

Kim, J.; Lee, J. Effects of COVID-19 on Preference for Private Dining Facilities in Restaurants. *J. Hospital. Tour. Manage.* **2020,** *45*, 67–70.

Kussmann, H. *Will the Coronavirus Crisis Force Hoteliers to Implement New Technologies Faster?* 2020. Hotel Online: www.hotel.online.com/press_releases/release/will-the-coronavirus-crisi-force-hoteliers-to-implement-new-technologies-faster/

Lemon, K.; Verhoef, P. Understanding Customer Experience Throughout the Customer Journey. *J. Market.* **2016,** *80* (6), 69–96.

Loh, D. *Robots in, Buffets Out: Asia Hotels Try Socially Distant Hospitality*, 2020. Nikkei Asia: https://asia.nikkei.com/Spotlight/Coronavirus/Robots-in-buffets-out-Asia-hotels-try-socially-distant-hospitality

Neuman, L.W. *Social Research Methods: Qualitative and Quantitative Appraoches*; India: Pearson, 2007.

Ministerie van Landbouw, Natuur en Voedselkwaliteit. *Vietnam Food & Beverage industry*, 2018. agroberichtenbuitenland.nl: https://www.agroberichtenbuitenland.nl/actueel/nieuws/2019/08/01/vietnam-food--beverage-industry#:~:text=With%20a%20forecasted%20annual%20growth,2020%20according%20to%20EVBN%20report.

Movetovietnam.com. *Must Have and Top Phone Apps for Vietnam: From Transportation, Food Delivery, Services and More*, 2020. Movetovietnam.com: https://movetovietnam.com/top-phone-apps-to-use-in-vietnam/

Nankervis, A. Dreams and Realities: Vulnerability and the Tourism Industry in Southeast Asia: A Framework for Analyzing and Adapting Tourism Management Towards 2000. In *Tourism in Southeast Asia: A New Direction*; Chon, K., Ed.; New York: The Haworth Hospitality Press, 2000; pp 49–63.

Nhan Dan. *Nhan Dan Online*, 2020. Hospitality Businesses Strive to Overcome Difficulties as COVID-19 Spreads: https://en.nhandan.org.vn/travel/item/8555202-hospitality-businesses-strive-to-overcome-difficulties-as-COVID-19-spreads.html

Pang, X.; Zhu, Z.; Xu, F.; Guo, J.; Gong, X.; Liu, D. et al. Evaluation of Control Measures Implemented in the Severe Acute Respiratory Syndrome Outbreak in Beijing, 2003. *JAMA* **2003,** *290* (24), 3215–3221.

Pizam, A. Terrorism and Tourism. *Int. J. Hospital. Manage.* **2003,** *3* (1), 1–3.

Roos, R. *WHO: Vietnam May Defeat SARS, and World Has Chance to Follow Suit*, 2003. Centre for Infectious Diseases Research and Policy: https://www.cidrap.umn.edu/news-perspective/2003/04/who-vietnam-may-defeat-sars-and-world-has-chance-follow-suit

Savills. *How Do Cleaning Costs Increase in a Pandemic?* 2020. Savills: https://www.savills.com.vn/insight-and-opinion/savills-news/186190/how-do-cleaning-costs-increase-in-a-pandemic-

Shin, H.; Kang, J. Reducing Perceived Health Risk to Attract Hotel Customers in the COVID-19 Pandemic Era: Focused on Technology Innovation for Social Distancing and Cleanliness. *Int. J. Hospital. Manage.* **2020,** *91*, 102664.

Sigala, M. Tourism and COVID-19: Impacts and Implications for Advancing and Resetting Industry and Research. *J. Busi. Res.* **2020,** *117*, 312–321.

Skytrax. *Vietnam Airlines COVID-19 Safety Policies*, 2020. Skytrax: https://skytraxratings.com/airlines/vietnam-airlines-COVID-19-safety-policies

Statista. *Food & Beverage Vietnam*, 2020. Statista: https://www.statista.com/outlook/253/127/food-beverages/vietnam

STR. *Asia Pacific Hotel Review: Industry Performance for the Month of March 2020*; Nashville, TN: STR, 2020a.

STR. *Asia Pacific Hotel Review: Industry Performance for the Month of April 2020*; Nashville, TN: STR, 2020b.

Thanh, D. *Quang Nam Reopens Entertainment Services with COVID-19 Under Control*, 2020. VnExpress: https://e.vnexpress.net/news/news/quang-nam-reopens-entertainment-services-with-COVID-19-under-control-4157469.html?fbclid=IwAR2Hn7FJdXoUb44Ig1s-bCA8BnG2N7LJfbYVj9I7Fy2JrnvOUXVMCeZ8Qs8

Thong, V. *Food Industry Considers Survival Plans as COVID-19 Spreads*, 2020. Vn Express: https://e.vnexpress.net/news/business/industries/food-industry-considers-survival-plans-as-COVID-19-spreads-4138452.html

Thu, T. *Startup Develops App That Allows Diners to Book Tables with Discounts*, 2017. Vn Express: https://e.vnexpress.net/news/business/startup-develops-app-that-allows-diners-to-book-tables-with-discounts-3606091.html

Tran, X. *Self Servioce Kiosk Check in Wink Hotel 2020*. Comq.vn: https://www.comq.vn/blogs/news/sefl-service-kiosk-check-in-khach-san-wink-hotel

Tu, A. *Coffers Close to Empty, Vietnam Airlines Seeks Government Credit*, 2020. Vn Express: https://e.vnexpress.net/news/business/companies/coffers-close-to-empty-vietnam-airlines-seeks-government-credit-4115007.html

UNWTO. *UNWTO World Tourism Barometer* **2020,** *18* (2).

Viet Nam News. *HCM City Develops 3D-Printed Robot to Disinfect Rooms During Pandemic*, 2020. Viet Nam News: https://vietnamnews.vn/economy/653896/hcm-city-develops-3d-printed-robot-to-disinfect-rooms-during-pandemic.html

Vietnam Airlines. *Travel Advisory*, 2020a. Vietnam Airlines: https://www.vietnamairlines.com/nz/en/travel-alerts/2020/0611-travel-advisory

Vietnam Airlines. *Complusory Pre-Departure Declarationfor All Passemgers Travelling on Vietnam Airlines' Domestic Flights*, 2020b. Vietnam Airlines: https://www.vietnamairlines.com/vn/en/vietnam-airlines/press-room/travel-advisory/2020/0324-EN-Compulsory-Pre-departure-Health-Declaration-on-Domestic%20Flights

Vietnam Airlines Heritage Magazine. Our Commitment to Safety. In *Heritage*; Vietnam Airlines Magazine, Ed.; Vietnam Airlines: Hanoi, 2020; pp 110–111.

Vietnam Insider. *Hanoi Asks All Restaurants, Cafes to Measure Customers' Body Temperature*, 2020. Vietnam Insider: https://vietnaminsider.vn/hanoi-asks-all-restaurants-cafes-to-measure-customers-body-temperature/

BIBLIOGRAPHY\l 1033 Vietnam Investment Review. *Using Lessons from SARS to Combat New Epidemic* 2020. Vietnam Investment Review: https://www.vir.com.vn/using-lessons-from-sars-to-combat-new-epidemic-73989.html

Vietnam+. *MoH Issues Urgent Instructions Over Chinese Disease Outbreak*, 2020. Vietnam+: en.vietnamplus.vn/moh-issues-urgent-instructions-over-chinese-disease-outbreak/167679.vnp

VnExpress. *Vietnam Confirms 17th COVID-19 Patient*, 2020. VnExpress: https://e.vnexpress.net/news/news/vietnam-confirms-17th-COVID-19-patient-4065517.html

World Health Organisation. *Viet Nam SARS-Free*, 2003. World Health Organisation: https://www.who.int/mediacentre/news/releases/2003/pr_sars/en/

CHAPTER 14

TOWARD AN INTEGRATED ACCOUNT OF COMPETENCE-BASED THEORY OF COMPETITION FOR COMMUNITY-BASED MICE TOURISM SEGMENT TRANSFORMATION EFFORT

CHAI CHING TAN

Ratttanakosin International College of Creative Entrepreneurship, Rajamangala University of Technology Rattanakosin, Salaya, Phutthamonthon, Nakhon Pathom, Thailand

ABSTRACT

Community-based tourism (CBT) is a feasible avenue to help revive the tourism markets during and in post-COVID-19, and in particular, due to social distancing and safety measures requirements, communities can pivot toward the MICE (Meetings, Incentives, Conventions, and Exhibitions) tourism markets. This paper contributes in twofold domains. First, the theoretical clarifications and an easily implementable framework that integrates the oft-confused terms, such as competencies, competences, and capabilities, is proposed, based on a systematic literature review. Second, the Tribes Experience, Tea, and Coffee communities, located in Chiang Rai, Thailand, who are qualified to extend their CBT projects to the MICE market, for cushioning the negative impact of COVID-19 and as a part of strategic pivoting, are selected for the illustration and structural equation modeling validation.

Post-COVID Tourism and Hospitality Dynamics: Recovery, Revival, and Re-Start.
Umendra Narayan Shukla & Sharad Kumar Kulshreshtha (Eds.)

14.1 INTRODUCTION

In recent news, it states that if both direct and indirect contributions are included, the tourism industry in Thailand could represent some 22% of Thai GDP, or as much as 25% of the nation's employment (Spindler, November 5). Judging by the significant role of tourism industry in Thailand, various measures during COVID-19 period are implemented by both the government and the private sectors. Procedurally, tourists or visitors to Thailand have to provide certification of COVID-19 free 72 hours before boarding and a booking at a quarantine-partnered hotel (Bangkok Post, 2020, December 18). Another tourism segment that has been widely and negatively impacted is the MICE (Meetings, Incentives, Conventions, and Exhibitions) markets. The mandatory 14 days of quarantine has brought the inbound and outbound tourism to a complete stop (Worrachaddejchai, 2020, October 8), leaving to domestic tourism as a feasible avenue for stimulating the tourism markets. To stimulate the domestic tourism markets, Thailand relies mainly on liquidity-supportive national policies, such as the domestic tourism incentives through subsidizing room charges and expenditures of the tourists (Taylor, 2020, December 29), while ensuring the pandemic safety measures. The Thai government's measures for the domestic tourism are well-aligned with the systematic reviews of the literature, such as by pushing forward the rural or community-based tourism (CBT) rather than the urban versions, as the former is believed to be more resilient in reviving the tourism industry in post-COVID 19 period (Sharma et al., 2021).

In view of the above context, and in favor of small-group domestic travels, this article aims to propose a competence-based model for responsively adapting the existing CBT activities for the domestic MICE markets, which will open up the corporate customers from the neighboring provinces. The shifting of the existing CBT competences to MICE segments matches also with the finding of the World Economic Forum (De Cuper et al., 2020, Jun 19), that is in search for industry revitalization through pivoting, such as by re-purposing and re-directing the existing competences to new needs. Pivoting of business pursuit is always challenging as it involves the matching between demand and supply. At the least, the pivoting toward MICE segment helps cushioning the current struggles in the travel and event industries during COVID-19 pandemic (Ionescu-Somers and Tarnawa, 2020). Nevertheless, it is vital that the pivoting has to project a perception that the CBT products and services are "essential" to the MICE market (Charm et al., 2020). From the social distancing and safety perspectives, the engagement of the tourists to the experiences of CBT activities, rather than putting them in crowded space for large indoor events, seems to be a preferable choice.

In order to convince the communities and the government's execution committees for accepting the quality and responsive expansions of the existing CBT activities to the MICE markets, this paper adapts the socio-psychological theories, in particular, relating to the roles of confidence and perceived behavior control (Ajzen, 1991), through competence development (Chuang et al., 2018; Wang et al., 2020), in influencing the intentions (Lima et al., 2020; Vaz et al., 2020) and the immediate investments and behaviors of the communities (Tan et al., 2020). Thus, specifically, the research objective is set as follows:

> *The purpose of this study is, first of all, through literature review, to clarify the competence-related terms, and organize them in strategy-mapping structure that enables the practitioners (the communities in this study) and the local and national policy makers (the government), as implications, to strategically and operatively understand the roles and practical usages of competences and, to test, using community-based cases, and by the use of structural equation modeling (SEM) statistics, to provide the theoretical-empirical model-matching support for the proposed model.*

14.2 LITERATURE REVIEW

Literature is an important world 3 of the three worlds when it is presented in the knowledge context of Professor Karl Popper (1963, 1970), known of the cognitive mental model, which reflects the world 1's physical world, whereas world 2 connotes the states of consciousness, mental states, and behavioral dispositions to act (Nagano, 2020) that serves to bridge the gaps between world 1 and world 3. These three worlds, thus, form the epistemic problem solving and research process, shown in Popper (1972) as P1 (the original problem, as revealed in the world 1) → TT1 (tentative or temporary theory, the World 3) → EE1 (the error elimination) → P2 (new problem arises) → TT2 (new proposed tentative theory) → EE2 (the next level of error elimination), which is a continuing process of research execution and attitude, that also serves to facilitate the growth of knowledge. Nagano (2020) illustrates how the Popperian epistemic process is used to develop and implement resource-based view (RBV) so as to prevent rigidity in RBV applications, which is applicable to competence-based approach of this study. The three worlds of Karl Popper are shown interactively in Figure 14.1, and the interactions between world 2 and world 3 establishes the domain of literature review, and the world 1 → world 2 → world 3 the inductive research approach, and world 3 → world 2 → world 1 the deductive research approach.

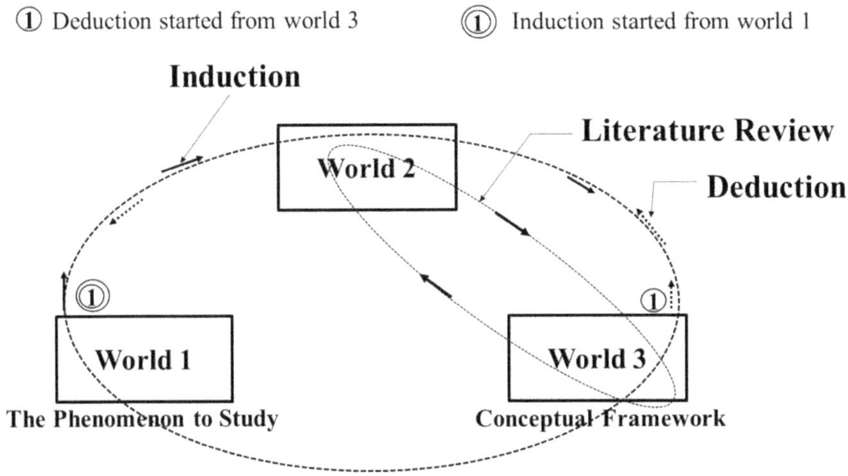

FIGURE 14.1 The Karl Popper's three worlds structure for literature review process.

Source: Developed for this Study.

The literature review surveys the theoretical and empirical studies in leading journals on topics relating to competences, competencies, capabilities, and capacities, and incorporates also the various dominant theories that have these variables as their strategic themes. To limit the searches, only Emerald and Science Direct databases were used, by using variables and their related theories of competition as the key words in the searching.

Driven by the keywords, the literature review searching is not limited to any set of specific journals. In particular, the literature review method used is "integrative" in nature, which aims to "assess, critique, and synthesize the literature on a research topic in a way that enables new theoretical frameworks and perspectives to emerge" (Snyder, 2019: 335), and thus, this study also serves the purpose to overview the knowledge base of competences, leading to a theme-based conceptualization of concepts and theoretical framework development. Specifically, the synthesis is aimed to meet customer-centric and business performance requirements. That is, competence is fundamentally reckoned as serving to create a basis for enabling and supporting business strategies (Simon et al., 2015).

At the elemental level, competence connotes the composite states of knowledge, skills, and judgement, known as competencies (Chen, 2009;

Garside and Nhemachena, 2013), including personal characteristics that are required to perform a task that is subjected to continuing changes (Axley, 2008). Competency is often a result of learning and growing based on experiences (Kaplan and Norton, 2004; Tan et al., 2020), relating to specific task assignments, such as the mastering of resources and innovation. A careful literature review concludes that competencies can be classified into 2X2 broad categories, namely conceptual and operational on the task domain, and the personal and occupational on the influential spaces, for which personal and occupational are accounted. Figure 14.2 depicts the competency requirement on the far left, which gets coordinated and applied to meet the business goals through business model execution in the middle, and the overall competence-based applications will, finally, result in the so-called capability to meet any possible emerging customer needs. Fundamentally, capability connotes the capacity for firms to render specific outputs (goods and services) to the market, and thus, implies the current and potential readiness for actions (Freiling et al., 2015). Without using the capability (the "flow"), capability is only a "stock" at best (ibid, p. 94). When competences and capabilities are executed to maturity (Vaz et al., 2018), to the impression of the customers and the market, it would lead to the formation of organizational character (Selznick, 1957). In short, the capability "flow" (such as enabled by boundary-spanning capabilities, interorganizational relations and connectedness, and integration of competencies, Prevot, 2015) enables one to utilize the capability "stock."

In addition, when an organization or a profession can demonstrate the necessary competence, it is said the organization has gained the capacity and capability to accomplish the tasks needed in unfamiliar and familiar situations (Mocke et al., 2013), and it is exactly this competence maturity that delivers the requisite sociopsychological confidence (Ajzen, 1991; Chuang et al., 2018; Lima et al., 2020; Tan et al., 2020; Vaz et al., 2020; Wang et al., 2020) for enabling the responsive shifting of the CBT competences to the MICE segments, as articulated in the introductory section.

14.2.1 MODEL CONCEPTUALIZATION

While competency is judged at the individual, task level, competence prioritizes at the business level, and thus valid illumination of the construct "competence" would need to connote an ability to influence the business goals such as the evidential ability of executing the business model to success. Based on Figure 14.2 as the structural linkages between competency, competence, and capability, the strategy mapping configuration of Kaplan and Norton (2004, 2008) becomes very practically useful. Another supporting rationality is that competencies are heuristic result of learning and experiences (Macpherson and Holt, 2007), and when the functional or task-based competencies are strategically integrated and coordinated for a business purpose, it leads to the formation of business model-enabling competences (Lee, 2010). Figure 14.3 depicts the 5Cs strategy map, which is an important theoretical and practical contribution of this paper, and the component understandings are given as follows:

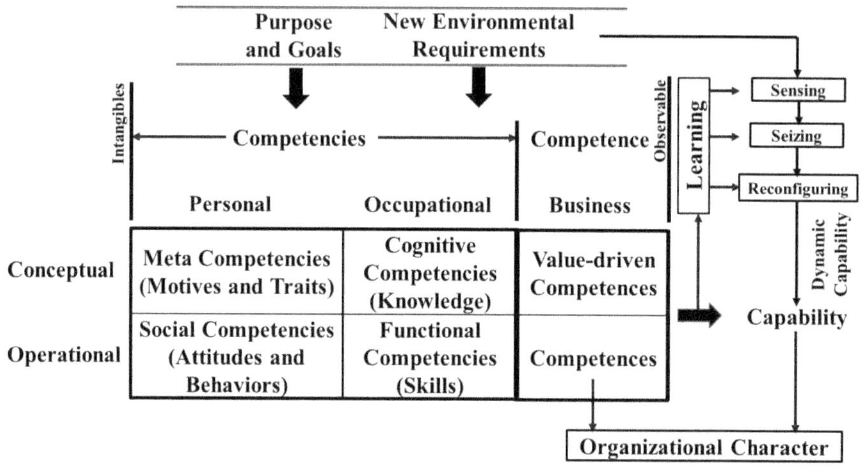

FIGURE 14.2 The interrelationship structure of competencies, competence, and capability constructs.

Source: Developed for This Study.

- *Competency level and domain*—Competencies identified to be important and are used in the study include "attitude and belief toward CBT, human capital development, innovation, resource acquisition and development, and tourism resource management." These competencies need to be integrated for the unifying ability to develop and execute the competences to execute the business model to success. Statistically, it is evidenced by the interrelated positive correlations of the various aspects of competencies. For instance, by gaining the human capital, innovative ways to develop the business model are improved and made possible (Danquah and Amankwah-Amoah, 2017). In addition, as noted in sociopsychological theory, such as the theory of planned behavior (Ajzen, 1991), the competencies form the sociopsychological forces to motivate the community members to continue to exert efforts for competence development and capability achievements. To have stronger ability to impress visiting guests and tourists, communities would also need to stress on innovation, which is particularly in need when the highland communities are far from the cities and coastal towns (Sanchez-Canizares and Castillo-Canalejo, 2014).

- *Competence level and domain, including character of the destination*—The business model-deliverable ability is reflected by the ability to make possible the supply-side activities (tourism activities and products development), the demand-side activities, that is, demand identification and marketing strategy development, and the values offered to tourists, which also reflect the destination personality and identity (the so-called character). In its simplest sense, based on an activity-based view advocated in Taylor (1945) and Barnard (1938), by breaking down into component parts as most efficient means (Schwartz, 2007), business model can be considered as the integration of activities consisted of value design, value creation, value manufacture, value delivery, and value capture.
- *Competitive advantage level and domain*—The level of sustainability achievements as epoxy base for competitive advantage. Competitive advantage can be directly a result of competency resources or business process competences. VRIN (valuable, rare, inimitable, and nonsubstitutable) characteristics (Eisenhardt and Martin, 2000) are often used to characterize the resources, competencies, and competences that have the potential for establishing competitive advantage. To be exact, RBV (resource-based view of competition) is not really about resources (Shafeey and Trott, 2014) but about the resource attributes that can contribute to sustain the competitive advantage of firms (Barney and Clark, 2007: 249–250). Criterion for competitive advantage or competitive position of firms in the market, on the pragmatic sense, is judged by the ability of firms to create more economic value (which is the total surplus beyond the economic cost, and is the summative customer value [customer surplus] and economic rents [producer surplus] than other players in the market in the course of providing a good or service (Peteraf and Barney, 2003). Thus, centralized whether on competitive advantage, or RBV, or DBV (Dynamics based View), the concept of "value" is the key. Scanchez (2008) goes to the extent to clarify the possible misconceptions in VRIN or VRIO (where "O" stands for organized) in that RIO dimensions are not the characteristics of a resource to be valuable, but

as the characteristics for the strategically "valuable" resource to be a possible source of competitive advantage.

- *Capability domain*—which is manifested by the capability of the entire competence-based strategy map execution outcome. Concept as such is known as dynamic capability (Teece, 2007; Teece and Pisano, 1994), which connotes the ability to execute the business model to success and in turn, yield competitive advantage. The dynamic capability puts the organization in constant "alertness," which possesses the ability to capitalize on the possibilities for profitable actions (Freiling et al., 2015). Putting into Winter's (2000) term, capability is "a high-level routine (or collection of routines) that, together with its implementing input flows, confers upon an organization's management a set of decision options for producing significant outputs of a particular type" (p. 983).

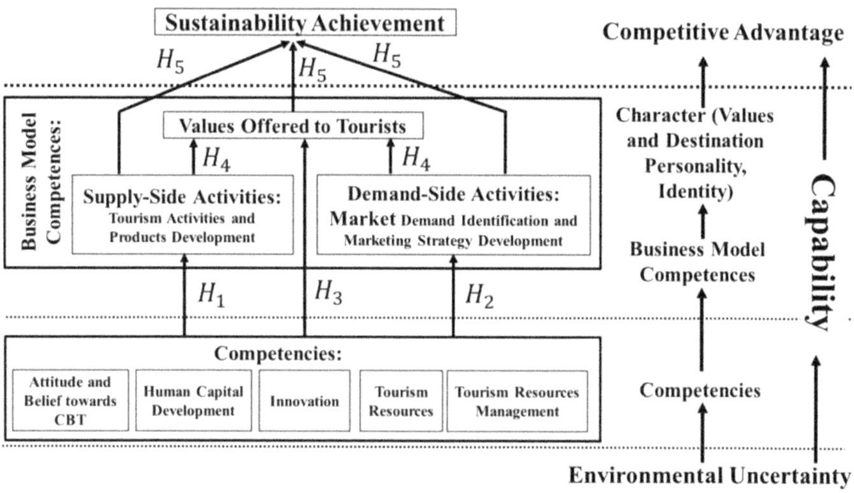

FIGURE 14.3 The proposed 5Cs strategy map—a competence-based approach

Source: Developed for This Study.

The five hypotheses of the 5s competence-based strategy map structure are listed as follows:

H1. The growth-oriented competencies, constituting of attitude and beliefs toward CBT, human capital development, innovation, tourism resources and tourism resource management, can significantly predict tourism activities and product development competences (representing the supply-side activities).

H2. The growth-oriented competencies, constituting of attitude and beliefs toward CBT, human capital development, innovation, tourism resources, and tourism resource management, can significantly predict market demand identification and marketing strategy development competences (representing the demand-side activities).

Both H1 and H2 illuminate how the growth-oriented competencies are aggregated into business competence. The growth-oriented competencies incorporate not only the resource-advantage view of competition, but also sociopsychological variables in terms of attitudes and belief. H1 and H2 configure the first two layers of the BSC (Balanced Scorecard) and strategy mapping structure, and establishes the linkage between organizational competencies and business competences. In addition, the use of business model concept, which incorporates both the supply-side and the demand-side activities of the business, and the values it offers, has been empirically shown as effective resource and competencies aggregator needed for establishing competitive advantage (Lu et al., 2020).

To operationalize the customer perspective, the concept of customer value proposition (CVP, Yi et al., 2020), which bridges the supply-side and the demand-side activities of the business model, is adapted. CVP is at the heart of "competitive strategy" as customer-centric value can effectively be used to define strategic vision, for instance, in a digital retail environment (Ziaie et al., 2021), and is also a key element of business competence in which the organization masters the exploitative use of the cognitive acceptance, emotional acceptance, normative acceptance and behavioral acceptance of customers for service advantages (Shulga and Busser, 2020). The following hypotheses H3 and H4 interconnect the growth perspective and the internal business process perspective to the customer perspective, and form the structural shape of business model, and enacts the *competitive-advantage* base to realize sustainability goals and vision.

H3. The growth-oriented competencies, constituting of attitude and beliefs toward CBT, human capital development, innovation, tourism resources, and tourism resource management, can significantly predict the values offered to tourists.

H4. Both the supply-side and the demand-side activities of CBT and their competences can significantly explain the values offered to the tourists, which is an outcome of the CVP (customer-value proposition)-deliverable competence.

The sustainability goal and vision are aimed to maintain the well-being of the economy, society, and ecological environment simultaneously (Formentini and Taticchi, 2016). The sustainability evidence, together with destination personality and identity (the "*character*" of the destination), and the distinctive values offered to visiting guests and tourists, would mold a socially responsible image of the destination (Ritala et al., 2018), leading to H5.

> **H5.** The business model competences of CBT, constituting of the values offered to the tourists, and the supply-side and the demand-side activities and competences, can significantly predict the sustainability impact.

14.3 METHOD

14.3.1 THE COMMUNITY PARTICIPANTS

The Tribes ETC (Experience, Tea, and Coffee) communities, who are qualified to extend their CBT projects to the MICE market, for cushioning the negative impact of COVID-19 and as part of strategic pivoting, are selected as the sampling population. They are located in Chiang Rai, such as Mae Chan Tai village, Doi Chang, Huai Nam Guen village, and Ban Pha Mee village. Other Tribes ETC communities include Huai Ma Kliang village, Khiang Mae Loi village, Phyapai Litu village, Phayapao Mae Mhor village, Pang Khon village, Mae Mon village, Khun Lao village, Pang Ton Phung village, Ban Thai Smakkhi village, and Ban Huai Khrai village, are excluded. A total of eighty-seven sets of statistics-analyzable questionnaires were returned and were all useful for statistical analysis.

The participating communities regularly hold cultural and social responsibility events to exhibit their shared identity and belief system (Figure 14.4) that, ultimately, can support the sustainable competitive advantage of their fertile lands (see Figure 14.5), and they also hold regular discussions with the stakeholders from the university, government, and private sectors, to improve their competencies (Figure 14.6). The communities integrate their competencies for making products and CBT services, such as food and entertainment, agricultural field practices, and homestays, as shown in Figures 14.7 and 14.8. These author-captured pictures illuminate the general activities and the community outlooks of the 5Ss strategy mapping structure of this study.

14.3.2 THE CONSTRUCTS OPERATIONALIZATION

The relevant constructs are referred to the hypothetical 5Cs model in Figure 14.3 and the literature. In this section, the questionnaire variables are stated.

FIGURE 14.4 Cultural dance and CSR activities of the community.

Source: Photos by author.

FIGURE 14.5 Ecological health of the community soil and environment.

Source: Photos by author.

14.3.3 COMPETENCIES

Competency requirements, as have already discussed in aforementioned sections, are to fulfill the execution needs of the elements of competences, and by expressively outlining the observable performance standard (i.e., as depicted by the five-stage process of Dreyfus and Dreyfus, 1980, and the underlying attributes, competency can serve to improve work performances (Cizel et al., 2007). In this regard, competency can be defined and measured

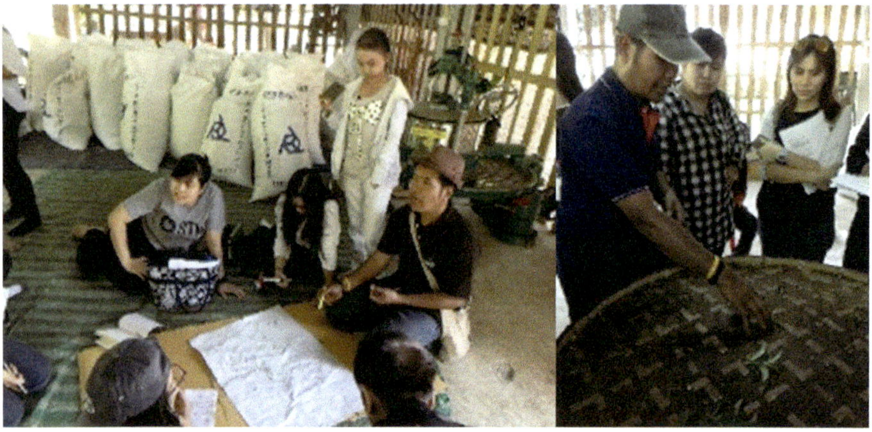

FIGURE 14.6 Community-based learning.
Source: Photos by author.

FIGURE 14.7 Products of the communities and the community-based tourism services.
Source: Photos by author.

FIGURE 14.8 Community homestays.
Source: Photos by author.

by the measurable or observable characteristics that are "causally related to effective and/or superior performance in a job" (Audrey, 1989: 21), which based on extensive research, identify knowledge, skill, and ability (Cizel et al., 2007). Because of the ability to identify these competence determinants to relate to and cause high performance to tasks and organizations, it has brought academic and practical interests toward competence-based HR processes (Heffeman and Flood, 2000). The learning and growth mechanism enables tasks-based competencies to be developed, improved, and maintained (Kaplan and Norton, 2004; Tan et al., 2020). The operationalized variables are:

- Attitude and belief toward CBT delineate the nature of the attitude and beliefs of the community members, and are operationalized by items such as committing on CBT expansion program without diluting the identity of the community, and by accentuating on the community's uniqueness and its livelihood, and lifestyle.
- Human capital development explains the extent to which the communities make efforts in developing leadership of the community members as a thrust for the CBT initiatives.
- Innovation measures the effort of innovation of the community toward CBT in terms of innovative services and reaching-out innovation.
- Tourism resources present the nature of ecological and man-made resources available to the communities, manifested, for instance, in terms of the lifestyle, abundancy of fertile land, conservation measure, culture, and heritages.
- Tourism resources management articulates the extent of efforts of the community in managing the various aspects of resources, and activities measured involve waste management, tourism carrying capacity, monitoring and continuous development, network leveraging of tribe members, and design of rules and regulations for appropriate resource management purposes.

14.3.4 COMPETENCES AND CHARACTER OF DESTINATION EMBEDDED

Three competence components are taken into consideration as guided by the concept of business model:

- Tourism activities and product developments are the supply-side activities, which involve developing the experiential activities for the tourists, such as eco-trekking, agricultural practice learning, local culinary tasting, leisure and recreation, homestay experiences, stories and interpretation of local attractions, enjoyment of natural landscapes, local lifestyle understanding, and exhibiting souvenir production.
- Market development identification and marketing strategies are the two market-side or demand-side activities that the community focuses, in the design and implementation of the business model, which focuses on the level of efforts in pricing, promotion strategies, and market development and fulfillment.
- The values offered to the tourists are operationalized by the degree of efforts made in the various domains or features of tourist values, such as functional value, educational value, cultural value, ecological value, which provides a bridging to brand studies in another aspect of the author's research with the communities.

14.3.5 COMPETITIVE ADVANTAGE-BASED OUTCOMES

Competitive advantage is measured by the degree of performances in each aspect of the sustainability goals, namely:

- Economic impact (i.e., family income increase, job creation, better earning, profit distribution),
- Cultural impact, which touches on cultural awareness increase of the community members, cultural maintenance, and new knowledge and valuable experiences gained as a result of open interactions with the visitors,
- Social impact, which focuses on infrastructure development, medical and educational improvement, and quality of life and sense of belonging of the community members, and
- Environmental sustainability, which includes measurement items such as systematic waste and water management benefits, and sense of love and care for the environment.

14.4 RESULTS AND DISCUSSION

This section presents the statistical analysis, which justifies, first, the discriminant and convergent validity, and the reliability of the measurement instrument. While reliability, which measures dependability or consistency of the measurement in defining the construct, is acknowledged by the Cronbach Alpha presented in Table 14.1, overpassing the recommended threshold of 0.70 at the minimum, the construct validity is represented by the assessments of both the discriminant validity and convergent validity. As explained in Neuman (2006), convergent validity means that "multiple measures of the same construct hang together or operate in similar ways" (p. 194), and discriminant validity is the opposite, which means that "the indicators of one construct hang together or converge, but also are negatively associated with opposing constructs" (p. 194).

With the square root of average variance extracted (AVE) shown exceeding the cross-correlations terms in Table 14.1, as highlighted by the diagonal value exceeding the off-diagonal in each column, then, and as suggested in Fornell and Larker (1981), divergent validity is established. Furthermore, with each factor loading (not shown) of the measurement items exceeding the 0.70 threshold, and with AVE, over 0.5, and with reliability index beyond 0.70 threshold, the convergent validity is also established.

Table 14.2 below is a creative combination of both the mean descriptive of the constructs and the ANOVA and t-test of important variables. Table 14.2 shows significant differences between the CBT head and nonhead. Apparently, there is a perception gap, which highlights that the CBT heads, in general, perceive at a higher level than the nonhead across the constructs studied and captured in the theoretical model. In addition, although coffee shops seem booming at an accelerated rate much higher than the tea-based beverages in Chiang Rai, the communities who rely on tea as the main livelihood show better perceptions across the board when compared with the communities planted coffees in majority. The perceptual mean that crosses four scale (of the five Likert scale, with 1 = strongly disagree, 2 = disagree, 3 = neutral, 4 = agree, and 4 = strongly agree) belongs to only two constructs, namely attitude and belief toward CBT, and tourism activities and product development, which highlights a more collective synergy among the members, and also shows relatively homogeneous, higher level of commitment in the production-side. In general, the communities lack the skills, competencies, and knowledge in the demand-side and the sustainability domains of goals.

The five hypotheses raised in the literature review section basically can be, fundamentally, confirmed by the structural equation models (SEM) shown in separate sustainability domains, namely Figure 14.9 for economic

TABLE 14.1 Discriminant and Convergent Validity, and Reliability Assessments.

Latent construct	1	2	3	4	5	6	7	8	9	10	Cronbach Alpha	AVE
1	0.854										0.814	0.729
2	0.354	0.831									0.943	0.691
3	0.508	0.691	0.801								0.921	0.631
4	0.553	0.519	0.553	0.811							0.892	0.659
5	0.519	–	0.364	0.505	0.897						0.757	0.804
6	0.402	0.612	0.657	0.471	0.325	0.842					0.794	0.709
7	0.735	0.504	0.612	0.661	0.626	0.565	0.847				0.905	0.717
8	0.435	0.658	0.729	0.671	0.461	0.596	0.594	0.799			0.823	0.638
9	0.514	0.552	0.603	0.649	0.333	0.377	0.567	0.669	0.872		0.895	0.760
10	0.758	0.575	0.763	0.675	0.593	0.611	0.754	0.761	0.719	0.872	0.969	0.759
Discriminant Validity Status	OK	OK	OK	OK	OK	OK	OK	OK	OK	OK		

Note: All the correlations coefficients presented are significant at the 0.01 level (two-tailed), and the diagonal value represents the square root of the AVE (Average Variant Extracted) for the corresponding construct. 1 = Attitude and belief toward CBT. 2 = Human capital development. 3 = CBT management. 4 = Innovation. 5 = Tourism resources. 6 = Tourism resource management. 7 = Tourism activities and products. 8 = Market demand identification and marketing strategy development. 9 = Values to tourists. 10 = Sustainability impact.

TABLE 14.2 Mean Descriptive and T- or ANOVA Test.

Latent construct	Mean	Std. Dev.	Agricultural livelihood type—main income earner							CBT role	
			Mean value			Test of homogeneity of variance		ANOVA		Mean value	
			Tea	Coffee	Both	Leven static	Sig.	F	Sig.	CBT head (11.4%)	Not CBT Head (88.6%)
1	4.14	0.62	4.40	3.89	4.24	1.494	0.2222	4.681	0.001	4.43	4.09
2	3.44	0.72	3.86	3.11	3.42	1.085	0.36	5.976	0.000	3.96	3.37
3	3.66	0.59	3.91	3.35	3.84	0.337	0.799	7422	0.000	4.12	3.59
4	3.73	0.61	3.90	3.60	3.62	2.47	0.06	2.909	0.039	4.04	3.69
5	3.77	0.52	3.88	3.70	3.70	2.228	0.091	1.522	0.215	3.92	3.74
6	3.45	0.72	4.04	3.00	3.00	0.772	0.513	15.003	0.000	3.81	3.39
7	4.02	0.56	4.37	3.80	3.80	0.604	0.614	6.334	0.001	4.32	3.97
8	3.78	0.55	3.98	3.62	3.62	0.692	0.559	2.38	0.075	4.15	3.72
9	3.91	0.61	3.94	3.78	3.78	0.663	0.577	2.37	0.086	4.10	3.88
10	3.95	0.48	4.20	3.72	3.72	1.019	0.388	6.737	0.000	4.22	3.91

Note: 1 = Attitude and belief toward CBT. 2 = Human capital development. 3 = CBT management. 4 = Innovation. 5 = Tourism resources. 6 = Tourism resource management. 7 = Tourism activities and products. 8 = Market demand identification and marketing strategy development. 9 = Values to tourists. 10 = Sustainability impact.

impact, Figure 14.10 for cultural impact, Figure 14.11 for social impact, and Figure 14.12 for environmental impact. There are minor additions of the direct relationship between attitude and belief toward CBT and the various facets of the sustainability achievements, signifying a significant value and role of collective sociopsychological variable.

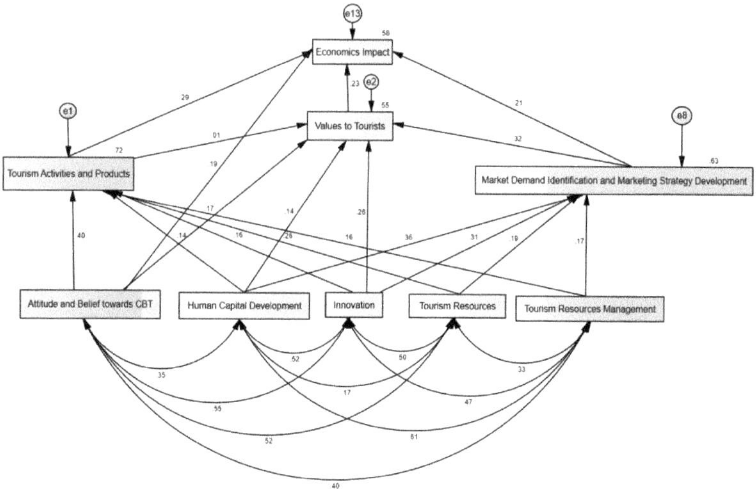

FIGURE 14.9 Economics impact.

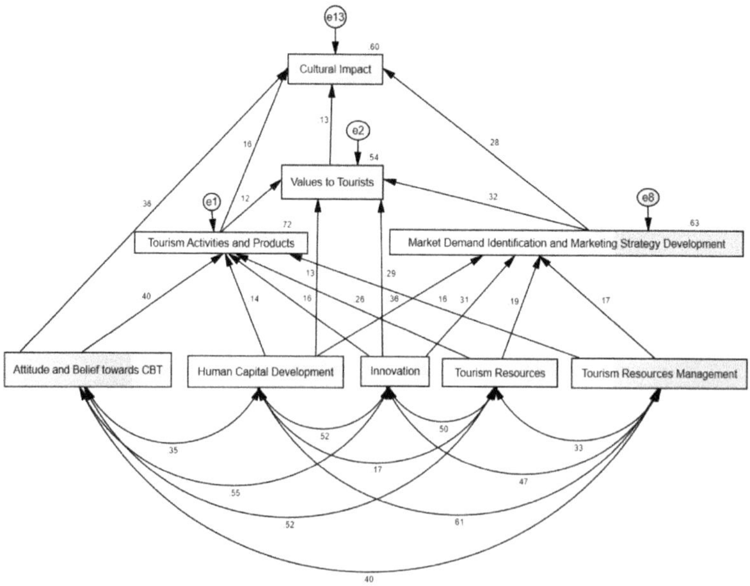

FIGURE 14.10 Cultural impact.

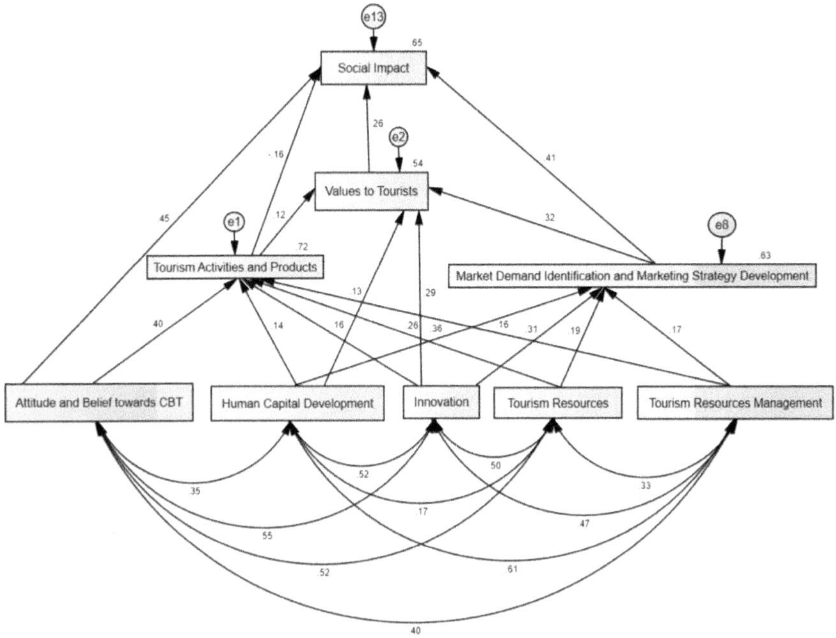

FIGURE 14.11 Social impact.

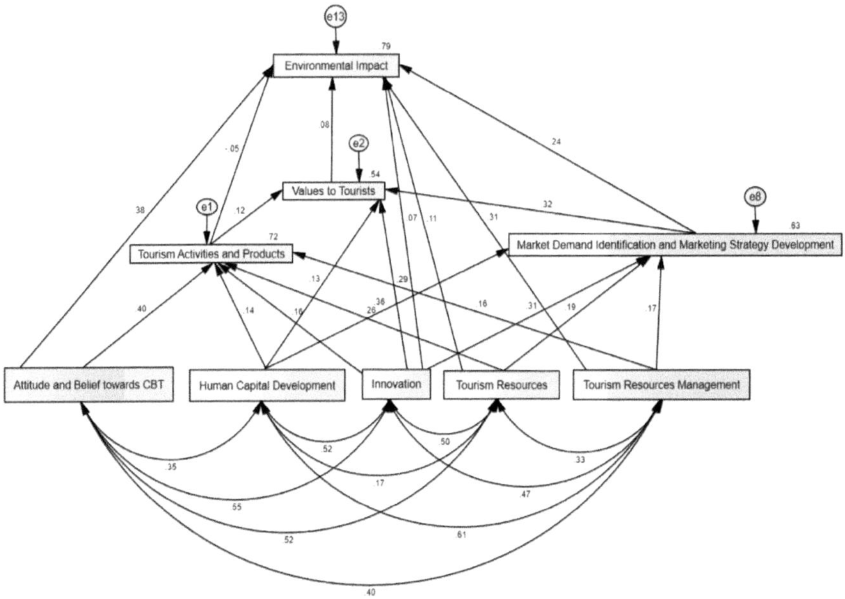

FIGURE 14.12 Environmental impact.

The SEM fit statistics are determined as follows:

- Chi-square/df = 1.446 (p = 0.172, model matched), NFI = 0.977, RFI = 0.897, IFI = 0.993, TLI = 0.966, CFI = 0.992, and RMSEA = 0.072 for economic impact,
- Chi-square/df = 1.047 (p = 0.40, model matched), NFI = 0.982, RFI = 0.927, IFI = 0.999, TLI = 0.997, CFI = 0.999 and RMSEA = 0.023 for social impact,
- Chi-square/df = 1.566 (p = 0.119, model matched), NFI = 0.972, RFI = 0.889, IFI = 0.990, TLI = 0.989, and RMSEA = 0.081 for cultural impact, and
- Chi-square/df = 1.45 (p = 0.192, model matched), NFI = 0.984, RFI = 0.907, IFI = 0.995, TLI = 0.969, CFI = 0.995, and RMSEA = 0.072, for environmental impact.

RMSEA represents the root mean square error of approximation, which indicates how well the SEM fits the population, not just a sample used for the estimation, and acceptable value should be below 0.10 (Hair et al., 2006: 748). A good fit is also shown in Chi-square/df (goodness of fit, GOF) in the range of 3:1 or less (Hair et al., 2006: 748). While RMSEA and GOF are measures of absolute fit, in quantifying the differences between the observed and the estimated covariance matrices (Hair et al., 2006), the incremental fit statistics are represented by NFI (Normed Fit Index), CFI (Comparative Fit Index), TLI (Tucker Lewis Index), and RFI (Relative Fit Index), which should exceed 0.90 threshold, or approach 1, for a good model fit.

The SEM configurations conclude good incremental and absolute model fits, which highlights that the values offered to tourists are a result of two types of activities typically understood in a business model concept (Aung and Tan, 2016; Tan and Anomasiri, 2017), namely the supply-side competence (tourism activities and product development) and the demand-side competence (market demand identification and marketing strategy development). Apart from the roles of business model competences played in contributing to environmental sustainability, it is also shown that the four types of competencies also have directly significant impacts, namely the attitude and beliefs toward CBT, innovation use of the community, the tourism resources, and their management.

Both human capital development and innovation competencies are shown significantly able to impact positively on the degree of efforts made on tourist value proposition design and development. The strong positive interrelationship between innovation and customer value design is the central theme of the blue ocean strategy, in which Leavy (2005: 13) states that "In blue ocean strategy, value and innovation are inseparable. Value innovation places equal emphasis on value and innovation." As to human capital aspect of competency, its role has the knowledge spillover effect, which is often regarded as the engine of sustained growth and development (Chang et al., 2016), being realized through the contribution of human capital in the various aspects of CBT business model, whether in the tourism activities and products, values offered to tourists, or market demand identification and marketing strategy development.

The SEM also shows the significant linkages between the extent of efforts in tourism resources and their management competencies, and in the developments and uses of both the supply-side and demand-side aspects of the CBT business model competences. The role of resources is significantly important for the communities, who make use of the available natural, cultural and livelihood resources as potential sources of attractions and tourist organism stimulations. In Ryan (2010), it is noted that the multifaceted resources, in combination, can provide the sources of stimuli in stimulating the formation of images, and could also help the tourists formulate the criteria by which they evaluate the success of otherwise of their stay (Chen et al., 2016).

The fact that a significant percentage of variance of the sustainability achievements, in the ranges between 0.58 and 0.73 as shown in Figures 14.9–14.12, can be explained by the business model competences and the growth-oriented competencies of innovation and human capital development, is well aligned with the definition of CBT as follows:

> *CBT is a sustainable, community-owned and community-based tourism initiative that enhances conservation and in which the local community is fully involved throughout its development and management are the main beneficiaries through community development.* (Manyara and Jones, 2007: 637)

Environmental or ecological management competency (i.e., resource management) ensures that resources are maintained to the best productive levels possible to favor the production activities, which is also in line with the guidelines advocated in UNEP and UNWTO (2005: 11–12), namely "sustainable tourism should make optimal use of environmental resources that constitute a key element in tourism activities, maintaining essential ecological processes and helping to conserve natural heritage and biodiversity."

14.5 CONCLUSIONS

Competences are very important strategic terms, which are often used to demonstrate the resource-based advantages of firms in competitive environment. Competences are also terms of great confusion, and are often interchangeably used with other competitive constructs such as competencies and capabilities. This study provides a clarification and illustrates a strategy map that integrates competencies, competence, character, capability, and competitive advantage, which is shown to parallel to the structure of BSC and strategy mapping of Kaplan and Norton (2004; 2008), and as such, due to the conceptual clarity and the wider generalizability of the concepts, can easily convince the practitioners and the policy makers in adapting for applications. This serves the purpose of this study, which is to support the theoretical-empirical validity of the proposed competence-based strategy mapping framework. Having rooted the explanatory base in the theories of sociopsychologies, such as the theory of planned behavior (Ajzen, 1991), the competence-based 5C strategy mapping framework would serve to help the communities to extend their original CBT services to the MICE segments responsively.

CBT is a unique feature of the alternative tourisms offered by Chiang Rai, Thailand, and as such, can be better used for the MICE (Meeting, Incentive, Conference, and Exhibition) segments, which are of high customer expectation and revenue-per-customer level (Mehta et al., 1991). MICE is also emerging as one of the fastest-growing aspect of tourism worldwide (Chiang et al., 2012), with visitor expenditures of around US & 43 billion back in 2005 (World Tourism Organization, 2006). The competence model shown in this study is systemic in nature, which embraces the competence continuum that spans the competency at the tasks level to competence level in delivering business model performances, and thus, it fills the available literature gaps relating to not only competence-related concepts, but also to the MICE applications.

KEYWORDS

- **competency**
- **competence**
- **organizational character**
- **competitive advantage**
- **community-based tourism**
- **MICE**
- **COVID-19**

REFERENCES

Ajzen, I. The Theory of Planned Behavior. *Org. Behav. Human Decis. Process.* **1991,** *50* (2), 179–211.

Audrey, C. Managers' Competence: Rhetoric, Reality and Research. *Personnel Rev.* **1989,** *18* (6), 20–26.

Aung, Z. M.; Tan, C. C. Using Business Model Concept as a Broad-Based S-O-R (Stimulating-Organism-Response) Consumer Behavior Model for Tourism Industry: The Case of Bagan, Myanmar. *J. Mekong Soc.* **2016,** *12* (2), 21–44.

Axley, L. Competency: A Concept Analysis. *Nurs. Forum* **2008,** *43* (4), 214–222.

Bangkok Post. Thailand Eases Curbs on Travel. *Bangkok Post* December 18, 2020.

Barnard, C. I. *The Functions of the Executive*; Harvard University Press, 1938.

Barney, J. B.; Clark, D. N. *Resource-Based Theory: Creating and Sustaining Competitive Advantage*; Oxford University Press, 2007.

Bharwani, S.; Jauhari, V. An Exploratory Study of Competencies Required to Co-Create Memorable Customer Experiences in the Hospitality Industry. *Int. J. Contemp. Hospital* **2013,** *25* (6), 823–843.

Bitner, M. Servicescapes: The Impact of Physical Surroundings on Customers and Employees. *J. Market.* **1992,** *56* (2), 57–71.

Brunner-Sperdin, A.; Peters, M. What Influences Guests' Emotions? The Case of High-Quality Hotels. *Int. J. Tour. Res.* **2009,** *11*, 171–183.

Chang, C. F.; Wang, P.; Liu, J. T. Knowledge Spillovers, Human Capital and Productivity. *J. Macroecon.* **2016,** *47,* 214–232.

Charm, T.; Coggins, B.; Robinson, K.; Wikie, J. *The Great Consumer Shift: Ten Charts That Show How US Shopping Behavior Is Changing*; McKinsey & Company, 2020.

Chen, C. M.; Chen, S. H.; Lee, H. T.; Tsai, T. H. Exploring Destination Resources and Competitiveness—A Comparative Analysis of Tourists' Perceptions and Satisfaction Toward an Island of Taiwan. *Ocean Coastal Manage.* **2016,** *119*, 58–67.

Chen, G. M. Competence Theories. In *Encyclopedia of Communication Theory*; Littlejohn, S. W.; Foss, K. A.; Eds.; SAGE Publications, Inc, 2009; pp 149–155.

Chiang, C. C.; King, B.; Nguyen, T. H. Taiwan's MICE Visitors: Business, Leisure and Education Dimensions. *Int. J. Cult. Tour. Hospital. Res.* **2012,** *6* (1), 21–33.

Chuang, Y. H.; Lai, F. C.; Chang, C. C.; Wan, H. T. Effects of a Skill Demonstration Video Delivered by Smartphone on Facilitating Nursing Students' Skill Competencies and Self-Confidence: A Randomized Controlled Trial Study. *Nurse Educ. Today* **2018,** *66*, 63–68.

Cizel, B.; Anafaarta, N.; Sarvan, F. An Analysis of Managerial Competency Needs in the Tourism Sector: The Case of Turkey. *Tour. Rev.* **2007,** *62* (2), 14–23.

Danquah, M.; Amankwah-Amoah, J. Assessing the Relationships Between Human Capital, Innovation and Technology Adoption: Evidence from Sub-Saharan Africa. *Technol. Forecast. Soc. Change* **2017**. http://dx.doi.org/10.1016/j.techfore.2017.04.021.

De Cuper, L.; Kucukkeles, B.; Reuben, R. Discovering the Real Impact of COVID-19 on Entrepreneurship. *World Econ. Forum* Jun 19, 2020.

Dreyfus, S. E.; Dreyfus, H. L. *A Five-Stage Model of the Mental Activities Involved in Direct Skill Acquisition*; Stoming Media, 1980.

Eisenhardt, K. M.; Martin, J. A. Dynamic Capabilities: What Are They? *Strategic Manage. J.* **2000,** *21* (10/11), 1105–1121.

Formentini, M.; Taticchi, P. Corporate Sustainability Approaches and Governance Mechanisms in Sustainable Supply Chain Management. *J. Clean. Product.* **2016,** *112*, 1920–1933.

Fornell, C.; Larcker, D. F. Evaluating Structural Equation Models with Unobservable Variables and Measurement Error. *J. Market. Res.* **1981,** *18* (1), 39–50.

Freiling, J.; Gersch, M.; Goeke, C.; Sanchez, R. Fundamental Issues in a Competence-Based Theory of the Firm. In *A Focused Issue on Fundamental Issues in Competence Theory Development*; Sanchez, R.; Ed.; 2015; pp 79–106. http://dx.org/10.1016/S1744–2117 (08)04002–4.

Garside, J. R.; Nhemachena, J. Z. Z. A Concept Analysis of Competence and Its Transition in Nursing. *Nurse Educ. Today* **2013,** *33* (5), 541–545.

Hair, J. F.; Black, W. C.; Babin, B. J.; Anderson, R. E.; Tatham, R. L. *Multivariate Data Analysis*; 6th ed.; Prentice Hall, 2006.

Heffeman, M. M.; Flood, P. C. An Exploration of the Relationship Between the Adoption of Managerial Competencies, Organizational Characteristics, Human Resource Sophistication and Performance in Irish Organization. *J. Eur. Ind. Train.* **2000,** *24* (2), 128–136.

Ionescu-Somers, A.; Tarnawa, A. *Diagnosing COVID-19 Impacts on Entrepreneurship: Exploring Policy Remedies for Recovery*; Global Entrepreneurship Research Association, 2020.

Kaplan, R. S.; Norton, D. P. Linking the Balanced Scorecard to Strategy. *CA Manage. Rev.* **1996,** Fall, 53–79.

Kaplan, R. S.; Norton, D. P. *Strategy Maps: Converting Intangible Assets Into Tangible Outcomes*; Harvard Business School Press, 2004.

Kaplan, R. S.; Norton, D. P. *The Execution Premium: Linking Strategy to Operations for Competitive Advantage*; Harvard Business School Press, 2008.

Leavy, B. Value Pioneering—How to Discover Your Own Blue Ocean: Interview with W. Chan Kim and Ranee Mauborgne. *Strategy Leadership* **2005,** *33* (6), 13–20.

Lee, C. Y. A Theory of Firm Growth: Learning Capability, Knowledge Threshold and Pattern of Growth. *Res. Policy* **2010,** *39*, 278–289.

Lima, P. G. L.; Damasceno, J. C.; Borges, J. A. R.; Santos, G. T. D.; Bankuti, F. I. Short Communication: Socio-Psychological Factors Influencing Dairy Farmers' Intention to Adopt High-Grain Feeding in Brazil. *J. Dairy Sci.* **2020,** *103 (*11), 10283–10288.

Lu, X. X.; Li, K. P.; Xu, H. C.; Wang, F.; Zhou, Z. Y.; Zhang, Y. G. Fundamentals and Business Model for Resource Aggregator of Demand Response in Electricity Markets. *Energy* **2020,** *204*, 117885. https://doi.org/10.1016/j.energy/2020/117885.

Manyara, G.; Jones, E. Community-Based Tourism Enterprises Development in Kenya: An Exploration of Their Potential as Avenues of Poverty Reduction. *J. Sustain. Tour.* **2007,** *15* (6), 628–644.

Mehta, S. C.; Loh, J. C. M.; Mehta, S. S. Incentive-Travel Marketing: The Singapore Approach. *Cornell H. R. A. Quart.* **1991,** 67–74.

Morcke, A.; Dornan, T.; Eika, B. Outcome (Competency) Based Education: An Exploration of Its Origins, Theoretical Basis, and Empirical Evidence. *Adv. Health Sci. Educ.* **2013,** *18* (4), 851–863.

Nagano, H. The Growth of Knowledge Through the Resource-Based View. *Manage. Decis.* **2020,** *58* (1), 98–111.

Neuman, W. L. *Social Research Methods: Qualitative and Quantitative Approaches*; Pearson Education, Inc, 2006.

Peteraf, M. A.; Barney, J. B. Unraveling the Resource-Based Tangle. *Managerial Decis. Econ.* **2003,** *24* (4), 309–323.

Popper, K. R. *Conjectures and Refutations: The Growth of Scientific Knowledge*; Routledge and K. Paul, 1963.

Popper, K. R. *Objective Knowledge: An Evolutionary Approach*; Clarendon Press, 1972.

Porter, M. E. *Competitive Advantage*; Free Press, 1985.

Prevot, F. *The Management of Competences in the Context of Interorganizational Relations: In Competence Building and Leveraging in Interorganizational Relations*; Published online: 09 Mar 2015; pp 7–35. http://dx.org/10.1016/S0749–6826 (07)11001–5

Pullman, M. E.; Gross, M. A. Ability of Experience Design Elements to Elicit Emotions and Loyalty Behaviors. *Decis. Sci.* **2004,** *35*, 551–578.

Rahman, S. A.; Amran, A.; Ahmad, N. H.; Raghizadeh, S. K. Prospective Entrepreneurial Competencies to Ensure Subjective Wellbeing of the Entrepreneurs at the Base of Pyramid. In *Proceedings Book of ICBSSS*, Malaysia Handbook on Business Strategy and Social Sciences, PAK Publishing Group, 2014; pp 56–64.

Ritala, P.; Huotari, P.; Bocken, N.; Albareda, L.; Puumalainen, K. Sustainable Business Model Adoption Image S&P 500 Firms: A Longitudinal Content Analysis Study. *J. Clean. Product.* **2018,** *170*, 216–226.

Ryan, C. *The Competitive Destination—The Future Role of the World City*. Keynote Paper for the Conference Hosted by Chinese Cultural University, May 6, 2010.

Sanchez, R. A Scientific Critique of the Resource-Based View (RBV) in Strategy Theory, with Competence-Based Remedies for the RBV's Conceptual Deficiencies and Logic Problems. In *Fundamental Issues in Competence Theory Development (Research in Competence-Based Management*; Sanchez, R., Ed., Vol. 4; Emerald Group Publishing Limited, 2008; pp 3–78.

Sanchez-Canizares, S. M.; Castillo-Canalejo, A. M. Community-Based Island Tourism: The Case of Boa Vista in Cape Verde. *J. Culture Tour. Hospital. Res.* **2014,** *8* (2), 219–233.

Schwartz, M. The Business Ethics of Management Theory. *J. Manage. Hist.* **2007,** *13* (1), 43–54.

Selznick, P. *Leadership in Administration*; Harper and Row, 1957.

Shafeey, T. E.; Trott, P. Resource-Based Competition: Three Schools of Thought and Thirteen Criticisms. *Eur. Busi. Rev.* **2014,** *26 (*2), 122–148.

Sharma, G. D.; Thomas, A.; Paul, J. Reviving Tourism Industry Post-COVID-19: A Resilience-Based Framework. *Tour. Manage. Perspect.* **2021,** *37*, 100786.

Shulga, L. V.; Busser, J. A. Customer Acceptance of Four Types of Hospitality Value Propositions. *Int. J. Hospital. Manage.* **2020,** *87*, 102388. https://doi.org/10.1016/j.ijhm.2019.102388.

Simon, M. C.; Welling, M.; Freiling, J. Internationalization Strategies—A Competence-Based Framework. In *A Focused Issue on Fundamental Issues in Competence Theory Development*; Sanchez, R., Ed.; 2015; pp 315–330. http://dx.org/10.1016/S1744–2117 (08)04009–7.

Spindler, J. Opening Up Is the Only Way to Rescue Thai Tourism. *Bangkok Post*, November 5, 2020.

Snyder, H. Literature Review as a Research Methodology: An Overview and Guidelines. *J. Busi. Res.* **2019,** *104*, 333–339.

Tan, C. C.; Anomasiri, S. Attention-Based View Approach to the Use of Performance Measurements to Drive Organizational Performance. *J. Acc. Prof.* **2017,** *13* (39), 58–82.

Tan, C. C.; Damnoen, P. S.; Phrapanyaprattanakorn, Phrakrupipithvarakijanukam, & Phumphongkhochasorn, P. Theory of Planned Behavior in Support for Post COVID-19 New Normalization Responses of Teachers Towards Online and Blended Learning. *Solid State Technol.* **2020,** *63* (5), 2166–2178.

Taylor, F. W. *Scientific Management: Comprising Shop Management, the Principles of Scientific Management, Testimony Before the Special House Committee*; Harper & Brothers Publishing, 1945.

Teece, D. J. Explicating Dynamic Capabilities: The Nature and Micro-Foundations of (Sustainable) Enterprise Performance. *Strategic Manage. J* **2007,** *28* (13), 1319–1350.

Teece, D. J.; Pisano, G. The Dynamic Capabilities of Firms: An Introduction. *Ind. Corporate Change* **1994,** *3* (3), 537–556.

Taylor, M. Domestic Tourism Subsidy Extended 200,000 Rooms Booked in First 12 Hours. *Thaiger* Dec 29, 2020. https://thethaiger.com/hot-news/tourism/domestic-tourism-subsidy-extended-200000-rooms-booked-in-first-12-hours.

UNEP & UNWTO. *Making Tourism More Sustainable: A Guide for Policy Makers*; World Tourism Organization Publications, 2005.

Vaz, C. R.; Selig, P. M.; Viegas, C. V. A Proposal of Intellectual Capital Maturity Model (ICMM) Evaluation. *J. Intell. Capital* **2018**. https://doi.org/10.1108/JIC-12–2016–0130.

Vaz, E. D.; Gimenes, R. M. T.; Borges, J. A. R. Identifying Socio-Psychological Constructs and Beliefs Underlying Farmers' Intention to Adopt On-Farm Silos. *NJAS-Wageningen J. Life Sci.* **2020,** *92*, 100322.

Walls, A.; Okumus, F.; Wang, Y.; Joon-Wuk Kwun, D. Understanding the Consumer Experience: An Exploratory Study of Luxury Hotels. *J. Hospital. Market. Manage.* **2011,** *20* (2), 166–197.

Wang, D.; Petitt, C. E.; Goel, N. S.; Ash, M. M.; Mervak, J. E. Confidence and Competency in the Use of Dermoscopy Among New First-Year Dermatology Residents: A Repeated-Pairs Pre/Post-Assessment Study of an Online Learning Module. *J. Am. Acad. Dermatol.*, 2020. https://doi.org/10.1016/j.jaad.2020.11.028.

Winter, S. G. The Satisficing Principle in Capability Learning. *Strategic Manage. J.* **2000,** *21*, 981–996.

Worrachaddejchai, D. TCEB Lines up Future MICE Arrivals Next Year. *Bangkok Post* Oct 8, 2020.

World Tourism Organization (WTO). *Measuring the Economic Importance of the Meetings Industry: Developing a Tourism Satellite Account Extension*; World Trade Organization, 2006.

Yi, Y.; Wang, Y.; Shu, C. Business Model Innovations in China: A Focus on Value Propositions. *Busi. Horiz.* **2020,** *63*, 787–799.

Ziaie, A.; ShamiZanjani, M.; Manian, A. Systematic Review of Digital Value Propositions in the Retail Sector: New Approach for Digital Experience Study. *Electr. Com. Res. App.* **2021,** *47*, 101053. https://doi.org/10.1016/j.elerap.2021.101053.

CHAPTER 15

PUBLIC POLICIES FOR TOURISM AND INTERGOVERNMENTAL COORDINATION FOR POST-COVID-19: ANALYSIS OF THE BRAZILIAN CONTEXT

FÁBIA TRENTIN[1], CLAUDIA CORRÊA DE ALMEIDA MORAES[2], and ISABELA DE FÁTIMA FOGAÇA[3]

[1]*Professor at the Faculty of Tourism and Hospitality at Universidade Federal Fluminense and the Master's Program in Tourism, Coordinator of the Policy, Governance, and Tourism Laboratory (LabPGTUR), Rua Noronha Torrezão, Santa Rosa, Niterói/RJ/Brazil*

[2]*Professor at the Faculty of Tourism and Hospitality and Coordinator of the MBA in Service Management at the Universidad Federal Fluminense (UFF), Vice-leader of the Research Group Tourism, Management and Territory, São Paulo/SP/Brazil*

[3]*Professor at the Tourism Course (Bachelor's degree) at the Federal Rural University of Rio de Janeiro (UFRRJ) and the Postgraduate Program in Heritage, Culture, and Society (PPGPACS) at UFRRJ, Rio de Janeiro/RJ/Brazil*

ABSTRACT

In 2020, the new SARS-CoV-2 coronavirus, which causes COVID-19 disease, is globally disseminated, causing a pandemic. Tourism is strongly affected by the pandemic since social isolation is adopted as a prophylactic measure

Post-COVID Tourism and Hospitality Dynamics: Recovery, Revival, and Re-Start.
Umendra Narayan Shukla & Sharad Kumar Kulshreshtha (Eds.)

and displacements and, consequently, tourist flows are interrupted. Public tourism policies, which commonly seek to attract visitors, are interrupted, and citizens' well-being and health security are placed while the economic crisis is installed in the tourism sector. In this context, this chapter adopts bibliographic and documentary research of a qualitative nature to analyze how governments, and consequently the tourist trade, acted to reduce the spread of COVID-19, the intermediate conflicts generated by divergent or converging interests, and to incorporate the issues arising from this situation (health and economic crises in tourism) in order to maintain the well-being of its citizens and the tourism sector.

15.1 INTRODUCTION

The pandemic caused by the new coronavirus SARS -CoV-2 fits, which brought COVID-19 to the world population, changing the environmental, economic, and socio-cultural dynamics today. In Brazil, COVID-19 began spreading in February 2020, and in about 1 year and 4 months, 17,375,000 people were infected and 486,300 died (Brazil, 2021). When the pandemic started, the governments in all their bodies (federal, state and municipal) including supranational organizations to act swiftly in order to contain the propagation of the disease and the hardness of its consequences, by adopting restrictive measures on people's movements and even banning it completely.

With the need to stretch the restrictions, the concern with the economy and with the specific productive sectors appeared, as well as with its workers. Maintaining the balance and intermediating conflicts generated by divergent or convergent interests, became a challenge, primarily in societies driven by neoliberal politics and characterized by regional socioeconomic differences, as is the Brazilian reality.

In this context, tourism, as a socio-cultural and environmental phenomenon that has within its primary characteristics the dislocation of people to the exterior of the territories in which they reside, was deeply affected by the restrictive measures imposed as a result of the pandemic.

The present chapter aims to study the implementation of the World Tourism Organization's (UNWTO, 2020) recommendations for the resumption of tourism, post-COVID health crisis, from the national, state, and municipal plans, with regard to the strategies for the resumption of the tourism sector, as well as for the protection of visitors and residents of tourist sites. For this purpose, an analysis was made of the plans elaborated by the Ministry of Tourism (MTur) (2020a), by the Secretary of Tourism of the State of Rio de

Janeiro (SETUR-RJ) (2020) and by the municipal governments of Armação dos Búzios (PMAB) and of Paraty (PMP).

Based on the criteria of economic dependence on tourism, and the relevance in the national tourist context, concerning the attraction of Brazilian and foreign visitors, the destinations of Armação dos Búzios and Paraty, both belonging to two distinct regions and strategic for the sectorial planning trajectories of the sectors of the state of Rio de Janeiro (Brazil): Costa do Sol and Costa Verde (Figure 15.1).

Map of Brazil with the location of the state of Rio de Janeiro

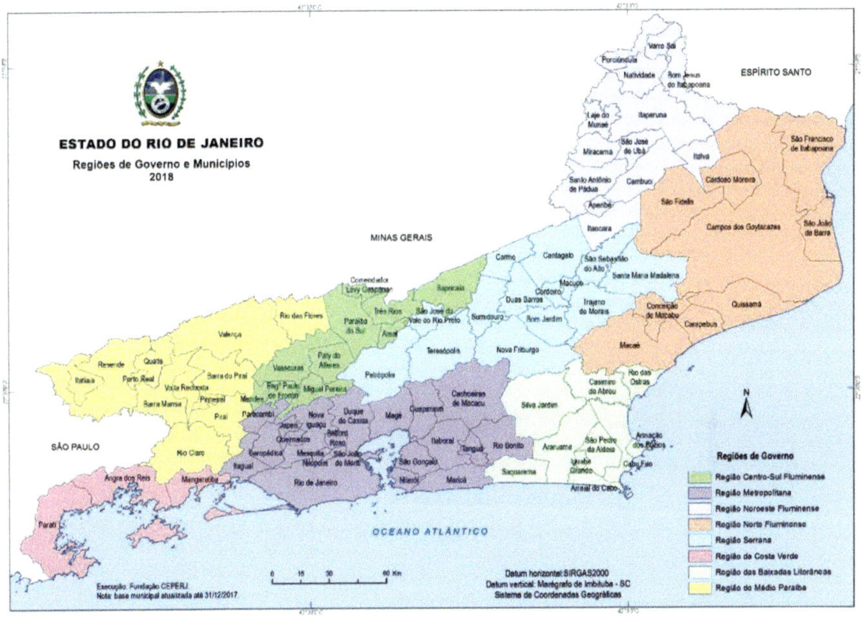

FIGURE 15.1 Localization of the municipalities of Armação dos Búzios and Paraty in Rio de Janeiro state, Brazil.

Source: Reprinted from Governo do Estado do Rio de Janeiro = Centro de Estatísticas, Estudos e Pesquisas, 2018, CC BY-SA 2.5, 2018.

Armação dos Búzios has 27,500 residents (IBGE, 2020) distributed in a territory of approximately 70 km^2, divided into a peninsular and a continental part. The peninsula has undergone an accentuated touristification process, "understood as a complex process in which various stakeholders interfere, transforming a territory through tourist activity" (Ojed & Kieffer, 2020) and now concentrates a significant part of the tourist infrastructure such as hotels, restaurants and other touristic services. It is located on the north coast of the state of Rio de Janeiro, a touristic region with relevance in the segment sun and beach (PMAB, 2020).

Its touristic flow is the second largest in the state of Rio de Janeiro and the fifth most sought-after destination by international tourists in Brazil. The primary issuing markets are Argentina, United States, Chile, and Peru. As for traveling motivations, Ecotourism ranks like Cultural Tourism, both with 16% in motivation, although less than Sun and Beach Tourism with 53% of the reasons for travel (Rede de Inteligência de Mercado [RIM)] (2019). The leading national flow is composed of residents of the state's capital (Rio de Janeiro) and residents of São Paulo and Minas Gerais's neighboring states.

Paraty has a population of 43,000 residents (IBGE, 2020), distributed in a territory of approximately 928,000 km^2, divided between a continental part and 65 islands in its Bay. It is located in the extreme south of the state, in a tourist region with many luxury resorts and the presence of nautical tourism. The Historic Centre, considered a National Monument (Instituto do Patrimônio Histórico e Artístico Nacional [IPHAN] (2020), concentrates the historical-cultural tourist attractions and other touristic services. In other areas of the city live the rest of the population and 30 traditional communities that contribute to its cultural diversity. It is a Creative City of Gastronomy and declared the Cultural and Natural Heritage of Humanity by the United Nations (UN) (PMP, 2019).

According to the research on international tourist demand in Brazil (MTur, 2019), made between 2014 and 2018, Paraty occupied the 12th position. Between Brazilians, the city of São Paulo is its primary issuer center (Serviço de Apoio às Pequenas e Micro Empresas [Sebrae]), (2019) and between foreign nationalities there are danish, French, Norwegian, English, German, Canadian and Mexican (MTur, 2019). Paraty's economy is 90% dependent on tourism, and its events are essential to maintain its regular tourist flow.

15.1.1 REVIEW OF THE LITERATURE

For a better understanding of intergovernmental coordination, it is essential to highlight that Brazil is a federative republic, composed of the Union (National State), states, municipalities, and the Federal District (subnational entities) (Brazil, 1988). A federalist situation presumes conjunction of characteristics related to the existence of territorial heterogeneities—such as great territorial extension and/or significant physical, cultural—ethnic, linguistic, socioeconomic—diversity, regional diversities, and political ones—differences in the constitution of the elites inside a country and/or an intense rivalry between them (Abrucio, 2005).

According to Abrucio (2005), a federalist country stands for unity in diversity, safeguards local autonomy, and searches for ways of maintaining territorial integrity in a highly heterogeneous context. A federalist country is established to account for the heterogeneities, maintain social stability, and preclude the nation's fragmentation. A federalist arrangement has the prerogative of autonomy of power and interdependence. The parties' integrity is mutually recognized to promote a particular unity among themselves, for better sharing of decisions and interaction between federated entities (Elazar, 1987).

Therefore, interdependence evokes coordination and intergovernmental relations (Silva & Cabral Neto, 2018). Federative coordination involves the sharing of objectives to reduce the duplication and superposition of federated entities' actions (Bakvis & Juillet, 2004) and territorial inequalities (Abrucio et al., 2020). Coordination evokes collaboration (Castro & Young, 2017; Trentin, 2016).

Federative coordination must dispose of action mechanisms that respect subnational entities' autonomy and competence regarding cooperation and collaboration relative to public policies. Coordination of public policies is necessary between organizations and political and bureaucratic actors that make up the same sphere of government, such as sectoral ministries at the federal level (Organization for Economic Cooperation and Development [OECD], 2013), and between different levels of government (Souza, 2019; OECD, 2013), as a way to obtain better results in meeting the collective interest (Trentin & Fonseca Filho, 2020).

However, Silva (2012) identifies some challenges in the social area that are also present in tourism. Amongst them, stand out:

- The challenge of coordination between management units and productive agents (coordination with the market).
- The challenge of coordination between management units and civil society (coordination with society).
- The challenge of coordination between management units and different powers—Executive, Legislative, and Judiciary (republican coordination).
- The challenge of coordination between management units and distinct federative levels of government—Union, States, and Municipalities (federative coordination).

- The challenge of coordination within the management unit itself (horizontal coordination).
- The challenge of coordination between management units and international institutions (international coordination).

The multiple challenges around coordination have been put under strain in the face of the pandemic's seriousness. This scenario evoked speed on the part of the public authorities in its actions, primarily aiming to prevent the spread of COVID-19, and posteriorly, regarding the necessity of dealing with the adverse effects in public health and in the economy, demanding strong coordination between the different levels of government, productive sectors, international institutions, civil society and amid public policies, including in tourism. This socio-cultural and environmental phenomenon has among its primal characteristics the mobility of persons outward from their territories.

In tourism, due to the pandemic, the UNWTO (2020) estimated a 60–80% reduction in the number of tourists landing worldwide, resulting in 850 million to 1.1 billion fewer international tourists in circulation in its member countries: also, losses of 100–120 million jobs in the sector and a 910–1.2 trillion of dollars reduction in exports.

With this scenario in place, the UNWTO launched the Global Guidelines to Restart Tourism, a document prepared with the contribution of the International Civil Aviation Organization (ICAO), the International Maritime Organization (IMO), the International Labour Organization (ILO), World Bank (WB), Airports Council International (ACI), Cruise Lines International Association (CLIA), International Air Transport Association (IATA), World Travel and Tourism Council (WTTC), and the Government of Spain, to guide the restart of touristic activities worldwide.

To the UNWTO (2020), the recuperation of tourism should be based on priorities such as providing liquidity and protecting jobs; recovering confidence through safety and security; public–private collaboration for an efficient reopening; opening of borders with responsibility; harmonization and coordination of protocols and procedures; addition of value jobs through new technologies; and innovation and sustainability as the new normal. Thus, the directives presented in the guide to the gradual recovering of tourism worldwide were organized in eight axes: (i) safe and seamless border management; (ii) private sector—cross-cutting measures; (iii) safe air travel; (iv) hospitality; (v) tour operators and travel agencies; (vi) meetings and events; (vii) attractions and thematic parks; and (viii) destination planning and management.

Starting from the categories found in the Global Guidelines to Restart Tourism document, this study identifies the actions undertaken by the Ministry of Tourism, the government of Rio de Janeiro State, and by the municipalities of Armação dos Búzios e Paraty, taking as a reference the challenges of intergovernmental coordination in Silva (2012).

15.1.2 METHODOLOGICAL PROCEDURES

This chapter adopts bibliographic and documentary research of a qualitative nature to analyze how governments, and consequently the tourist trade, acted to reduce the spread of COVID-19, the intermediate conflicts generated by divergent or converging interests and to incorporate the issues arising from this situation (health and economic crises in tourism) in order to maintain the well-being of its citizens and the tourism sector.

Thus, a cross-sectional approach is applied, combined with empirical observation and research in the scientific and documentary literature on public policies, intergovernmental coordination, norms and legal acts and recommendations from international organizations and institutions, in order to understand how the government is fulfilling its role, coordinated or not their actions in an articulated way and how were the results concerning society and the tourism sector.

15.2 RESULTS AND DISCUSSION

Brazil, a member country of the UNWTO, developed, in the context of the COVID-19 pandemic, two guiding documents. The manual of elucidation about measures adopted in the fight against the pandemic, maintenance of employment and liquidity of companies and the Recovery Plan to be developed in four stages: protection, resumption, incentive, and promotion, which guiding motto would be "Security for the Recovery of Tourism and Employment for the Recovery of the Economy" (MTur, 2020b), which was published in November 2020, instituting four axes of action and the parameters for the development of programs, projects, and actions aimed at the recovery of tourism in Brazil, in order to mitigate the negative effects caused in the sector: (i) preservation of companies and jobs in the tourism sector; (ii) improvement of the structure and qualification of destinations; (iii) implementation of biosecurity protocols; and (iv) promotion and incentive to travel.

The municipalities of Armação dos Búzios and Paraty, based on the recommendations of the UNWTO, MTur and the documents "Rio de Janeiro Conscious Tourism—10 commandments to the conscious tourism and the "Rio de Janeiro State Social Pact for Health and Economy," organized the regulations and its protocols in order to attend to the necessities of the sector in the pandemic crisis context. In Paraty, the orientations are inscribed in the document "Booklet of Guiding Principles of the New Tourism of Paraty" (PMP, June/July 2020). The documents elaborated by the WTO and by the three government levels in Brazil are the basis to the analysis henceforth unfolded.

15.3 SAFE AND SEAMLESS BORDER MANAGEMENT

Among the actions suggested by the WTO is the flexibilization of travel visas, adopted by Brazil even before the COVID-19 pandemic. Within the virtual platform "Rio de Janeiro Turismo Consciente," the state government created the "Tourist Destination Monitoring Report," in which it presents the tourist facilities that are open, partially open, closed and lacking information and, where there are municipal barriers to circulation, providing access to reliable information on travel restrictions and protocols for the private sector and travelers.

Both the municipalities of Armação dos Búzios and Paraty created links on their official websites. They provide information to the local population, entrepreneurs, and tourists about the administrative decrees that regulate the pandemic's response in their territories. Crisis committees were created, integrating several sectors such as tourism, health, transport, among others. Tourism protocols were revised according to the accumulated knowledge on the subject.

15.4 PRIVATE SECTOR CROSS-CUTTING

The private sector cross-cutting measures, led by MTur, focus on the preservation of companies and jobs in the tourism sector and on the promotion and incentive of travel through action split into three axes:

1. *Concession of fiscal credits and benefits*: The federal government has created lines of credit for small- and medium-sized enterprises to maintain employee payments, in addition to having facilitated and extended credit for tourism enterprises impacted by the pandemic

and the granting of exemption from fines for travel postponement and the right to use credit for a period of 12 months.

2. *Reduction of working hours and wage flexibility*: Working hours and wages were made more flexible as a means of maintaining job vacancies, as result of reduced economic activity.
3. *Emergency aid for workers in the tourism sector and outplacement in the labor market*: Includes autonomous and informal workers and individual microentrepreneurs affected by the COVID-19 pandemic. A platform was created with the purpose of supporting the reallocation of professionals that have lost (or not) their jobs as result of the pandemic. Establishment of the Aldir Blanc Law for the culture sector's aid.

As a method of supporting local enterprises and establishments, and within the constitutional competence of the municipalities, Armação dos Búzios's local government suspended the collection of the "Territorial and Urban Property Tax," the billing of taxations and fines during the emergency period and Paraty's government suspended the payment of basic service bills, such as water and electricity, besides working with the "Restart Project," liberating emergency credit for microentrepreneurs and microenterprises, through an emergency fund created by Paraty's prefecture.

Rio de Janeiro's state government adopted measures for the recovery of tourism business and events sector, such as fairs, congresses and conventions and made it possible for tour guides to register their own automobiles for professional activities, combining them with the provision of tourist transport services, to be able to develop their activities and at the same time avoid agglomeration in the itineraries.

In Armação dos Búzios and in Paraty, amid the policies for the private sector stand out protocols that contemplate health, security and hygiene for the collaborators and tourists that were trained in the new conducts of service provision during the pandemic. The adopted protocols were publicized internally in establishments and in some cases, released digitally.

Procedures related to the redesign of operations into contactless digital operations at all possible points, creation of internal COVID-19 committees, registration of statistical information for planning and management of tourism in plain coordination with local and national tourism and the statistical authorities were compromised since they already were fragilities inherent to the sector, regardless of the pandemic.

Dialog between the public and private sectors occurred to establish protocols and adopt protective measures. In Paraty, a poll was made by the municipal government asking the population about whether the touristic activities should return or not, in which 58.8% (in august) of the residents responded positively.

15.5 SAFE AIR TRAVEL

Airlines adapted their operational procedures in order to meet World Health Organization (WHO), UNWTO, and MTur's recommendations on providing safer services in terms of consumer's health. Hygiene protocols on operations contemplate, especially, limits on movements inside the cabin area during flights, simplifying buffet services, and other procedures that reduce the crew's movement and interaction between passengers.

Measures such as reducing items in the cabin, limitation of carry-on baggage and control of proximity between people by scattering seats, for example, are not being respected. Although there is social distancing control at boarding and landing, there is no effective control and passengers tend not to embrace the orientations.

In terms of partnerships and of decrease in airport service's costs, including financial assistance, the federal government announced working capital credit for airlines, postponement of payment of fares for the right to operate airports, and 12 months for reimbursement of tickets purchased until December 31, 2020.

15.6 HOSPITALITY, TOUR OPERATORS AND TRAVEL AGENCIES, MEETINGS AND EVENTS AND ATTRACTIONS

The MTur recommends that the public and private sectors to intensify personal hygiene, establish a crisis management mechanism, set up biosafety protocols, and to incorporate to Responsible Tourism Seal, a declaratory seal.

It also recommends that communication with establishments about the disease, symptoms and behavior to prevent contagion is reinforced. In case of those infected, MTur orients on treatment plans and on self-isolation protocols: 7 days after domestic flights and 14 days for those traveling internationally.

The State of Rio de Janeiro created sanitary protocols, which resulted in the "Conscious Tourism Seal," which allows business owners of the sector to subscribe to their establishments in an online portal where they auto-declare suitable for service provision.

The municipalities organized biosafety protocols and directly involved segments such as hospitality, operators and travel agencies, events, conventions, and tourist attractions.

Armação dos Búzios developed sanitary protocols in consonance with WHO recommendations for the resumption of activities in the corporate hospitality segment, free fairs, parks, entertainment attractions, water activities, physical activity outdoors, and events and land tours (Bugres, Troller). It trained the population through courses and lectures about each sector's protocols for the return of the activities, being the establishment obliged to expose a certificate of such training sessions.

Paraty proceeded in the same way, following WHO's proposed guidelines, and establishing a dialog with the private sector and organized civil society, through the Municipal Council for Tourism (COMTUR) and the Municipal Council for Cultural Policies (CMPC) and of identified tendencies in a postpandemic scenario on tourism. It also adopted more strict measures regarding the prohibition of tourist vehicles (collective).

Both municipalities reinforced the oversight of public streets, establishments, and businesses according to different segments that belong to the tourist sector. During weekends, attendance capacity was limited for each type of business, and implemented sanitary responsibility terms to be signed by tourism establishments.

In this sense, in the comparison between the actions recommended by the UNWTO (2020) and federal, state, and municipal protocols, there are recommendations for the hospitality, operations, travel agencies, events, and tourist attractions segments aiming at the increase of cleaning and hygiene frequency of spaces, communication to employers and tourist about protocols and measures to be followed targeting joint health and protection. The use of masks was compulsory in all establishments. This study did not verify new job posts such as hygiene manager and the guest tutor.

From the point of view of product and marketing, federal and state documents have foreseen the diversification of tourism offer, although there is no specific focus on nature, rural environment, or culture. At a federal level, were developed the advertisement campaigns "Brazil waits for you," "The pandemic will pass, tourism will not," and "Do not cancel, RESCHEDULE!," to highlight for tourist consumers the importance of not canceling, but instead postponing its travel plans, as a way of maintaining the sector alive.

At the state level, the current policy is focus on the training of municipal managers and on actions regarding the pandemic such as the biosafety protocols. There is also the campaign "Rio, your best gift" which encourages tourism beyond the homonymous capital after the end of the pandemic.

In Armação dos Búzios, no strategic planning actions with new products were observed, but on the contrary, there were actions in terms of adequacy of existing products to safety and recovery protocols. Besides recommending the use of technology necessary for proper operational procedures in lodging organizations, Paraty is proposing that the municipality change its tourism to more sustainable, inclusive, with more experiences and fewer tourists.

15.7 DESTINATION PLANNING AND MANAGEMENT

MTur, in terms of planning and management actions of destinations in Brazil, includes connectivity, campaigns to encourage travel to support local commerce, and an online intelligence platform for Brazilian tourism with a portfolio of business opportunities in strategic tourist routes. The improvement of mobility, connectivity, and the infrastructure of touristic destinations which must, for example, avoid connections and stopovers in flights. However, so far, in terms of reducing passengers and reorganizing flights due to the decrease in demand, more connections with a shorter duration between flights connecting different regions of the country have been observed.

Following UNWTO's recommendations, the coordination between tourism, health, and transport policies was verified in Armação dos Búzios and Paraty and the definition of roles and responsibilities for governments, the private sector, and for travelers.

In Armação dos Búzios, the Ministério Público (Public Prosecution Service), an independent control organ, considering the intensive care units (ICU) occupancy rate, determined that all tourists must leave Armação dos Búzios within 72 h from December 17, 2020, when the city entered the Red Flag stage. However, after business owners and workers' protests, the Rio de Janeiro's Court of Justice reversed the decision.

Regarding safe and continuous frontiers management, in the pandemic context, the federal government is responsible for the international frontiers. Simultaneously, the states and municipalities can use sanitary control barriers and limit the entrance of people in their territories. This policy was admitted after actions were filed at the Supremo Tribunal Federal (Federal Supreme Court) against the national executive.

Thus, states and cities had more clearly the limits and attributions of their roles amid the omission of the federal government (UNWTO, 2020, OECD, 2020) and, hence adopted restrictive measures primarily in terms of essential activities and services. Posteriorly, they could better organize lifting the restricted measures on nonessential activities, judging from contamination, and ICU occupation rates.

Comparing the actions described in the Global Guidelines to Restart Tourism (UNWTO, 2020) and the Recovery Plan (MTur, 2020b), a conversion was verified in terms of the improvement of structure and qualification of destinations implementation biosafety protocols. Regarding the preservation of jobs and businesses in the tourism sector and promotion of travel, the Brazilian document presents different, albeit complementary, approaches. The Global Guidelines for Restarting Tourism (OMT, 2020) suggest new products and destinations within the pandemic context and, while Brazil recommends travel after the pandemic has calmed or passed.

As mentioned earlier, the seals are declaratory, and there is no verification in terms of compliance or not with their terms. In both cases, tourists may serve as the confirmation as it may endorse or report establishments through channels such as the ombudsman services guaranteed by the Constitution of 1988 or through social media. However, there is little clarification about this possibility for tourists and consumers.

Lastly, when analyzing municipal, state, and federal documents and the Global Guidelines to Restart Tourism, based on the intergovernmental coordination, a framework was elaborated that summarizes the results revealed by this chapter (Framework 15.1).

15.8 KEY CONSIDERATIONS

The negationist posture, the strategies, and ideology behind the central government have undermined intergovernmental coordination led by it, enabling a rearrangement, in which subnational governments have strengthened themselves through the judicialization of several demands that have been presented to the Federal Supreme Court, the results of which have historically been contrary to the federal government and in favor of states and municipalities.

After the analysis of the documents named above, it was verified that, as in member countries of the OECD (2020), Brazil provided financial support to protect businesses, families, vulnerable populations, and small and micro-enterprises within areas directly affected by the crisis, by promoting the flexibilization of fiscal rules and norms.

Despite the intergovernmental lack of coordination in health, there seems to be a coordinated dialog between the Ministry of Tourism and representatives of the tourism trade. About intergovernmental coordination on the part of the federal government, in the sector of tourism amid COVID-19, there is a gap involved in the rhetoric of discontinuity of actions since, during the crisis, the Minister of Tourism was replaced by an interim one.

FRAMEWORK 15.1 Syntheses of intergovernmental coordination of tourism in the pandemic context (march to December 2020).

Intergovernmental Coordination	MTUR	Rio de Janeiro's State Government	Armação dos Búzios	Paraty
Coordination between management units and productive agents (market coordination)	**Market coordination occured in all levels (federal, state and municipal)**			
	The federal government consulted class entities on the elaboration of its recovery plan.	The state government consulted the trade sector on elaborating the booklet with the ten commandments of Conscious Tourism.	Armação dos Búzios's local government was pressured by commercial associations and business owners of the tourist sector to elaborate on the pandemic's contingency measures.	The planning of the gradual reopening of tourism was made with the prefecture's collaboration through the Tourism, Culture, Environment and Health public offices, the Public Attorney's Office, Communication Department, and COMTUR.
Coordination between management units and civil society (coordination with society)	–	–	The municipal levels coordinated actions with civil society, especially in Paraty, where the prefecture made a public survey with the local population regarding the restart of touristic activities in August 2020. Armação dos Búzios had the collaboration of public power and the tourist sector.	
Coordination between management units and different powers—Executive, Legislative and Judiciary (republican coordination)	The public powers changed the cooperative coordination arrangement, structured since 1988 with intense federal action, giving the subnational entities, mainly the states, more autonomy.		Occasionally via the judiciary power.	Coordination between local public authorities and the Public Prosecutor's Office to monitor compliance with health protocols.
Coordination between management units and distinct federative levels of government—Union, States and Municipalities (federative coordination)	Occasional, with a highlight to the Aldir Blanc Law, disposes of emergency actions destined for the cultural sector to be adopted during the public emergency state aimed at states and cities.			

FRAMEWORK 15.1 *(Continued)*

Intergovernmental Coordination	MTUR	Rio de Janeiro's State Government	Armação dos Búzios	Paraty
Coordination within the management unit itself (horizontal coordination)	Visible in the advertisement of investments in infrastructure, which results until December 18th, 2020, was worth 16%.	It was made between different secretaries that involve tourism, health, transport, and social services.	They were conducted with the secretaries of Tourism, Culture, and Historical Patrimony, Social Development, Employment and Revenue, Education, Science and Technology and the Special Secretary for Communications.	It was achieved by the secretaries of Health, Education, Tourism, Culture, Environment, Gabinet, Attorney General of the City and Construction and Transportation of Paraty.
Coordination between management units and international institutions (international coordination)	It was achieved, however, not in all principals and axes.	It was identified in some actions.	Paraty has the Global Guidelines to Restart Tourism available on its official website. Protocols of both cities present correlation with the content of their biosafety and tourism restart protocols.	

Source: Elaborated by the authors based on NWTO (2020); MTur (2020); SETUR-RJ (2020); PMAB (2020); PMP (2020).

While OECD member countries (2020) have responses to introduce, activate or redirect multilevel coordination bodies, supporting cooperation between cities and regions seeking to minimize uncoordinated responses and competition for resources, Brazil experiences a (dis) coordination at the federal level, which violates the cooperative model by deconstructing and/or dismantling participatory structures.

In this study, coordination between the three government levels with the market concerning pandemic control measures and the restart of tourism, a lot of the times tensioned by economic interests, stand out. These interests questioned the restrictions and rigidity of measures to control COVID-19, sometimes in collaboration with actions.

Municipalities played fundamental roles in the horizontal and society coordination's—this last one, with occasional conflicts that have been addressed. The republican coordination is currently in a tension period undergoing structure alterations defined by the 1988 Federal Constitution. The federative coordination was verified, mostly, by the implementation of the Aldir Blanc Law intended for the cultural sector. International coordination was identified primarily in the municipalities through compliance with OMT's directives. It is observed that Brazil and its federative entities have not yet surpassed the federative coordination challenges, nor have they been proposing alternative new models.

An absence of coordinated intergovernmental policy structures for the restart of tourism beyond the usual measures (infrastructure, training, promotion, and events) is noticeable. Once more, Brazil loses an opportunity to (re)plan tourism and promotes profound changes to potentialize the local development of tourism based on the guarantee of criteria guided by strategic regional priorities by integrating sustainability's social objectives into the recovery plans prepared by all levels of government.

KEYWORDS

- **COVID-19 pandemic**
- **UNWTO**
- **public policies**
- **tourism recovery**
- **Brazil**

REFERENCES

Abrucio, F. L.; Grin, E. J.; Franzese, C.; Segatto, C. I.; Couto, C. G. Combating COVID-19 Under Bolsonaro's Federalism: A Case of Intergovernmental Incoordination. *Braz. J. Public Admin.*, Rio de Janeiro **2020,** *54* (4), July/Aug.

Abrucio, F. L. A Coordenação Federativa no Brasil: a Experiência do Período FHC e os Desafios do Governo Lula. *Revista de Sociologia Política,* Curitiba **2005,** *24*, 41–67. DOI:10.1590/S0104-44782005000100005

Bakvis, H.; Juillet, L. *The Horizontal Challenge: Line Departments, Central Agencies and Leadership*; Canada School of Public Service: Ottawa, 2004.

Brazil. Portaria nº 754 de 10 de novembro de 2020. Institui os eixos de atuação e os parâmetros para desenvolvimento de programas, projetos e ações para retomada do turismo no Brasil, com vistas a mitigar os efeitos negativos causados no setor, em decorrência da pandemia do COVID-19. Diário oficial da União. Brasília, 2020. http://www.turismo.gov.br/2020/14005-portaria-n%C2%BA-754,-de-10-de-novembro-de-2020.html.

Brazil. *Constituição da República Federativa do Brasil*; Senado: Brasília, DF, 1988.

Castro, B. S. de; Young, C. E. F. Os problemas de coordenação de políticas públicas: desafios para a gestão ambiental no Brasil. *Revista TCE—RJ* **2017,** *12* (1), 32–53. Jan/jun.

Elazar, D. Exploring Federalism. Tuscaloosa: University of Alabama.

Governo do Estado do Rio de Janeiro = Centro de Estatísticas, Estudos e Pesquisas—CEEP = http://www.ceperj.rj.gov.br/ceep/info_territorios/divis_politico_administrativo.html, CC BY-SA 2.5, através da wiki Wikimedia Commons.

Instituto Brasileiro de Geografia e Estatística [IBGE]. 2020. https://cidades.ibge.gov.br/brasil/rj/rio-de-janeiro/pesquisa/48/48980

Instituto do Patrimônio Histórico e Artístico Nacional [IPHAN] Paraty (RJ), 2020. http://portal.iphan.gov.br/pagina/detalhes/381/

Ministério da Saúde [MS] Boletim Epidemiológico Especial—doença pelo Coronavírus COVID-19 23, 2021. https://www.gov.br/saude/pt-br/media/pdf/2021/junho/18/boletim_epidemiologico_covid_67.pdf.

Ministério do Turismo [MTur] Dados e Fatos. Demanda Turística Internacional. Ficha Síntese Destinos **2018,** 2019. http://www.dadosefatos.turismo.gov.br/2016-02-04-11-54-03/demanda-tur%C3%ADstica-internacional.html.

Ministério do Turismo [MTur]. *Manual sobre coronavírus*. Brasília, 2020a. http://www.turismo.gov.br/%C3%BAltimas-not%C3%ADcias/13428-mtur-apresenta-manual-sobre-o-coronav%C3%ADrus-aos-viajantes-e-entidades-do-setor.html.

Ministério do Turismo [MTur]. Em Ouro Preto (MG), o ministro anunciou o Plano Nacional de Retomada do Turismo: Durante a agenda, Álvaro Antônio apresentou ações para o fortalecimento do setor em todo o Brasil. Brasília, 2020b. http://www.turismo.gov.br/%C3%BAltimas-not%C3%ADcias/13744-em-ouro-preto-mg-ministro-anuncia-plano-nacional-de-retomada-do-turismo.html.

Ojed, A. B.; Kieffer, M. Touristification. Empty Concept or Element of Analysis in tourism geography? *Geoforum* **2020,** *115*, 143–145. DOI: https://doi.org/10.1016/j.geoforum.2020.06.021

Organization for Economic Co-operation and Development (OECD) The Territorial Impact of COVID-19: Managing the Crisis Across Levels of Government, 2020. https://read.oecd-ilibrary.org/view/?ref=128_128287-5agkkojaaa&title=The-territorial-impact-of-covid-19-managing-the-crisis-across-levels-of-government.

Prefeitura de Paraty. Paraty espera por você. Protocolos Sanitários, 2020. https://www.paraty.rj.gov.br/paraty-espera-por-voce.
Prefeitura de Paraty [PMP] Cartilha de Princípios Norteadores do Novo Turismo de Paraty–junho/julho, 2020. https://www.paraty.rj.gov.br/conteudo/downloads/paraty-espera-por-voce/Cartilha_ Retomada_Economica.pdf
Prefeitura de Armação dos Búzios [PMAB]. Protocolos para Retomada de Atividades, 2020. https://buzios.aexecutivo.com.br/publicacoes.php?pagina=1.
Serviço de Apoio às Micro e Pequenas Empresas [Sebrae-RJ] Resultados da pesquisa sobre o perfil do turista em Paraty são surpreendentes, 2019. http://www.paratyonline.com/jornal/2018/05/pesquisa-perfil-do-turista-na-costa-verde/.
Secretaria de Estado do Turismo do Rio de Janeiro [SETUR] Rio de Janeiro Turismo Consciente, 2020. https://turismoconscienterj.com.br/segmentos/
Silva, L. A. L. da. Gestão de políticas sociais: desafios contemporâneos de cooperação e coordenação. Temas de Administração Pública, Araraquara, SP **2012,** *4* (7).
Silva, R. R. de O. da; Cabral Neto, A. Relações intergovernamentais e a coordenação federativa no contexto do plano de ações. *Revista Eletrônica de Educação* **2018,** *12* (3), 621–634, set./dez.
Souza, C. Coordenação, uniformidade e autonomia na formulação de políticas públicas: experiências federativas no cenário internacional e nacional. Cadernos de Saúde Pública. Rio de Janeiro **2019,** *35* (supl. 2).
Trentin, F. Governança turística em destinos brasileiros: comparação entre Armação dos Búzios/RJ, Paraty/RJ e Bonito/MS. *PASOS Revista de Turismo y Patrimonio Cultural* **2016,** *3* (14), 645–658.
Trentin, F.; Fonseca Filho, A. S. Gestão Pública em Turismo: Coordenação e Comunicação no Processo de Inventário da Oferta Turística. *Caderno Virtual de Turismo* **2020,** *20* (1), 1–17. DOI: 10.18472/cvt.20n1.2020.1768.
World Tourism Organization [UNWTO] Global Guidelines to Restart Tourism. Madrid, 2020. https://webunwto.s3.eu-west-1.amazonaws.com/s3fs-public/2020-05/UNWTO-Global-Guidelines-to-Restart-Tourism.pdf.
World Health Organization (WHO). Coronavirus Disease (COVID-19) Advice for the Public, 2020. https://www.who.int/emergencies/diseases/novel-coronavirus-2019/advice-for-public

CHAPTER 16

REVIVING AND RECOVERING DOMESTIC TOURISM IN INDIA POST-COVID-19 PANDEMIC: CHALLENGES, PROSPECTS, AND STRATEGIES

SUNILDRO L. S. AKOIJAM[1], T. R. BANDARU[1], and GINO A. SANGMA[2]

[1]*Department of Management, North Eastern Hill University (NEHU), Tura Campus, Tura, Meghalaya, India*

[2]*Department of Agri-Business Management & Food Technology, North Eastern Hill University (NEHU), Tura Campus, Tura, Meghalaya, India*

ABSTRACT

In spite of the tremendous potential of the tourism industry, the scenario is going to be different post-COVID-19 in India. There will be a tough time ahead, and the industry needs to adopt new strategies and business models to attract tourists. India is the Asian country with the highest number of corona cases in the world and stands second rank in the world with the highest number of corona cases (Worldometers, as on November 25, 2020). As per a recent report by UNWTO, 100% of all destinations worldwide continue to have some form of COVID-19-related travel restrictions and will continue further, especially for foreign tourists. As a result, the number of foreign tourist arrivals in India will also decline drastically, which calls for the industry to focus on domestic tourism. Domestic tourism contributed significantly in Indian tourism industry in past years. According to the Ministry of Tourism, Government of India, domestic tourist arrivals in different states of India were 1854.93 million which was almost 180 times as that of foreign tourist

Post-COVID Tourism and Hospitality Dynamics: Recovery, Revival, and Re-Start.
Umendra Narayan Shukla & Sharad Kumar Kulshreshtha (Eds.)

arrivals in the country (10.56 million) in 2019. However, after the pandemic, the main challenge will remain as how to create confidence in the minds of domestic tourists to visit various tourist destinations in the country. In this regard, all stakeholders of the tourism industry particularly state government and central government have to discharge their roles effectively in framing protourist policies that will rebuild the confidence of safety from COVID-19 among domestic tourists. This chapter is an attempt to identify and analyze the challenges, prospects, and strategies in reviving and recovering domestic tourism in India post-COVID-19 pandemic.

16.1 INTRODUCTION

In India, the first case of COVID-19 was reported on January 30, 2020. India currently has the largest number of confirmed cases in Asia and the second largest number in the world after the USA (Worldometers). As on November 25, 2020, India has crossed the mark of 92 lakhs confirmed cases and surging to cross 1 crore confirmed cases in a couple of days (Arogya Setu app). Since the confirmation of COVID-19 case in India, the economy started shrinking due to the fear psychosis generated in the minds of people across the industry. But the dramatic downfall in the economy started when the Prime Minister of India, Shri Narendra Modi, announced for nationwide lockdown phase 1 from March 24, 2020, for a period of 21 days due to the immense spike in the number of confirmed cases of COVID-19 across the country. Even before the completion of 1st nationwide lockdown, on the advisory of various state governments, the Government of India extended the lockdown further in phase 2 (April 15 to May 3, 2020) followed by phase 3 (May 4 to May 17, 2020), phase 4 (May 18 to May 31, 2020), and so on.

It was during the initial phase of nationwide lockdown for almost 3 months, the economy collapsed with complete closure of almost all industries except few which fall under essential commodities. Among all the industries, tourism is the one which is affected the most due to complete restrictions of movements of people across the nation, region, state as well as district. Indian tourism industry is facing its biggest economic challenge with larger and combined effect of 9/11 and economic recession in 2009 and estimated bigger effect due to the pandemic COVID-19. The central as well as state governments had officially closed all the tourist spots across the country due to the pandemic and till today the restrictions are still not lifted fully due to the surge in positive cases of COVID-19. According to the Federation of Associations in Indian Tourism and Hospitality (May 2020), COVID-19 has posed an economic risk of the country estimated around Rs 10 lakh crores which was doubled from estimated value of Rs 5 lakh crore in March 2020. Over 95% MSMEs of 53,000 travel agents, 115,000 tour operators, 15,000 adventurers, 911,000 tourist transporters, 53,000 hospitality, and 5 lakh restaurants are facing the heat due to lack of cash flows, said the Federation. Unlike other industries, tourism industry is the one which is affected the most but will have longer time to recover.

Realizing the status of economic collapse as well as the industry downfall, a stimulus economic package of Rs 20 lakh crores was announced on May 12, 2020, by Hon'ble Prime Minister to pave the way for postpandemic recovery and unleash the wave of economic growth. The economic recovery stimulus is estimated to be around 10% of the GDP of India and it is one of the largest economic recovery stimuli announced by any country in the world during this pandemic. The Prime Minister stressed on a self-reliant India which will rely on five pillars as the foundations—economy, infrastructure, system, democracy, and demand. However, tourism industry people were shocked and showed disappointment and claimed that the tourism industry has been ignored in this recovery package as there is not any scheme or package meant exclusive for this hard-hit industry. This will impact those states of the country which depend extensively on tourism as a major source of state revenue. The states include Kerala, Rajasthan, Uttarakhand, Goa, Sikkim, Himachal Pradesh, Tamil Nadu, Maharashtra, Meghalaya, Assam, and so on. In this regard, it is very important to analyze how domestic tourism can be revived and recovered on priority basis during and immediately after the pandemic so that the loss and the negative impact of the pandemic can be minimized. Hence, this chapter attempts to analyze the challenges, prospects, and strategies to revive and recover Indian domestic tourism post-COVID-19 pandemic.

16.2 INDIAN TOURISM INDUSTRY

Indian tourism industry is one of the dominant and fastest growing industries in the country that has a significant contribution in the GDP and employment generation. It has tremendous potential of growth because of its rich culture and numerous places of historical importance. Moreover, it is blessed with natural beauties and landscapes and biodiversity hot-spot reserves which are popular across the globe. The tourism industry generated US$ 240 billion which is 9.2% of GDP in 2018. The sector is predicted to grow at an annual rate of 6.9% to US$450 billion by 2028 (9.9% of GDP). As of 2019, 4.2

crores jobs were created in the tourism sector in India which was 8.1% of the total employment in the country. The number is expected to increase by 2% per annum to 52.3 million jobs by 2028. According to World Travel and Tourism Council, India ranked third among 185 countries in term of travel and tourism's total contribution to GDP in 2018. India ranked 34th in Travel and Tourism Competitiveness report published by the World Economic Forum. Indian tourism industry handled over 10.5 million foreign tourists, more than 5 million visiting NRIs, 1.8 billion domestic tourists, and over 26 million outbound travelers.

From Table 16.1, we can see that number of domestic tourists visiting different parts of the country had been increasing continuously since 2000 till 2018. But the number of foreign tourist arrivals fluctuated during the same time period which is evident from the negative annual growth rates in some of the years. Moreover, if we compare the number of domestic tourists and foreign tourists arrivals for the same period, domestic tourist arrival was more than 60 times of foreign tourists arrivals in India. This clearly indicates that domestic tourism in India has a huge potential in our country for which the Government of India has already initiated some of the schemes which focused only on promotion and development of domestic tourism.

In India, domestic tourists have been the major driving force for tourism industry. As a result, various stakeholders including travel companies are designing customized tour packages with attractive prices for different categories of domestic tourists across the country. In order to promote domestic tourism, the Directorate of Civil Aviation, Government of India, is expanding air connectivity across corners of the country with several schemes. This is evident with the establishment of several airports even in small towns and cities in different parts under the scheme UDAN, Regional Connectivity Scheme which was initiated in 2017. The objective of the scheme was setting the common citizen of our country "fly," aimed at making air travel affordable and widespread, to boost inclusive national economic development, job growth, and air transport infrastructure development of all regions and states of India. Similarly, the government also came up with a scheme called "Dekho Apna Desh" to promote domestic tourism and motivate domestic tourists to explore the country. Under the scheme, the Ministry of Tourism, Government of India, will bear the travel expenses of any Indian citizen who visits and documents 15 domestic tourist destinations within a span of 1 year. The target participant for this scheme is the rising middle-class group with rising disposable income. These initiatives of the government clearly indicate the importance of domestic tourism in the country, and as result, the government is also focusing more on promoting and developing domestic tourism.

TABLE 16.1 Domestic and foreign tourist arrivals in India (2000–2018).

Year	Tourist visits		Annual growth rate	
	Domestic	Foreign	Domestic	Foreign
2000	220,106,911	5,893,542	15.44	1.05
2001	236,469,599	5,436,261	7.43	**–7.76**
2002	269,598,028	5,157,518	14.01	**–5.13**
2003	309,038,335	6,708,479	14.63	30.07
2004	366,267,522	8,360,278	18.52	24.62
2005	392,044,495	9,949,671	7.04	19.01
2006	462,439,634	11,747,937	17.96	18.07
2007	526,700,493	13,267,612	13.90	12.94
2008	563,034,107	14,380,633	6.90	8.39
2009	668,800,482	14,372.300	18.79	**–0.06**
2010	747,703,380	17,910,178	11.80	24.62
2011	864,532,718	19,497,126	15.63	8.86
2012	1,045,047,536	18,263,074	20.88	**–6.33**
2013	1,142,529,465	19,951,026	9.33	9.24
2014	1,282,801,629	22,334,031	12.88	11.94
2015	1,431,973,794	23,326,163	11.63	4.44
2016	1,615,388,619	24,714,503	12.81	5.95
2017	1,657,546,152	26,886,684	2.61	8.79
2018	1,854,933,384	28,872,384	11.91	7.39

Note: The bold fonts indicate negative growth.

Source: Ministry of Tourism, Government of India (2019).

16.3 CHALLENGES FACED BY THE TOURISM SECTOR IN INDIA THAT MAY ACT AS A HURDLE IN THE REVIVAL AND RECOVERY OF DOMESTIC TOURISM POST-COVID-19

The Indian tourism industry is one of the fastest growing and significant contributors in the economy of India. The importance of domestic tourists is more significant than that of foreign tourists in the Indian tourism sector and is at the forefront of making a significant contribution to the development of the tourism sector. However, there are a few challenges that are acting as hurdle in the process of growth and development of the industry. These challenges will also disrupt the revival and recovery of tourism industry, particularly the domestic tourism in India post-COVID-19. The heritage and culture of India are fascinating and very diverse from one region to another and from one state to another. But there is lack of proper preservation of this rich culture and heritage of India. Only a few ancient monuments and historical sites have been identified in the tourism sector. However, countries like Malaysia and Italy, which are much smaller than India, are attracting a much larger number of foreign tourists as well as domestic tourists by focusing and leveraging on the unique diversity of their culture and heritage. Many tourists are encouraged only to those areas (Bhatia, 2019). This is a big challenge for the tourism industry to leverage on India's rich culture and heritage for tourism promotion and development for drawing competitive advantages over other tourist destinations. Another challenge is lack of effective coordination between central and state governments, and among state governments for integrated promotion, advertisement and marketing of Indian tourism products. This acts an obstacle in designing effective tourism product mix as well in devising marketing strategies for attracting tourists from across the world.

For an area to thrive as a tourist destination, it must have a unique attractiveness, ease of access to the area, and basic facilities like accommodation, food, electricity, internet, and a tour guide. Many parts of India have a unique appeal but undeveloped tourist destinations due to lack of infrastructures such as roads, electricity, banking system, transport facilities, and basic amenities such as food, accommodation, and tour guide. Statistics of WEF (2019) on India's travel and tourism competitiveness confirm these things. Identifying tourist destinations and providing infrastructure and minimum tourism facilities is a challenge in India (Sonawala, 2019).

Atithi Devo Bhava is an ancient Indian tradition of treating a guest or a tourist as equal to God. But, due to the social, economic, technological, and political changes taking place in the society over the last century, the uniqueness of the guest or tourists has been neglected in the present times. As a result, tourists in some areas are being deceived or robbed. In some cases, passengers traveling alone may feel insecure (ASKIHM, 2019; WEF, 2019). This creates a sense of insecurity among prospective tourists who like to visit different parts of our country. In this regard, local tourism development agencies and authorities should create awareness among the local people about the protection of tourists and need for tourism development and the economic development that will be brought to the local people in the long run.

The availability of skilled human resources is low in the tourism sector in India. Comparatively, the availability of human resources specializing in service sectors such as banking, insurance, hospitality management, and telecommunications and other manufacturing sectors is very high. This leads to hiring people who are mostly less skilled in the tourism sector due to the lack of skilled human resources. Due to this, the tourism industry is unable to provide full-quality services (ASKIHM, 2019; WEF, 2019). Developing skilled human resources is costly. The use or awareness of new tourism technology for small tourism companies can be financially burdensome. Due to this, small tourism companies are lagging behind with big companies in competition and growth (Sonawala, 2019).

Due to COVID-19, the desire of tourists to travel to far areas with the priority to protect themselves is being temporarily postponed. Traveling to distant places has almost stopped due to lack of travel facilities as well. The number of people recovering from COVID-19 is increasing nowadays as people are flocking to local tourist destinations in their respective areas. Controlling this unexpected tourist influx has become a challenge for tourism companies and local governments. Tourists want to dine or live temporarily in an area where there are no or fewer COVID cases. As a result, hotels are displaying COVID-free/COVID clean certificate. Adopting new technologies to clean hotel rooms and surroundings while trying to be as human touch-free should be encouraged as much as possible. In the process, an additional charge is also being levied that adds up to the price of the tourism products for tourists. This is likely to increase dissatisfaction among many tourists and some may not be able to get those services due to increased prices or may be reluctant to do so. The income of many people has been reduced and the cost of essentials has increased due to COVID-19. The increased cost may discourage many tourists to undertake any tour in the near future. This is the biggest challenge facing by tourism sector due to COVID-19 in India which is likely to continue for few years post-COVID-19 pandemic also.

16.4 PROSPECTS IN REVIVING AND RECOVERING DOMESTIC TOURISM IN INDIA POST-COVID-19

Despite having numerous challenges that will act as a stumbling block in the process of revival and recovery of domestic tourism in India post-COVID-19, there are several prospects that will act as a catalyst in the revival and recovery process. Some of them are:

16.4.1 INCREASED INVESTMENT IN INFRASTRUCTURE

The Government of India has initiated two schemes to boost the tourism infrastructure to promote domestic tourism across the country—Swadesh Darshan and PRASHAD. Under Swadesh Darshan, 15 themed circuits were outlined to be developed into 77 projects. A cumulative of $863.6 million was sanctioned out of which $526 million was already released in December 2019 for the development of these circuits. Under the PRASHAD scheme, a total of 51 sites were identified across 28 states. At an initial phase, $120.2 million has been sanctioned for the development of 28 projects in 17 states. Until December 2019, $72.3 million has been released for the same. The Ministry of Tourism, Government of India, has been collaborating with and supporting multiple central agencies such as the Archaeological Survey of India, the Port Trust of India, and the Indian Tourism Development Corporation to develop potential destinations under their purview to increase tourism. In order to promote the regional tourism, the Ministry of Tourism released an amount of Rs 74.7 million to multiple states and union territories for the purpose of fairs, festivals, and tourism-related events. In order to promote tourism even in remotest corner of the country, the Airport Authority of India will be spending $2.5 billion to enhance regional connectivity, especially under service regions in the country.

16.4.2 PROTOURISM GOVERNMENT POLICIES AND MEASURES

In order to promote domestic tourism in the country, central as well as state governments have taken up measures and framed tourism supportive policies. For instance, the Ministry of Tourism launched a 24*7 toll-free tourist helpline available in 12 languages to support travelers in obtaining information and lodging-related complaints. Similarly, many of the state tourism departments have instated tourist police to support travelers and ensure their safety. In order to educate stakeholders in the tourism industry, the Ministry of Tourism signed an MoU with the Eco-Tourism Society of India to conduct workshops on sustainable and responsible tourism practices.

16.4.3 GROWTH IN DOMESTIC TOURISM DEMAND

India is a country with a majority of the people being young and their disposable incomes relatively higher. This group of young minds prefers experience-based tourism rather than a typical destination-based tourism. There is increased number of domestic tourists who prefer specialized tourism products such as adventure tourism, wellness tourism, sports tourism, and spiritual tourism.

16.4.4 FAVORABLE FISCAL CONDITIONS

The tourism industry raised USD 13 billion dollars in capital investment in 2019 (till June). In order to promote this industry, the Government of India approved and allows 100% FDI investment through automatic route including road infrastructure, tourism projects, tourist activities, and construction of hotels. During 2019, many global companies in this industry have invested a huge amount of dollars to capture the lucrative industry in India. The companies include Airbnb, Radisson Hotel Group, and Hilton Group. Indian government under the GST council reduced the GST slab from 28% to 18% for room tariffs over Rs 7500, 18–12% for room tariffs between Rs 1000 and Rs 7500, and no GST for room tariffs below Rs 1000.

From Table 16.2, it is visible that the Government of India has initiated and introduced several schemes for the promotion of tourism in India. Out of the 14 schemes introduced by the Central Government for development of tourism industry, most of the schemes are directed toward the promotion and development of domestic tourism. This shows that the Government is taking serious efforts to capture the prospects of tourism industry that provide huge employment opportunities to millions of people and contribute significantly to the Indian economy.

These are the factors driving the tourism industry that will enhance the revival and recovery of domestic tourism in India post-COVID-19 pandemic. COVID-19 pandemic is prevalent in the current scenario and is about to coexist for another few months; domestic tourism is the most viable option to be targeted at this juncture. The government needs to leverage on these driving forces and try to revive the highly prospective domestic industry to lure domestic tourists which account for 98% of tourist arrivals in the country.

16.5 STRATEGIES FOR REVIVING AND RECOVERING DOMESTIC TOURISM POST-COVID-19

Post-COVID-19, tourists and all stakeholders in the tourism industry will maintain a hygienic and healthy lifestyle. The stakeholders involved in marketing and developing tourism industry should focus on this aspect as the top priority for attracting reasonable number of tourists' movements across the states in the entire country. Tourism and related industries comprised huge number of casual and self-employed workers who are affected badly during the pandemic. Government and employers should protect and help these workers where possible. Government should provide assistance to many of the tourism enterprises such as hotels and airlines who are about to go bankrupt. Assistance can be in the form of low interest rate loans or grants in some special cases. This will boost the morale and confidence of these enterprises and start working again in the revival of domestic tourism in the country.

TABLE 16.2 Scheme-wise outlay of the Ministry of Tourism 2019–2020 and 2020–2021.

Sl. no.	Name of schemes	Rs (in crore)		
		2019–2020		2020–2021
		BE	RE	BE
1	Asstt. To Central Agencies for Tourism	91	70	80
2	Viability Gap Scheme for Revenue Generating Tourism Project	10	0	30
3	PIDDC	5	0	5
4	Swadesh Darshan	1106	566	1200
5	PRASHAD	160.5	145	207.55
6	Capacity Building for Service Provider	38	33	60
7	Asstt. To IHM/FCI/IITTM/NIWS/NCHMCT	82.89	61	70
8	Domestic Promotion and Publicity including Hospitality	129.5	100	140
9	Market Research	5	3.61	9.66
10	Incentive to Accommodation Infrastructure	0.01	0	0
11	Buddhist Centre	0.01	0	0
12	Champion Service Sector Scheme	1	0	0
13	Construction of Paryatan Bhawan	0.01	0	0
14	Overseas Promotion and Publicity Including MDA	446.20	312.39	450
	Total Scheme	2075.12	1291	2375.21

Source: Ministry of Tourism, Government of India (2019).

With the adoption of the unlock policy of the central government on July 1, 2020, some of the industries initiated opening up taking into consideration the scenario of pandemic across the country. As a result, the government also started opening up the domestic tourism slowly in a phase-wise manner.

Several countries have started opening up the tourism sector, especially for domestic tourists. They have lifted some of the strict requirements for visiting a particular place that will facilitate the tourists' movement across a particular country. Similarly, the government can initiate some of the integrated and coordinated approach and protourists' policies following standard operating procedures (SOPs) to revive domestic tourism in the country. The relevant authorities can plan out certain rules considering the sanitation requirements. For instance, the rules of departure, arrival procedures, booking of hotels, taxis, procedure for entering tourist destinations, and so on can be well framed and made available for tourists on various platforms or medium.

Revival and recovery of international tourism would take time and therefore focus of the government is to promote domestic tourism. According to the Ministry of Tourism, Government of India (June 2020), the "Dekho Apna Desh" initiative launched by the ministry before the lockdown would be continued. The ministry further emphasized that people need to get over their fear for traveling and asserted that the initiative will encourage tourists to travel to their neighboring tourist destinations. He also informed that the promotion of off-beat attractions and wildlife and wilderness destinations is also in the ministry's plan.

Some of the strategies that will revive and recover the pandemic-hit domestic tourism in India are as follows.

- Divide the Indian tourism sector into regions and make a calendar of the various tourism opportunities or services that take place or are found throughout the year and link it to the country calendar. All of this information should be made available to tourists on a travel website.
- Regional tourist destinations should be identified and village, mandal, or local revenue officers should be linked in the identification process. These identified tourist destinations should be advertised in the local media.
- Infrastructure development in tourist areas should be initiated. There should be monitoring of the status of facilities already available at the destinations.
- Local youth should be trained as a tour guide. Homestay practices should be promoted in regional or rural tourist areas. These details should be made available on various websites, and in small and large hotels in the area. Details of those providing homestay should be made available at the gram panchayat or mandal offices. This will reduce the burden on the government to create tourist accommodations in rural areas. Tourists also get a special understanding of the local culture and food.

- The local school and college students should be educated on local and regional tourist sites, arts, culture, and literature. This will enable them to know the richness of local and regional historical literary cultures and support their preservation.
- By creating awareness on tourism, more and more young people are embracing tourism as their profession and coming forward to learn and train in it. Thereby, the shortage of quality human resources in the tourism sector can be overcome to some extent.
- The problem of vehicle parking is high in tourist area. At present due to COVID-19, tourists mostly use their own vehicles to visit the tourist spots near them. This has led to increased traffic congestion in tourist areas and is creating parking problems. To avoid this problem, the local tourism authority and government administration should take appropriate steps to set up parking spaces.
- Information on tourist destinations around the area should be made available to tourists in tourist areas so as not to increase tourist congestion in a single area. This information should be made available even in small-to-large hotels.
- Every tourist destination should develop different types of places rather than places that give tourists the same tourist experiences. Examples are spiritual tourism, agricultural tourism, adventure tourism, cultural tourism, and gastro tourism. This makes it more likely that different types or segments of tourists will visit an area.
- Locals should be trained in packaging and selling hygienic food sold locally in tourist area in a proper, neat, and hygienic manner. This makes it easier for tourists to buy and receive food from locals.

16.6 CONCLUSIONS

Indian tourism industry is an important industry that contributes significantly to the economy of the country. It is one of the largest job providers in the entire country. However, due to the COVID-19 pandemic, the industry is badly affected and it is one of the worst hit sectors in the country. Due to pandemic, the industry was completely shut for almost 5 months affecting various others industries including airline, hotel, restaurant, booking and reservation, and transportation. Millions of people lose their jobs thereby affecting millions of families across the country. Realizing the drastic downfall of the economy, the Government of India decided to open up economy Unlock 1.0 in phase–wise manner after almost 4 months of complete lockdown of the entire country in

June 2020. But it is only during Unlock 5.0 in September 2020 when tourism industry started opening up in a real sense even though with certain SOP. In the Indian tourism industry, domestic tourism has been continuously proving to be dominant in terms of the number of tourist arrivals as well as the annual growth rate in comparison with foreign tourism. The tourism industry has numerous challenges that hinder the growth of the industry, particularly domestic tourism that has multifold impact during the pandemic and will continue even after pandemic. However, despite numerous challenges, the same industry also has various prospects that will definitely supplement the revival of the domestic industry after the pandemic. The Government of India as well as state governments have initiated some revival and recovery plans for the tourism industry, especially for domestic tourists. But all stakeholders of the tourism industry also need to adopt respective strategies and play their roles effectively in the process of reviving and recovering the hard-hit domestic tourism post-COVID-19.

KEYWORDS

- **challenges**
- **COVID-19**
- **domestic tourism**
- **revival and recovery**
- **strategies**
- **prospects**

REFERENCES

ASKIHM. *Challenges and Prospects of Indian Tourism*, 2019. http://studyask.com/blog/hospitality/challenges-and-prospects-of-indian-tourism/

Bhatia, R. *India Tourism: Challenges and Opportunities*, April 8, 2019. https://timesofindia.indiatimes.com/blogs/hummingbird/india-tourism-challenges-and-opportunities/

Sonawala, H. *5 Challenges That India's, Inbound Travel Industry Faces*, Dec 19, 2019. https://www.entrepreneur.com/article/344071

World Economic Forum. *India's Travel and Tourism Competitiveness Index*, 2019.

CHAPTER 17

OLD NORMAL TO NEW NORMAL: CULTURAL SHIFT IN TOURIST GUIDE-SERVICE PARADIGM

SAMIK RAY[1,2,3]

[1]*Department of Folklore, University of Kalyani, Kolkata, India*

[2]*Travel and Tourism Management in MPTI, Kolkata, India*

[3]*WTCC School of Trade and Commerce, Kolkata, India*

ABSTRACT

The growth of tourism slumped below zero level for the first time since 1950 due to the COVID-19 pandemic (UNWTO, 2021a). This health calamity disrupted the global economy, and thus posed the greatest challenge to the sustenance of tourism too. The world fought it, introducing physical distancing as the primary strategic tool, which logically extends to a worldwide complete lockdown strategy resulting in the rocket fall in traveling. It reveals that self-styled intelligent and powerful species of the globe are defenseless, powerless, and helpless amid microbial attacks and compels human beings to change the survival culture. Tourism got affected massively as tourist arrivals slump down by 75%, planned holidays by 90% resulting in the loss of a few trillion USD and millions of jobs (Economic Times, March 2020; UNWTO, 2020). Though the gradual unlocking process succeeds a bit to make the economy crawling, it is not able to build enough confidence among travelers except for a few accidental incidences of travel. Returning to the prepandemic scenario will depend on the availability of vaccines or treatments and gradual and linear lifting of travel sanctions at the

Post-COVID Tourism and Hospitality Dynamics: Recovery, Revival, and Re-Start.
Umendra Narayan Shukla & Sharad Kumar Kulshreshtha (Eds.)

outset but building travelers' confidence is the greatest challenge for tourism stakeholders.

Travel and tour is a kind of experience achieved through authentic gazing, memorable feeling, and meaningful interaction at the visitation. A balance between tourism experience and COVID-19 protocol and confidence-building among travelers traumatized by the COVID-19 pandemic will be then a challenge. The role of tourist guides will be crucial in the balancing and confidence-building process as tourists spent the longest period at the visitation in contact with them.

The present chapter will deal with the necessary cultural shifts in the tourist guide-service paradigm from old to new normal to make the travelers enough confident and comfortable to get travel experience amid any kind of global health calamity like COVID-19.

17.1 INTRODUCTION

The COVID-19 pandemic is the most crucial global calamity since 1950. There have been about 116 million confirmed COVID-19 cases, 2.59 million deaths, and 65.8 million recovered by the beginning of March 2021 (WHO, March 7, 2021). About 1.48% of the world's population got affected. Among the affected, 57% recovered and 2.24% died since the outbreak (WHO, March 7, 2021). It is a pandemic as it spread across 200 countries leaving bearings on the global economy, social activities, health, and human psychology. The world had no pre-experience of it. In the 21st century, the world becomes an integrated village. The mobility of everything including disease and its effects thereby becomes faster than ever before.

Like other economic pursuits, tourism industry also got affected due to the pandemic outbreak. At the outset, the lockdown and sociophysical detachment measures to control or reduce the spread of COVID-19 led to a sudden halt or complete cessation of tourism activities across the continents. This has a significant consequence on tourism and beyond. It becomes apparent in the sudden closures of millions of tourism–hospitality and allied firms; enterprises and in the loss of trillions of USD and millions of jobs. The present populace is never cognizant of either viral or nonviral realities of the pandemic.

Since its first appearance, COVID-19 viruses changed and mutated many times. New variants or strains developed over time and persist with greater intensity and much faster way. Multiple variants prolong the sufferings, fear psychosis, lockdown, and relockdown. Presently, the world is struggling to be normal amid the emergence of new variants like B.1.1.7 (UK strain), B.1.351 (South African strain), P.1 (Brazil strain), and others, resuming work and business from home with the help of smart technology or from office and business premise maintaining COVID-19 protocol, carrying on essential activities outside home and recommencing local transportation phase by phase. The process of loosening restrictions begins to make the world return back to the prepandemic lifestyle. The vaccination process begins too. All those efforts will assist the Globe to come out of restricted living amid the warfare between macro and microorganisms. The world is hopeful about normalization by the end of 2021 except for the tourism sector. Most tourism experts do not foresee a return to prepandemic status or 2019 situation before 2023 or 2024 (Tourism Economics, 2020; WTO, 2020). All those expectations are hypothetical as the extent of the pandemic is not yet known to us (UNWTO, 2021b).

17.2 METHOD

This chapter involves a descriptive study based on both naturalistic and participant overt observations drawn primarily from interactions with professional tourist guides of India and self-experience of the author who is a researcher and regional level tourist guide of India and also from secondary sources like published academic researches and current web media.

17.3 EARLY APPROACHES

COVID-19 pandemic consequences on tourism are crucial and contemporary. Thus, it draws the attention of academic and industry stakeholders. The majority studies its impact and recovery aspects. Researches on impact dealt with tourism degrowth aspects (Fletcher et al., 2020), sociopolitical implications (Jamal and Budke, 2020), economic and business decline (Deegan, 2020), and impact typology (Yeh, 2020; Sigala, 2020). Studies on impacts in tourism demand, supply, and destination management (Sigala, 2020), impact on demand curves and socio-economic costs of sudden cessation (Mandel and Veetil, 2020; Qiu et al., 2020), and effect intensity, depth, and scale (Sigala, 2020; Marinko et al., 2021) are also important and worth mentioning. Besides those, a few scholars aim to examine the regional and destination-level impacts. COVID-19 crisis impact studies on Indian tourism–travel–hospitality (Dash, 2020; Divya, 2020; Radhakrishna,

2020; PTI, 2020), Malaysian tourism and hospitality industry (Lee-Peng, 2020; Medina, 2020; Karim et al., 2020), Thailand tourism (Yuda and Tani, 2020), and Taiwan tourism (Yeh, 2020) are crucial. Studies on COVID-19 pandemic crisis impacts on sectors like aviation (AFP, 2020; Tan, 2020) and hotels (Mahalingham, 2020; Nesrine et al., 2020; Ranasinghe, 2020) also contributed important observations. Simultaneously, well-thought-out suggestions toward the COVID-19 crisis management in the field of tourism and hospitality is the need of the hour. While Shih-Shuo (2020) elicits the importance of tourism crisis and disaster management toward mitigating the crisis impact and facilitating the postcrisis recovery, Kaushal and Srivastava (2021) emphasized multiskilling and professional development and SOPs for hygiene and sanitation. Jamal and Budke (2020) offer some research directions about managing the postpandemic situation. The Malaysian government's stimulus packages for the tourism industry set a way how states could respond and combat the crisis (Medina, 2020). The resident-participated recovery strategy is quite a new approach in the present crisis management (Qiu et al., 2020). While Marinko et al (2021) thought effective contingency plans could be crucial in the unprecedented pandemic crisis management; Kirby (2020) emphasized economic assistance programs. Mandel and Veetil (2020) endeavored to estimate the recovery time concerning typological variants of lockdown. Marron (2020) developed a four-part economic strategy to manage the crisis. The strategy focuses on the acceptance of economic losses, protection of health, support people experiencing a sudden loss of income by broadening existing security network programs, and protect productive capacity and use economic production capacity to the fullest extent when the virus will completely disappear. Indeed except for a few, studies about impacts of COVID-19 on tourism largely ignored the aspect of pandemic effects on tourist guides (FEG, 2020; Grancay, 2020).

17.4 IMPACTS ON TOURISM

The world had experienced many political, economic, and medical emergencies before but none of those were like COVID-19 pandemic. The world turns out to be more integrated than what it was in the predigital era as the internet, social media, instant communication technology, and mass transportation make people, goods, money, and ideas move faster than ever before. Thus, COVID-19 virus, its related news and impacts, newly emerged health and sociocultural protocols, and fear psychosis spread at a rocket speed or with a high degree of swiftness across the globe.

Tourism plays the most crucial role in the international economy since it accounted for 29% of total service-exports and about 300 million jobs across the world (UNCTAD, 2020a). Thus, sudden cessation has a wide tangible impact. Quantitative or statistical data are nothing but a form of evidence that supports an idea or leads to a conclusion. The intensity of impact-reality is sociopsychologically so deep that intangible mere data cannot reveal the actual truth. Still to measure impacts, the world has to rely on quantitative data as there are no many alternatives to it. The direct impacts of the COVID-19 pandemic are evident in the declines of international arrivals by 74% (UNWTO, 2020, 2021a, 2021b), overnight visitors by 80%, and domestic tourists by more than 70%. These tourism crises possibly contributed to a loss of 3% in world GDP (UNCTAD, 2020b) and put 174 million jobs at risk globally (WTTC, 2020). Among the world regions, Asia and the Pacific incurred the highest loss in 2020 international arrivals. It was about an 84% decrease in comparison to that of 2019. In the Middle East and Africa, it was 75%. European territory and America had a decline of 70% (UNWTO, 2020, 2021b). COVID-19 crisis impacts on Indian tourism and hospitality are crucial too as it affects inbound, outbound, domestic, adventure, business, and cruise tourism (Dash, 2020) heavily. Indeed, the sudden closure of the industry due to pandemic outbreak contributed to over 38 million job loss, 70% of the total workforce used in tourism during the prepandemic period (Radhakrishna, 2020), and over Rs 10 lakh crore direct financial losses (PTI, 2020).

17.5 TOURIST GUIDE SERVICE

Tourist guide is a generic term used to denote a particular kind of tourism professionals. Thus, tourist guide service is not limited to destination attraction guiding. It includes services of accompanying individuals and groups of visitors/tourists/travelers/explorers to tourism and travel attractions, either on day trips or longer visits and assisting them while on tour, sharing information and insights about the destination and its tangible and intangible cultural and natural heritage in such a way that helps them to connect the place and people culturally (Govt. of India, 2018; FEG, 2019; CEN, 2020; Dept. of Tourism Republic of South Africa, 2021). This profession involves high physical and mental fitness with lots of energy, excellent interpersonal and communication skills. Thus, jobs of the tour manager, trip leader, tour guide, regional level guide, guide lecturer, museum guide, adventure guide, monument guide, state-level guide, forest guide, nature guide, tourist facilitators, mountain guides, and others come under this concept. Variations in nomenclature develop by subtle and minute differences in the job profile.

Generally, visitors/tourists/travelers/explorers spend maximum time with those professionals during visitation. Tourist guide service is highly customer focused, thus service is always imparted in high contact with tourists. Hence, tourist guides are highly exposed to the contagious diseases which customers carry. The greatest sense of professional risk comes from traffic safety while personal health risk comes from long and odd working hours, long physical activities, and no regular diet and rest conditions. The low professional social status, unstable and irregular income, and no social security make their career at risk. This chapter involves a descriptive study based on both naturalistic and participant overt observations drawn primarily from interactions with professional tourist guides of India and self-experience of the author and also from secondary sources like published academic research studies and current web media (Hu and Chen, 2015; Chang et al., 2018; FEG, 2019).

17.6 COVID-19 CRISIS AND TOURIST GUIDE SERVICE

The impact of the COVID-19 pandemic on tourist guides is quite crucial as their services are consumable only in close contact. Tourist guides across the world are self-employed freelancer and single-business professionals. About 80% are not having other sources of earnings (FEG, 2020). Prepandemic armchair theorists' perception about the professional risks comes true with the living and lifestyle deterioration of tourist guides in the COVID-19 pandemic perspective. It was evident in the loss of several hundred thousand wage units and service hours. Most governments or employers subtly ignored the issue except few states from developed worlds that announced compensation packages for individuals involved in this profession. Thus, recovery of the loss becomes difficult. Further, its perishability feature and zero or fewer opportunities to access government funding and support make the recovery quite challenging. Thus for sustenance, they depend only on their life-long savings. Besides financial instability, health risk factors also make tourist guides professionally vulnerable. They are highly open to contagious diseases carried by the guests as were evident in the incidences of corona transmission from the guests to the guides. Above all complete cessation of work for more than a year and uncertainty regarding work resumption leads many to develop hypertension symptom or related ailments.

17.7 NEW NORMAL

COVID-19 pandemic bearings are evident in all spheres of life, living, and lifestyle. The term "new normal" and its concept is the most crucial contribution of the COVID-19 pandemic to the world as it appears with high intensity or frequency in the cloud, printed format, and orally of late. The term refers to a condition or state of economic, physical, psychological, and social behavior, emerged to combat and acclimatize crisis and its consequences that disrupt the existing system. The newly emerging behavior differs from the precrisis pattern and sets the society and its activity well adjusted to the disrupted condition. The term was used in lesser frequency with the world wars and other world crises that occurred before the COVID-19 pandemic. In the present pandemic perspective, its usage refers to the behavioral changes in the society, among its members and in the existing intangible cultural traits. It has already been apparent in daily life and the work domain.

In the COVID-19 pandemic context, the concept of the new normal is interpreted as a kind of social and health-related behavioral protocol that makes the human world acclimatized with the COVID-19 crisis and its consequences. Primarily, it was a set of human social and health-concerned behavioral rules suggested and adopted by the WHO and world health experts to control the spread of COVID-19. The states compelled the subjects to be habituated with the new normal behavioral pattern introducing contact tracing, ban on travel and large gatherings, complete or partial lockdown, and other strategies. Eventually, those behavioral patterns specified as the new normal behavioral protocol since a greater part of human society out of COVID-19 phobia accepted and translated it into practice. The protocol aims to limit person-to-person physical contacts and encourages sociophysical distancing, mask; face shield; head-cover use, sanitization, washing hands with soaps at regular intervals, and many other health-protective habits. As a result, Indian "Namaste" culture replaces western handshake and hugging habits. Webinar culture becomes popular in place of seminars. Conference and meetings switch to web conference and meetings. Online shopping habit increases. Culture of work from home, the habit of getting online utility services, and online receipt-payment transactions escalate. The line of difference between personal and work time has been blurred. Thus, "New normal" surfaces as the nonviral reality of a viral pandemic (Bhattacharya, 2020).

17.8 SHIFT FROM OLD TO NEW NORMAL IN TOURISM

Both the exorbitant growth of tourism and COVID-19 pandemic are crucial phenomena of the 21st-century world. Naturally, the pandemic has bearings on tourism. When pandemic compels the travelers to be immobile, travel–tourism–hospitality experienced an unprecedented sudden halt. With the temporary decline of the pandemic intensity, the world struggles to resume all economic pursuits including tourism developing a new set of behavioral rules as proposed by the health experts for crisis consequence acclimatization, the sustenance in pandemic and postpandemic condition. Thus, health safety and psychological well-being are prioritized. It was assumed that COVID-19 spread by travel and through droplet transmission while human-to-human contact or interaction occurs at a distance less than 6 ft or 2 m. Most states prefer to be isolated, than putting sanctions on human mobility from and to the foreign land. New protocols like sociophysical distancing, mask; face-shield; head-cover usage, sanitization, washing hands with soaps at regular intervals, and many other health-protective measures come into force at health experts' recommendations. Bans on travel across the world, border closures, events cancellations, quarantine, contact tracing, and complete lockdown measures together with the growth of COVID-19 fear psychosis pushed the tourism–travel–hospitality industry to shut the business off all of a sudden (G¨ossling et al., 2020) and experience an unprecedented loss.

With the gradual lifting of lockdown measures, the world gets the opportunity to kick-start economic pursuits and its recovery though it is hindered by the COVID-19 fear psychosis and uncertainty about the large-scale vaccination. Industry stakeholders and states, then, developed a balancing act weighing resumption and recovery with public health and safety. Indeed, the introduction of the new normal protocols in the sociophysical interaction and work domain made the balancing. At this juncture, it is hard to say how long the balancing would continue or whether the shift from old to new normal would become permanent or set to be a temporary behavioral transformation.

Resumption of economic activities step by step in all the crucial sectors does not ensure tourism's return as its take-off is largely dependent on surplus time and money flow, health security and safety, vaccination process, and decline of COVID-19 phobia. As tourism is a complicated sociopsychological process, its revival thereby depends considerably on the complete disappearance of the phobia. The COVID-19 pandemic-led damage or loss recovery issue gets priority as soon as the world economy resumes rolling amid new normal condition, while the sense of recovery turns meaningless to the tourism–travel–hospitality sector as most tourism product and service offers are highly perishable.

At this point, tourism stakeholders become puzzled and bewildered as the industry hardly had any clue for a comprehensive revival plan. The idea of total damage that the industry already had and the knowledge about the future of the pandemic was missing from their perception as the post-1950 tourism industry never experienced a global crisis of that scale. Most post-1950 crises were either regional or local (Cetin, 2020). The impact was lesser than the present crisis. Today's global integration increases media mobility across the globe and inspires the state apparatuses to work in close contact. The worldwide media propaganda and states taken measures, thereby led the society to live with new normal protocol and COVID-19 fear psychosis. The tourism industry too perceives the need for developing SOPs (standard operating procedure) or new normal protocols to acclimatizing low and postpandemic condition. It is believed that SOPs or new protocols would be relevant toward recommencement and slow rolling of tourism even amid the emergence of new strains and confusion about the success of the vaccine.

The "Old normal" is referred to as a precrisis state of the tourism system and pattern the term "New normal" will denote a change in the system and behavioral pattern, making tourism meaningfully acclimatized any crisis consequence. Thus, the old normal signifies the prepandemic state and the new normal is the pandemic-led changed state. Tourism–travel–hospitality and its services are highly human-oriented as produced, offered, and consumed by humans. A competent, friendly, and face-to-face interaction brings tourist satisfaction. COVID-19 pandemic puts a serious challenge to this notion as the coronavirus spread and transmits through droplets and in close human contact and interaction.

It was found that the industry intended to be in touch with the market since the lockdown period as was evident in sporadic efforts of organizing webinars and context conversation on tourist destinations. Simultaneously, the habit of accessing and visiting the travel blogs and you tubes increases in rapidity during the pandemic. These technology-driven destination knowledge enhancing or virtual travel experience approaches was quite pragmatic to revive or keep the travel zeal alive among potential travelers or tourists. Indiatourism, Tourist Guides' Federation of India (TGFI), and some operators, agents, hoteliers too organized webinars and made video presentations using digital and social media platforms to promote India's destinations and increase the destination knowledge and information within the potential markets (Foodka, 2021; Indiatourism Kolkata, 2021; Ministry of Tourism, 2021; RLGwebinar, 2021; Safariplus, 2021). "Dekho Apna Desh" and "Ek Bharat Shrestha Bharat" series of Incredible Indiatourism (Incredible India, 2021; Indiatourism Kolkata, 2021, Ministry of Tourism, 2021) become the most successful since it has drawn the largest viewers and participants. Information and communication technology–aided promotion is cost-friendly and time saver, advocates contact-free faster interaction and communication, minimizes internal human resource involvement, and connects the maximum number of people just by one click. It could be considered as the leading acclimatizing initiative. As an indication of further digital advancement in the industry to meet the level of personalization that guests demand is apparent, industry initiates the technology-driven new normal protocols to make the return to health concerned and safe business pursuit successful. The industry opts to design and develop a human contact-free or technology-abetted new normal system, then. A shift from an old system or behavior pattern to a new normal set in motion to ensure the move from crisis management to recovery or acclimatize to the newly emerging situation. The process is mainly driven by principles like ensuring and delivering a safe experience, rebuilding trust and confidence in the market, enabling the industry to confront any crisis and acclimatizing newly emerging consequential conditions of any crisis. Thus, shift to new normal tourism includes implementation of human contact-free online booking and cancellation of services with the flexibility of accommodating benefits to counter any kind of crisis or emerging rules of local or international travel restrictions and introduction of e-check-in e-marketing, e-promotion, and e-payments along with computerized digital room key, online entry at the attraction sites, virtual visits at indoor museums and galleries, automatic doors in the public areas of the hotels; restaurants; airport; rail stations; any waiting areas, automatic sanitization, self-service kiosks, automatic body temperature checking system, and many more facilities. Besides technology abetted changes' shifting is evident in the behavior of both guest and hosts like physical distancing, wearing masks, washing hands with soaps or sanitization of uncovered body parts at regular intervals. The new normal protocol also includes sanitization of public areas where tourists and hosts have to pass or use frequently. Traveling by self-driven vehicles, online space blocking in the restaurants or sites of entertainments, use of drink machine, Namaste culture, and virtual or open-air business meetings becomes customary in the new normal condition.

17.9 CULTURE SHIFT IN TOURIST GUIDE SERVICE

In the prepandemic tourism world, tourist guides played a crucial role in shaping the experience of the tourist guests at the visitation, portraying and creating the destination image connecting the guests with local culture meaningfully and making the tourists satisfied turning a tour authentic and culturally meaningful. Tourists use the guide services for a bespoke experience. Tour operators, hoteliers, and travel agencies use this service to provide an authentic, safe, secure, satisfactory, mesmerizing, and knowledge govern experience to their guests. As pandemic features immobility, tour guides have had no choice but to suspend temporarily. A decline in pandemic intensity sets the world to unlock. The concept of "staycation" then emerged as the most preferred form of tourism since it involves traveling but to a day-trip distance, thus no need for overnight accommodation. Visitors from Athens to Amsterdam flock to nearby museums, bars, and beaches. Istanbul's tourist places resume the business. Beaches along 75 mi long coastline of California begin to experience the footfalls of the local visitors. Germans feel for sunbathing at local beaches. As time rolls enthusiast travelers start to spend weekend overnights at a nearby place that maintains the COVID-19 safety protocols. Most small and medium Indian hotels experience "sold out" or overfull weekends of late. Both staycation and weekend overnight footfall trend encourages self-driven travel. Indeed, sociophysical-distancing protocols, restrictions on long-distance travel, irregular public transportation, and COVID-19 fear psychosis cause the emergence of those new normal traveling trends and keep travelers away from having tour guide services.

There is a clear indication of the ongoing transformation in travel choice, pattern, demand, and supply. Transformation is inevitable to resume socio-economic pursuit acclimatizing a condition, emerged from the COVID-19 crisis consequence. Experts prefer to call it a new normal condition. Tourism's return to prepandemic status is not yet known as the emergence of new strains may prolong the pandemic. Although the return is uncertain, the transformation from old to new normal is definite as COVID-19 fear psychosis may continue for a significant period.

Tourist guide as a kind of community shares the same professional emotion; interest; beliefs, sense of professional relatedness, and the feeling of belonging to the community. Its survival is subject to a process of evolution and growth making the community relevant to any new emerging perspective that evolved out of either a societal crisis or a transformation, occurred within the greater society. Survival of the tourist guide profession amid the present pandemic thereby utterly depends on acclimatization of the guide service to emerging neo-normal condition. Acclimatization would be possible if a change in the guide-service paradigm or in the existing culture of imparting the guide service in agreement with neo-normal demands begins within the community at the will of the members. It is an ongoing phenomenon of the cultural shift in the belief, perception, norms and mode of practice, pattern of delivering service, and behavioral protocol related to guide service. This phenomenon continues until the acclimatization to the new environment completes with the institutionalization of shifts and adaptation of new beliefs, perceptions, norms as emerging traditions. In the end, tourist guide service will be relevant and positioned within the new normal context.

The shift is essential to make the guide service relevant and alive in the newly emerging context and also to bring the confidence of tourists traumatized by COVID-19 back on tour guide service. Change in the guide-service culture is both a priority and a challenge, then. Giving up a habit or habitual norms is often substantially difficult and challenging since the practicing community is comfortable in it. When old beliefs require to be made new the process of freezing, the existing beliefs and comfort will be effective. It would be possible organizing critical events with a significant role for guides relevant to new condition. It is already evident in the initiatives of turning tours into at-home seminars or webinars where the guides play the role of lecturer at virtual platform. Efforts of Indiatourism, TGFI, FEG, and Context travel are quite crucial in this context and appear to be a win-win initiative as it simultaneously generates earning opportunity for the tour guides, reconnects travelers with tour guide services making them feel the relevance of tour guide even in the new condition, and encourages the tour guides to be acclimatized to new condition comfortably and prompts them to be ready to change their old beliefs. The initiatives of the government, nongovernment, and within the community are quite pragmatic since it makes the tourist-guide community perceive the importance of acclimatization and the need of shifts from old to the emerging cultural context redefining the existing belief, the pattern of work, and sense of comfort. In the end, a new narrative for tour guides' survival will evolve. However, the tour guides who already participated in the critical events act as the creator of new narrative or the vanguard of changes at the belief, practice, and outcome level crafting a new service narrative relevant to the emerging condition. As a role model, they silently encourage the shift from the old to the new culture of imparting guide service communicating the need for acclimatization to the emerging condition and making the community perceive the inappropriateness of old belief and the service pattern in the neo-context. A new service narrative emerged, then, illustrating the benefit of changes from the old to new-fashioned service pattern and behavioral protocol.

The creation of a new narrative for the guide service involves reframing the existing culture of doing guiding, thus makes the tourist guide and their services relevant to the emerging condition. The reframing begins with the construction of the emerging condition-suited set of new SOPs. The COVID-19 pandemic consequence context-sensitive or context-adaptive SOP construction mainly features the health safety and protection principle-based protocols. Hence, a series of the cultural shift in the guiding service paradigm is quite logical. Shifts are evident in the norms of greetings, interactions, speech delivery, personal hygiene practice, preparation of tour programs or itinerary, sharing any authentic and hands-on experience, entertaining the guests, and mode of receiving feedback. Thus, handshake and hug culture begin to transform into Namaste culture. Tourist guide profession is highly prone to contagious diseases including COVID-19 as the service is to confer in close contact or at a close vicinity of receivers. Sometimes it is offered in the outdoor amid a big crowd like at a local market or fair. Thus, the risk of being infected by contagious diseases and the possibility to infect others is quite high. Thus, taking care of personal hygiene is not a new practice for them. COVID-19 phobia only makes them mentally prepared for regular self-sanitization, thus increasing the frequency of self-cleaning particularly while doing the guiding which makes tourists feel confident to use the tourist guide service. The existing habit of interacting with guests within four walls or in a crowded outdoor in close contact or from a near distance without mask demands a change in agreement with the newly emerged format of one-to-one and public interaction. As a result, the length of physical distance during an interaction with guests extends from close vicinity to a minimum of 6 ft or 2 m. The culture of interaction between the guests and the guides without mask contradicts the new normal norm. Thus, interaction with mask happens to be the accepted norm. A shift from the conventional way of outdoor speech delivery is also evident when a VOX (voice-operated exchange) tour guiding machine or public address system is introduced for lecture delivery in the open air and crowded place. The transformation of the nondigital individual guide to a technology-savvy professional is possibly the most important shift in the culture of guide service as it develops operational readiness among the tourist guides for resuming their services in the new emerging condition.

In the present pandemic perspective, the cultural shift in tourist guide service is quite pragmatic as it is linked to its survival within the neo-normal market. Indeed, acclimatization to the emerging market is in progress as is found in the ongoing increase of tourist guides' participations in guiding in the new-normal settings (Diagram 17.1), in the webinars and virtual tour programs as speaker (Diagram 17.2) and as viewer participants (Diagram 17.3).

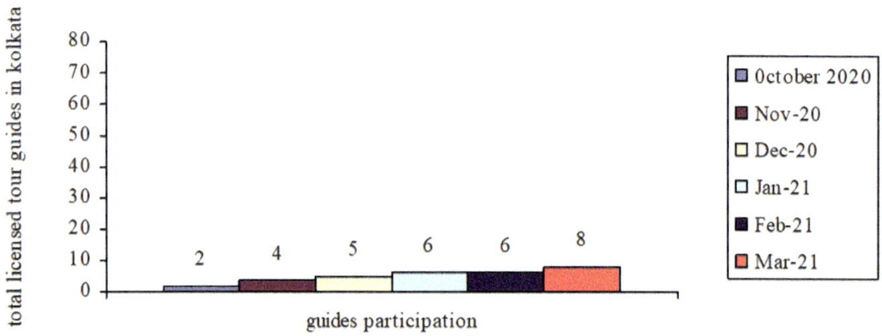

DIAGRAM 17.1 Participation of tour guides in new normal settings—new double-decker bus tour at Kolkata.

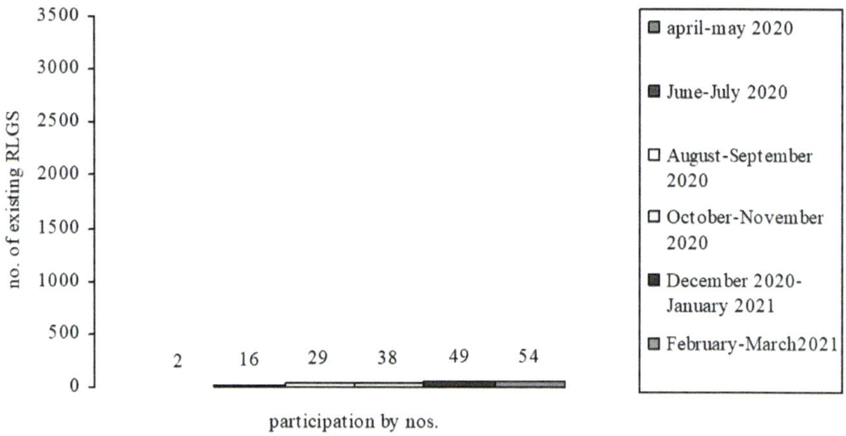

DIAGRAM 17.2 Participation of regional level guides as speaker in new normal settings—webinars and virtual tour seminars.

17.10 CONCLUSIONS

As an important component of the travel–tourism–hospitality industry, the profession of tourist guides has to experience an unprecedented sudden halt too when the ban on travel across the world imposed due to the spread of COVID-19. Complete cessation of tourism–travel–hospitality industry is the nonviral reality of a viral pandemic. The world had experienced many other crises but of different natures and scales that impacted differently too before the present pandemic. Crises emerge and disappear but left various impacts on the world and communities. Sometimes crisis consequence questions the survival of the world or its inhabitants or of a particular community like tourist guides. COVID-19 pandemic poses the greatest challenge to the sustenance of the world disrupting the global economy and health system. The tourism–travel–hospitality industry as a whole and particularly its crucial component tourist guide service also met the same challenge when tourism suddenly halted and its growth slumped far below zero level for the first time since 1950 due to the COVID-19 pandemic.

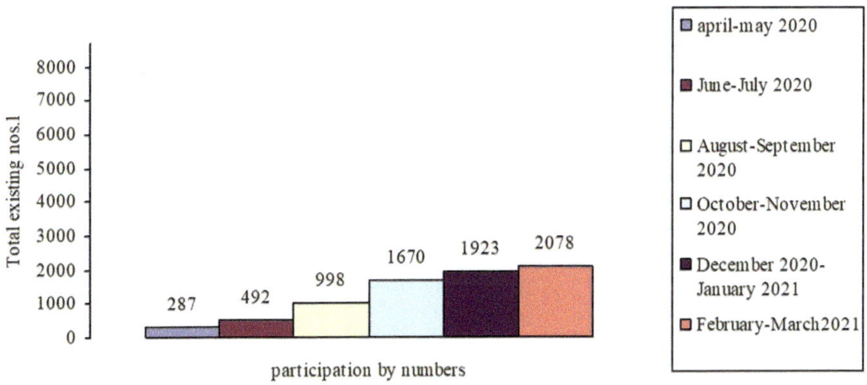

DIAGRAM 17.3 Participation of existing approved guides as viewer-participants in new normal settings—webinars and virtual tour seminars.

[All those diagrams prepared by the author based on information and data collected from WBTDCL guide assignment register and uploaded contents of Kolkata connect tours in social media and uploaded contents of Ministry of Tourism; Indiatourism Kolkata; Incredible India; Foodka, Safariplus websites, and social media (Debdut, 2021; Foodka, 2021; Incredibleindia, 2021; Indiatourism Kolkata, 2021; Ministry of Tourism, 2021; RLGwebinar, 2021; Shibaji, 2021; Wbtdcl, 2021; Sayantani, 2020; Safariplus, 2020)].

The gradual decline of pandemic intensity succeeds a bit to make the economy crawl but fails to build enough confidence among people for traveling. Thus, survival of tourism–travel–hospitality as a whole and particularly of the tourist guide service is crucial and challenging. Survival in any emerging and strange condition evolved out of a crisis like COVID-19 pandemic is the most difficult drive. In the present pandemic-consequential condition, the survival of tourist guides highly depends on both external and internal issues. Control on external issues like availability of a vaccine or treatment and gradual and linear lifting of travel sanctions is not in the hand of tourist guides though survival largely depends on those. At this juncture, the tour guide community has no choice but to make them prepared to be fit for rolling as soon as the condition becomes favorable. Indeed to become relevant in any disrupted, emerging, and strange condition, the community primarily need to be flexible to accept the disruption and changes that said condition demands. COVID-19 pandemic compels the world to accept new normal condition refers to a new way of living and going about lives, work, and interactions. Acclimatization to the new normal condition is a challenge to the community as they were habituated with an entirely different kind of professional culture or behavioral protocols. A series of shifts from habitual behavioral protocol to the new professional protocol is quite pragmatic to ensure cultural acclimatization of the community to the emerging work condition and the guide services to the emerging demands. Indeed, it assists the community to be repositioned and acclimatizes the service and the community to the new environment. Acclimatization is an ongoing process which continues until the cultural repositioning of the community and guide services in the new environment completes with the institutionalization of shifts and adaptation of new beliefs, perception, and norms as emerging tradition.

KEYWORDS

- **COVID-19**
- **new normal**
- **protocol**
- **shift**
- **tourist guide**

REFERENCES

AFP. Covid-19: Most of World's Airlines on Verge of Bankruptcy. New Straits Times, Mar 17, 2020. https://www.nst.com.my/world/world/2020/03/575499/covid-19-most-worlds-airlinesverge-bankruptcy (accessed on Mar 29, 2020).

Bhattacharya, S. The Social Impact of the COVID-19 Pandemic, *ORF Issue Brief No. 406*. Observer Research Foundation, 2020. ISBN: 978-81-948115-2-7.

CEN. 2020. https://www.guiasgranada.com/pdf/DefinicionGuiaTuristico.pdf

Cetin, G. Impact of Covid-19 on Tourism Industry. *J. Tourismol. Ed.* **2020,** *6* (1), editorial page.

Chang, T. Y.; Shen, C. C.; Li, Z. W. Establishing Tour Guide Work Safety and Risk Management Indicators System. *J. Tour. Hospital.* **2018,** *7* (2), 1000352. DOI: 10.4172/2167-0269.1000352

Dash, J. Covid-19 Impact: Tourism industry to Incur Rs 1.25 TRN Revenue Loss in 2020. 2020. https://www.business-standard.com/article/economy-policy/covid-19- impact-tourism-industry-to-incur-rs-1-25-trn-revenue-loss-in-2020-120042801287_1.html (accessed on July 11, 2020).

Debdut. 2021. #Kolkata #Doubledeckerbus #Doubledeckerbuskolkata; http://www.instagram.com/debdut_youtube; https://www.facebook.com/Debdutyoutube; https://www.youtube/debdutyoutube

Deegan, J. How the Coronavirus May Hit the Tourism Industry, March 3, 2020. RTE: https://www.rte.ie/brainstorm/2020/0303/1119857-coronavirustourism-ireland/ (accessed Mar 3, 2020).

Dept. of Tourism Republic of South Africa. 2021. https://tkp.tourism.gov.za/touristguide/Pages/What-is-Tourist-Guiding.aspx

Divya, A. Hospitality Sector Gears Up for a New Post-Lockdown Normal, 2020. https://indianexpress.com/article/business/hospitality-sector-gears-up-for-a-new-post-lockdown-normal-6482394/

Economic Times. March, 2020. https://economictimes.indiatimes.com/industry/services/travel/impact-of-coronavirus-on-indian-tourism-could-run-into-thousands-of-crores-of-rupees/articleshow/74592482.cms

FEG. 2019. https://www.feg-touristguides.com/cen_definitions.php

FEG. COVID-19 Pandemic and Next Steps in Tourism - the Tourist Guidesâ€™ Perspective, EuropeanFederationofTouristGuideAssociations,2020.https://www.feg-touristguides.com/post.php?i=covid-19-pandemic-and-next-steps-in-tourism-the-tourist-guides-perspective

Fletcher, R.; Murray Mas, I.; Blázquez-Salom, M.; Blanco-Romero, A. Tourism, Degrowth, and the COVID-19 Crisis. *Polit. Ecol. Netw.* Mar 24, 2020. https://politicalecologynetwork.org/2020/03/24/tourism-degrowth-and-the-covid-19-crisis/.

Foodka. 2021. https://www.youtube.com/hashtag/foodkaseason7

Gössling, S.; Scott, D.; Hall, C. M. Pandemics, Tourism and Global Change: A Rapid Assessment of COVID-19. *J Sustain. Tour.* **2020,** *29* (1), 1–20. https://doi.org/10.1080/ 09669582.2020.1758708

Govt. ofIndia. 2018. https://www.nqr.gov.in/sites/default/files/Curriculum%20-%20Tourist%20Guide.pdf

Grancay, M. COVID-19 and Governmental Measures to Support Tourist Guides—A Research Note MPRA Paper 102382. University Library of Munich: Munich, Germany, 2020.

Incredible India. 2021. https://www.incredibleindia.org/content/incredible-india-v2/en/events/dekho-apna-desh.html

Indiatourism Kolkata. 2021. https://www.facebook.com/pg/indiatourismkolkata/videos/? ref=page_internal#Indiatourismkolkata #foodka#Incredibleindiahttps://www.facebook. com/indiatourismkolkata/ https://www.youtube.com/channel/UCrLcuMkNeMDJ50vCbnm KeK whttps://www.instagram.com/indiatourism_kolkata

Jamal, T.; Budke, C. Tourism in a World with Pandemics: Local-Global Responsibility and Action. *J. Tour. Futures* 2020, *6* (2), 181–188. https://doi.org/10.1108/JTF-02-2020-0014

Hu, S. W.; Chen, J. H. Research on the Factors Influencing the Happiness Index of Tour Guide: Sample Data from Fujian and Network Complaints. *J. Guangxi Econ. Manage. Garde College* **2015,** 27, 81–86.

Karim, W.; Haque, A.; Anis, Z.; Ulfy, A. M. The Movement Control Order (MCO) for COVID-19 Crisis and Its Impact on Tourism and Hospitality Sector in Malaysia. *Int. Tour. Hospital. J.* **2020,** *3* (2), 1–07.

Kaushal, V.; Srivastava, S. Hospitality and Tourism Industry Amid COVID-19 Pandemic: Perspectives on Challenges and Learnings from India. *Int. J. Hospital. Manage.* **2021,** *92*, 102707. https://doi.org/10.1016/j.ijhm.2020.102707

Kirby, J. Will Coronavirus Cause a Global Recession? We Still Don't Know. Vox (2000). Mar 9, 2020. https://www.vox.com/2020/2/28/21153492/coronavirus-recession-china-stock-market-economy/

Foo, L-P.; Chin, M-Y.; Tan, K-L.; Phuah, K-T. The Impact of COVID-19 on Tourism Industry in Malaysia. *Curr. Iss. Tour.* **2020.** DOI: 10.1080/13683500.2020.1777951

Mahalingham, E. Hotel Sector Hit by Covid-19, Mar 26, 2020. The Star : https://www.thestar.com.my/business/business-news/2020/03/26/hotel-sectorhit-by-covid-19 (accessed on Mar. 28, 2020).

Mandel, A.; Veetil, V. The Economic Cost of COVID Lockdowns: An Out-of-Equilibrium Analysis. *Econ. Disasters Clim. Change* **2020,** *4*, 431–451. https://doi.org/10.1007/s41885-020-00066-z

Marinko, S.; Domingo, R. S.; Małgorzata, P. R. Impact of COVID-19 on the Travel and Tourism Industry. *Technol. Forecast. Soc. Change* **2021,** *163*, 120469. https://doi.org/10.1016/j.techfore.2020.120469

Marron, D. Macroeconomic Policy in the Time of COVID-19 Federal Budget and Economy, TaxVox (2020), Mar 17, 2020. https://www.taxpolicycenter.org/taxvox/macroeconomic-policy-time-covid-19

Medina, A. F. Malaysia Issues Stimulus Package to Combat COVID-19 Impact, March 19, 2020. ASEAN Briefing: https://www.aseanbriefing.com/news/malaysia-issues-stimulus-package-combat-covid-19- impact/ (accessed on Mar 28, 2020).

Ministry of Tourism. 2021. https://tourism.gov.in/whats-new/dekho-apna-desh-webinars-ministry-tourism

Nesrine, K.; Zoltan, L.; Ayoub, N. Pandemic and Tourism: Re-Preparation of Tourism Post COVID- 19. *J. Hotel Busi. Manage.* **2020,** 9, 198. DOI: 10.35248/2169-0286.20.9.198

PTI. COVID-19 Pandemic Hit India's Travel and Tourism, Immediate Survival Measures Required: FAITH, 2020. https://www.financialexpress. com/lifestyle/travel-tourism/covid-19-impact-loss-forecast-for-indias-tourism-sector-doubles-to-rs-10-lakh-crore-says-faith/1949030/

Qiu, R. T. R.; Park, J.; Li, S.; Song, H. Social Costs of Tourism During the COVID-19 Pandemic. *Ann. Tour. Res.* **2020,** *84*, art. no. 102994. https://doi.org/10.1016/j.annals.2020.102994

Radhakrishna, S. A. COVID-19 | Post-pandemic, India's Tourism Sector Stares at 70% Job Loss, 2020. https://www.thehindu.com/news/nationa l/kerala/coronavirus-tourism-sector-stares-at-70-job-loss/article31310234.ece.

Ranasinghe, R.; Damunupola, A.; Wijesundara, S.; Karunarathna, C.; Nawarathna, D.; Gamage et al. Tourism After Corona: Impacts of COVID 19 Pandemic and Way Forward for Tourism, Hotel and Mice Industry in Sri Lanka. 2020, 1-19. DOI: 10.13140/RG.2.2.27955.17442

RLGwebinar. 2021. https:www.facebook.com/RLGWebinarForum/

Safariplus. 2021. https://www.safariplus.co.in/indiatourism-story-telling-series https://www.safariplus.co.in/india-tourism-reconnect-organized-by-the-indiatourism-singapore

Sayantani. 2020. https://www.youtube.com/watch?v=mobC6HoSPVE https://www.getbengal.com/details/connect-with-a-kolkata-youve-never-seen-before

Shibaji. 2021. #kolkata#cityofjoy#kolkatacitytour#dalhousie#edengardens#BBDBag#greateasternhotel#kolkatahistory#princepghat#jamesprincep

Sigala, M. Tourism and COVID-19: Impacts and Implications for Advancing and Resetting Industry and Research. *J. Busi. Res.* **2020,** *117* (Sept.), 312–321. https://doi.org/10.1016/j.jbusres.2020.06.015

Tan, R. Airlines See Big Drop in Business, February 11, 2020. The Star Online: https://www.thestar.com.my/business/business-news/2020/02/11/airlinessee-big-drop-in-business (accessed on Mar 29, 2020).

Tourism Economics. Total Travel Industry Impact. Çevrimiçi, 2020. https://www.ustravel.org/toolkit/covid-19-travel-industry-research

UNCTAD. Covid-19 and Tourism: Assessing the Economic Consequences (UNCTAD/DITC/INF/2020/3), July 2, 2020a

UNCTAD. COVID-19 and Tourism Assessing The Economic Consequences, 2020b. https://unctad.org/system/files/official-document/ditcinf2020d3_en.pdf (unctad.org).

UNWTO. Impact Assessment of the Covid-19 Outbreak on International Tourism, 2020. https://www.unwto.org/impact-assessment-of-the-covid-19-outbreak-on-international-tourism

UNWTO. COVID-19 and Tourism | 2020: A Year in Review, 2021a. https://www.unwto.org/covid-19-and-tourism-2020

UNWTO. World Tourism Barometer and Statistical Annex, 19/1. UNWTO World Tourism Barometer (English Version) **2021b,** *19* (1) (e-unwto.org). DOI: https://doi.org/10.18111/wtobarometereng. October 2020.

Wbtdcl. 2021. https://www.wbtdcl.com/package

WHO. March 7, 2021. https://covid19.who.int/

WTO. International Tourism Highlights. Çevrimiçi, 2020. https://www.e-unwto.org/doi/ pdf/10.18111/9789284421152, Erişim tarihi: 12.04.2020.

WTTC. Press Release, Oct 29, 2020

Yeh, S-S. Tourism Recovery Strategy Against COVID-19 Pandemic. *Tour. Recreat. Res.* **2020,** 45. DOI: 10.1080/02508281.2020.1805933

Yuda, M.; Tani, S. Coronavirus Hits Thailand Hardest in Tourism-Reliant SE Asia: Hotel and Airline Stocks Sink as China's Travel Ban Kicks In, 2020. https://asia.nikkei.com/Spotlight/Coronavirus-outbreak/Coronavirus-hits-Thailand-hardest-in-tourism-reliant-SE-Asia (accessed on Jan 27, 2020).

CHAPTER 18

ADAPTING TO A NEW NORMAL: MOVE FOR A RE-START OF THE TOURISM AND HOSPITALITY INDUSTRY

SACHIN KUMAR BEHERA

Department of Tourism and Hotel Management, North-Eastern Hill University, Shillong, Meghalaya

ABSTRACT

Since the outbreak of the novel corona virus from the Wuhan city in China, authorities have stressed upon its population to maintain physical distancing in public places. Although a lot of restrictions have been lifted in many countries but everyone has to follow strict standard operating procedures issued by governments. An attitudinal change is set to follow in tourists when a lot of restrictions are put in place and fear grips in the minds as governments have asked its citizens to avoid nonessential travel. There is an obvious change seen in the behavior of travelers. Many prefer to stay home, avoid public transport, avoid crowded places, order food online instead of visiting a restaurant, etc. This article intends to showcase the impact of distancing on various service providers. There are both positive and negative impacts, which will be described in an elongated manner. Critical standard operating procedures and practices will also be discussed. Examples from practices being followed in movie theatres, shopping malls, events, etc. and case studies may be included to make the article interesting and understandable.

Post-COVID Tourism and Hospitality Dynamics: Recovery, Revival, and Re-Start.
Umendra Narayan Shukla & Sharad Kumar Kulshreshtha (Eds.)

18.1 INTRODUCTION

Vaccine has not been finalized and entered into market in India till date. Distancing is still been followed and taken as a new normal. Physical distancing is an important part of measures to control covid-19, but exactly how far away and for how long contact is safe in different contexts is unclear. In order to slow the spread of virus and save lives, citizens' voluntary compliance with social distancing rules and maintaining such measures for long play a pivotal role (Greenstone and Nigam, 2020). The major tool against Covid-19 will ultimately be masks, hand hygiene, and physical distancing that we all need to follow. Physical and social distancing is followed and encouraged but tourists are unwilling to travel the far distance as fear of infection has gripped their minds. Proximity tourism is gaining more importance as it is the kind of tourism that focuses on (re)discovering the places nearby, precisely because they are too close and familiar. Faraway destinations and exotic travels are usually avoided in this context. Traveling to long distance away from home may have become nonconformist these days. Traveling has reshaped the way in light of the pandemic outbreak.

The lack of spatial distance between tourists will be more common as there is a major shift in customer behavior and preferences after the dust of the virus settles. Tourists drop travel plans just because of social distancing requirements. Government officials and industry have predicted that in the postpandemic new normal, hygiene, health, and physical distance will be traveler's new main concerns. When there is higher concern about anything that becomes "the new normal." Here sanitation and distance are to be considered as the new normal and focus should be made on these principal things to anticipate the new tourism paradigm.

Better visitor management for busy destinations such as world heritage sites would be required to maintain physical distance between tourists. Previously tourists were ready to go to packed and crowded places but not anymore. Strategies such as avoiding transit flights to minimize contact, digital check-in, and contactless payment are being adopted in hotels and airports.

Let us understand briefly, how distancing has impacted and is being followed in different places of touristic interest such as concerts, cinema halls, hotels, eateries, destinations, etc.

18.2 CONCERTS/EVENTS

Large gatherings like musical concerts and events have been tremendously impacted since large gatherings may trigger the spread of virus (Eze et al., 2021). Musical events like Sunburn were made the most controversial Indian electronic dance music festival this year, as it was planned to be held in Goa this December. Prior announcements were made like: "Party People Might Have a Chance to Redeem Themselves this Year with Sunburn 2020."

A lot of controversies started when the government hinted of giving a green signal to the organizers of the musical festival but soon it was debunked because of widespread opposition. Government aimed to revive the two important state revenue sectors of Goa- Mining and Tourism, which was dysfunctional due to the COVID-19 related restrictions. Concerts like Sunburn could help in kick-starting tourism sector of the state. Although Goa is Fun's Own Country but had the event been given permission it would potentially prove as a super-spreader. India scaled back on Ganesh Puja and Dussehra, the Sunburn Festival is completely unavoidable. Ironically the tagline for the event this year was "Live Again" and India has already crossed one crore cases.

Since central government has encouraged a slow and phased reopening of tourism business, cultural events and musical concerts with huge crowd have either adapted to the virtual format, or have postponed to a faraway date. Musical festivals like NH Weekender and international electronic dance music festivals like Tomorrow land will be held virtually. Cultural and other festivals such as the Kochi Biennale, India Art Fair, and Jaipur Literature Festival have been shifted to 2021 and 2022. Hosting a physical event clearly seems like the more profitable option, and tourists are livid. The feel and enjoyment for which tourist comes to these concerts will not be normal even if it is held physically since government restrictions have to be followed. Mandatory download of the Aarogya Setu app, thermal screening, and bringing COVID-negative certificates to these concerts will further act as a deterrent. The fun lies in gathering and mixing up with different people, dancing together but restricting people to come in small groups and ensuring physical distancing between them will not let people have fun.

18.3 TRANSPORT

Transportation plays an important role in tourism. Restrictions on movement have affected people's lives and livelihoods (Mogaji, 2020). The negative effect of lockdown can be seen in the usage of public transport as well. Private transportation isn't much affected by the COVID-related regulations but changes can be seen in public transportation. Physical distancing is being given more importance. This result in fear among travelers, fall in ridership, loss of revenue by transporters, and mental stigma in the minds of travelers. Public transport users would be less likely to use public transport once the lockdown restrictions will be completely lifted as it is nearly impossible to follow distancing in public transport. The perception of the commuters/travelers toward hygiene will change which may result in preference to buy their own vehicles in the future. An important role is also played by the psychology of people, most will avoid traveling but those who cannot avoid will travel eventually.

The world came to a standstill when all modes of public transportation were stopped. Operators and rental services companies suffered huge losses due to suspension of services during the lockdown tenure.

In a country like India, implementing social distancing is a big challenge as there are various types of transport such as buses, railways, metro, rick-shaws, taxis, flights, etc. and always remain crowded and infrequent. Public transport fleet is limited and the number of commuters is huge which results in loss of weekly working hours of commuters due to traffic congestion (Obi, 2018). To ensure social distancing the fleet of public transport has to be increased with higher frequency and limited commuters. Most people use cash to buy tickets in public transport and although digital payments are being encouraged but still large population does not have access to digital payment.

Indian Railways has suffered a loss of Rs 6500 crore on revenue earned from ticket sales (during the first two phases of lockdown itself. In the first phases of lockdown only Shramik special trains were allowed to ferry migrant workers. The costs of tickets were borne by both central and state governments. Revenue of the national transporter was mostly generated by freight trains as regular passenger trains were suspended.

The Indian aviation industry has suffered a huge loss due to grounding of aircrafts. Companies were unable to pay their fixed costs in order to sustain in the market. All international flights were slowly reopened with air bubble agreements with different countries. Under air bubble agreement, limited flights were allowed to operate and was strictly under government super-vision. Distancing is also being followed in airlines by various ways such as keeping the middle seat vacant, mandatory usage of Personal Protective Equipment (PPE), discontinuation of in-flight meals, physical distancing while boarding and DE boarding, digital payment for in-flight purchases, self-check-in, etc.

Similarly road transport is also affected with less toll collections this year. People were discouraged to use even their private vehicles unless emergency. After planned reopening of business activities, taxi operators, and aggregators were allowed to ply with strict precautions. Distancing and sanitation have to be strictly followed during the ride. Both the driver and the commuters must wear mask. Temperature checks and all high-contact surface areas must be cleaned with sanitizer after every ride.

Metro rails were opened in selected cities and selected routes in order to cater the growing demand of urban population. Distancing measures such as 6ft distancing between commuters, zone-wise queuing, screening, and controlling crowds at the stations were taken. Masks and thermal screening are also made mandatory to enter the stations.

Covid-19 has taught a lesson to the governments to focus on transportation and particularly public transportation. It provides an alternative to private vehicles and plays an important role in sustainable transportation. People moving away from public transportation is not a good sign for the future as it leads to various problems. Congestion, pollution, traffic jam, high CO_2 emission, etc. will be the challenges if private transportation increases in future.

18.4 DESTINATIONS

Rural destination and noncrowded destinations are being preferred. Tourists do not prefer to move in groups. Destinations that pay attention to carrying capacity are being preferred. Tourists may prefer to go to quieter places with a little outdoor activity as we are inching toward the new normal. Instead of going to Instagram-worthy places, tourists would prefer visiting some place that offers nature-based activities and remains isolated. In a country like India, it was visible that as soon as the lockdown started most of the population started moving toward rural areas to their homes. This proves that rural tourism has a lot of potential if people start exploring places nearby to their home, which are also safer and more affordable. Domestic travel can be seen as an option for the time being, which is sustained by small locally owned businesses and usually ignored by the giant travel marketing machinery in favor of big-ticket international trips. Radical localism will prevail.

People will definitely crave for something that has been taken away from them—*Human Connection.* The lost human connection cannot be restored immediately until a vaccine arrives and people become fearless of contacting the virus. Till the time destinations must attract tourist and continue the business by following strict regulations. Stakeholders must take cognizance of environmental costs beyond economic costs. Destinations should move toward a zero carbon emission with higher levels of hygiene and sanitation, similarly tour operators and hoteliers should gravitate toward more responsible and meaningful experiences by showcasing the local for the global and positioning the host community as the centerpiece of the tourist experience.

18.5 SHOPPING MALLS/CINEMA HALLS/CONVENTION HALLS

The lockdown has severely devastated the entertainment industry (Moon, 2020). The pandemic has added enormous pressure to the already delicately-poised Indian retail sector. Since lockdown was announced in early March, malls and retail stores saw a huge dip in footfall with nearly 20–25% drop in revenue. The footfalls are also reduced due to thermal screening at the entry of malls and only asymptomatic visitors are allowed inside. People moving away from crowds and shifting to online platforms for purchase is the reason of reduced footfalls. In the pre pandemic world, people used to go to malls with family and friends to relax, watch movies, shop, dine and even enjoy gaming arcade in a safe and comfortable surroundings. Malls have the conducive atmosphere for spending as it is designed to soothe our nerves and make us feel good.

In the postpandemic, the new mall experience is unlikely to be good for mall economics. There will be change in consumer behavior. People will be repeatedly reminded of the virus threat which will lead to negative state of mood. Eventually, it will distract them from the buying process as the new distancing rules demand constant attention and alertness. In order to attain higher profits, fundamentally malls should capture and analyze consumer data to understand new behavior to build effective strategies.

Malls have many retail outlets, food, bar, gaming arcades; cinema hall, etc. and in the postlockdown, most malls will have to change their layout, and seating arrangements among other things.

Cinemas and food & beverage (F&B) are the two largest segments that attract footfall at malls and they will be last to revive at malls as till date cinema halls/food outlets are not allowed to run with normal occupancy. The timings of malls have also been tweaked, mall which were 24*7 open are now forced to work in single shifts. The reduced occupancy in the malls will lead to higher operation costs and lower revenue for few months. In order to implement distancing norms, malls have to employ more manpower and resources, which eventually leads to higher operation costs. The interval time in cinema halls will now be staggered to avoid crowding at food outlets.

People with higher disposable income will move toward other forms of leisure and entertainment but avoid visiting crowded places with lot of restrictions. At a time of surging demand for entertainment, pandemic disrupted both the demand side and the supply side of business in the entertainment sector (Nhamo et al., 2020).

The preventive measures that include distancing norms are to be followed to reduce the risk of COVID-19. These measures need to be followed by all the venue providers, exhibition organizers, service providers, cinema halls, malls, exhibitors, and business visitors/delegates, and are as follows:

1. Auditorium/hall occupancy shall not be >50%.
2. Vulnerable population which includes persons above 65 years of age, persons with comorbidities, pregnant women, and children below the age of 10 years are discouraged to attend any such events and are advised to stay at home, except for essential and health purposes.
3. Seating arrangements shall be made in such a way that adequate physical distancing is followed.
4. Every alternative seat shall be marked "Not to be occupied."
5. Hand wash and hand sanitizers are to be provided wherever necessary.
6. Everyone visiting these places shall be advised to install and use the Arogya Setu App.
7. In another way of discouraging people to come, only asymptomatic individuals shall be allowed to enter after thermal screening.
8. For different screens in a multiplex, staggered show timings shall be followed to avoid crowding.
9. Cash payment will be discouraged and digital modes of payments like UPI to be encouraged.
10. The box office and other nearby areas are to be regularly cleaned and disinfected.
11. The box office shall open sufficient number of counters to avoid long queues.
12. Audience shall be discouraged move during the intermission.
13. For better queue management at box office, the floors shall be marked with circles ensuring for physical distancing
14. Provision for tickets sale at box office to be open throughout the day, advance booking to be encouraged to avoid crowding.

15. Spitting in and around the abovementioned places shall be strictly prohibited.
16. Respiratory etiquettes to be strictly followed by everyone like covering nose and mouth while sneezing, use of mask, etc.
17. Only packaged food and beverages shall be allowed and no delivery inside hall
18. Food and beverages shall be sold at multiple counters.
19. The sanitization staff shall be provided with all safety measures such as gloves, boots, masks, PPE, etc.
20. Contact numbers of all visitors shall be taken to facilitate contact tracing.
21. COVID-19-related stigmatization or unruly behavior of anyone shall be dealt with strictly/asked to leave the venue.
22. Temperature setting of all air conditioning should be in the range of 24–30°C.
23. Before and after screening and during intermission, public service announcements on wearing masks, physical distancing, and hand hygiene shall be made.

18.6 EATERIES/RESTAURANTS

Nobody ever thought of a recession in this industry as everyone believed people will always eat and drink no matter whatever the circumstances are. But the pandemic has proved everyone wrong those who believed this industry as recession proof. After a short while of imposing nationwide lockdown, the Government ordered closing down of bars and restaurants and only allowed takeaway service after a while. The sales of thousands of restaurants plunged and this industry began to be counted as one of the most affected due to the pandemic (Bartik et al., 2020). Many restaurants and bars struggled to keep their heads above water as there was little or zero revenue and the fixed operating costs such as rent, utility expense and salaries were quite significant. Many fear the small businesses of the industry permanently closed (Brizek et al., 2021). In bigger cities, businesses were closed due to high cost of operations and in rural areas the proprietors chose other ways of earning livelihood.

People are afraid to dine out or to order food as the virus can be spread through surfaces. Fear has gripped the minds of people that outside food/nonveg food can contain the virus. During the lockdown period, as people spent more time stuck at home and discouraged from going to restaurants, they developed interest in home cooking. Cooking for self and loved ones became more common. Many people have changed their eating behavior and are now looking for healthy and simple one-pot meals that require less time and effort.

People have decreased their spending on dining out with the increase in COVID cases and some opted for stay-at-home orders (Yang et al., 2020). Large increase in order volumes is seen in places that are able to offer delivery and pick-up options. Meal kit delivery services are also on a major surge. Food aggregators like Zomato, Swiggy, etc. and direct restaurant channels have seen a surge in online orders since the lockdown.

Due to the pandemic, not just restaurants but its allied stakeholders like horticulture farmers, dairy producers, and food processing companies, suppliers, logistics companies and even delivery partners also saw a hit on revenue.

18.7 GOVERNMENT HAS RELEASED SOP FOR RESTAURANTS TO ENSURE PROPER DISTANCING

- Physical distancing of at least 6 feet to be followed as far as feasible.
- Use of face covers/masks to be made mandatory.
- Practice frequent hand washing with soap (for at least 40–60 s) even when hands are not visibly dirty. Use of alcohol-based hand sanitizers (for at least 20 s) can be made wherever feasible.
- Respiratory etiquettes to be strictly followed. This involves strict practice of covering one's mouth and nose while coughing/sneezing with a tissue/handkerchief/flexed elbow and disposing off used tissues properly.
- Dine-In shall be limited and takeaways to be encouraged. It is advised to the food delivery personnel to leave the packet at customer's door. The food packet should not be handed over to the customer directly.
- Before allowing home deliveries, the staff concerned shall be thermally screened and only asymptomatic staff and patrons shall be allowed to operate.
- All staff and patrons should use face cover/masks all times inside the restaurant.
- Posters/standees/AV media containing preventive measures about COVID-19 to be displayed prominently in such a place where it clearly visible to all.
- If possible, staggering of patrons to be done.

- Restaurant management must deploy adequate manpower for ensuring social distancing.
- Vulnerable employee's, that is, older employees, pregnant employees, and employees who have underlying medical conditions should be advised to take extra precautions. They should not be assigned any work requiring direct contact with the public (Front Office Management).
- Crowd should be efficiently ensured in parking lots and outside the premises.
- Waiting areas should be seated with additional patrons to follow social distancing.
- Valet parking shall be operational with operating staff wearing face covers/masks and gloves as appropriate. Disinfection of frequently used surfaces like steering, door handles, keys, etc. of the vehicles must be ensured.
- Encircling markings on the floors must be made with enough distance to manage the queue and ensure social distancing in restaurants.
- Separate entry and exits for patrons, staff, and goods/supplies should be provided.
- Extra precautions should be taken while handling supplies, inventories, and goods in the restaurant.
- Those visiting restaurants should maintain physical distance of a minimum of 6 feet, when queuing up for entry and inside the restaurant.
- In restaurants, not >50% of seating capacity is to be permitted. Tables should be placed in such a way that adequate distance in maintained.
- Instead of permanent menu book, disposable menus can be used.
- Disposable paper napkins should be encouraged instead of cloth napkins.
- Patrons involved in buffet service should also follow social distancing.
- Use of elevators by large number of people shall be restricted.
- In order to maintain distance while using escalators, one person should stand on alternate steps.
- For air-conditioning/ventilation, the temperature setting of all air conditioning devices should be in the range of 24–30°C, relative humidity should be in the range of 40–70%, intake of fresh air should be as much as possible and cross ventilation should be adequate.
- Large gatherings/congregations should be prohibited.
- Frequent sanitation of common-use places like lavatories, drinking, and hand washing station shall be maintained.

- Frequently touched surfaces like door knobs, elevator buttons, hand-rails, benches, washroom fixtures, etc. must be cleaned and regularly disinfected (using 1% sodium hypochlorite).
- Arrangements must be made for proper disposal of face covers/masks/gloves leftover by patrons and/or staff.
- At regular intervals, washrooms must be deeply cleaned.
- In order to ensure social distancing norms adequate crowd and queue management must be practiced in and out of restaurants.
- All precautionary measures like wearing of mask and hand gloves must be taken by all Staff/waiters.
- People should be encouraged to pay bills/order using Contactless mode and digital mode of payment (using e-wallets).
- As soon as the customer leaves, tables must be sanitized.
- Social distancing norms are to be followed by the staff in the kitchen area as well and it must be sanitized after regular intervals.
- Gaming arcades/children play areas in restaurants shall remain closed.

18.8 ALL HOTELS SHALL ENSURE THE FOLLOWING DISTANCING ARRANGEMENTS

- Hotels must mandatorily provide (sanitizer dispenser) and make provision for thermal screening at the entrance. Guest should be asked to sanitize their hands before and after filling of necessary forms at the reception including arrival and departure register.
- Guests should provide details of their (travel history, medical condition, etc.) along with ID and self-declaration form at the reception.
- Only asymptomatic guests shall be allowed to check-in.
- Guests must be asked wear face cover/masks all times inside the hotel except inside the room.
- Guests and staff of hotel should be made available appropriate personal protection gears like face covers/masks, gloves, hand sanitizers, etc.
- For ensuring social distancing norms, adequate manpower shall be deployed by hotel management.
- As a precautionary measure hotel staff should additionally wear hand gloves.
- Places like parking lots/entrance in the hotel as well as outside the hotel should not be overcrowded and proper crowd management should be ensured.

- Large gatherings/congregations shall continue to remain prohibited.
- The operating staff working for valet parking, must be wearing face covers/masks and gloves as appropriate. Disinfection of steering, door handles, keys, and other frequently used surfaces of the vehicles should be taken up in a proper manner.
- Separate entry and exits for guests, staff, and goods/supplies must be there in hotels. A minimum of 6 feet of distancing has to be maintained, when queuing up for entry and inside the hotel.
- To manage the queue, specific encircling markings on the floor may be made with sufficient distance.
- If hotels have elevators, the number of people using it at a time shall be restricted.
- In case of escalators, each guest should stand on alternate steps.
- Posters/standees/Audio Visual media about COVID-19 preventive measures to be displayed prominently in hotel.
- Contactless processes like QR code, online forms, digital payments like e-wallet, etc. for both check-in and check-out must be adopted by hotels.
- Only disinfected luggage must be sent to rooms.
- Guests should be discouraged to visit areas falling within containment zone.
- Only disinfected supplies, inventories, and goods should be allowed inside hotel.
- Gaming arcades/children play areas shall remain closed.

18.9 CONCLUSIONS

The definition of "normal" all over the world is changed, most industries are still scrambling and some are in the process to adjust with COVID-19 having the new normal. The effect of the pandemic on the tourism industry has been particularly dramatic. Many lasting consequences will be seen in the society like people's ability to focus, to feel comfortable around others; even to think of mingling out with others, may diminish. Durante et al. (2021) explored the relationship between social distancing and social trust attitudes among people. Distancing in every sphere will be given priority. Societies that have controlled their outbreaks want to keep travel restricted as they do not want the new ones to break out. Earlier humans were trapped in a hostile environment and thrived when they cooperated, typically in groups of a few hundred. Now again we are moving toward isolation and closer to nature.

There is re-emergence of our survival instincts. Amid widespread crisis there is a great psychological shift toward a "pro-social behavior"—checking in on neighbors, caring for the needy, cooking for friends but keeping distance in mind.

Changes will be seen in how we think, behave and relate to one another as for the last few months we have been living in a new and acute state of fear. Some deliberate changes will be seen but many will be unconscious, some temporary but others potentially permanent. The pandemic has taught people new tricks of living, pushing people to come up with new ways of shopping, working, learning, socializing, queuing, praying, playing, and even how to move and interact with one another. The enduring social and psychological impact of maintaining distance based on space will be long lasting. How we act and how we think is directly impacted by the way our bodies occupy space. There will be negative effect on subjective wellbeing as it is expected that due to social distancing—people will travel less, avoid public transport and travel places for recreation to short distance using own car/bike. The out –of- home participation of people will decline.

There will be drastic changes on how tourism will look like in the future. Automation will play a very important role in the industry as it decreases the fear of human contact. Tourism service providers and technology suppliers are working hand in hand to introduce automation (Alexis, 2017). It will also force companies to focus more on the human to human connection which is highly necessary in service based industries (Naisbitt and Philips, 2001). Fear may overpower everything. In order to transform the room cleaning and room service operations many hotels in India have already adopted robotics. The new beginning of guest hospitality will see a rise in service robots. Robots assure guests of utmost safety, hygiene and comfort and many are excited to take these services. In the coming years, hotels may deploy robot butlers to deliver snacks, toiletries and other amenities to the rooms. These robots will be able to use Wi-Fi for navigation of floors and will also be able to handle front office management like answering questions and chatting with guests. They can provide basic information about best places to hangout, spa, gym, tourist destinations, shopping hubs etc. This change will lead to elimination of some tourism and hospitality jobs while changing the nature of task performed with creation of brand new job positions. Overall service processes and the job requirements will changes (Ivanov, 2020).

Some examples from over the world adopting the new normal of distancing—In Italy, to monitor crowding at beach resorts, electronic bracelets are being used, along with one-way flow systems, umbrellas are fixed at five-meter distance, and booking apps assign time slots for sunbathing. Sardinia's La Pelosa beach plans to implement a limited admission ticketing system.

Another example is of Greek island of Santorini where Plexi glass screens have been installed around beach chairs and sun loungers at a beach bar.

Hardware businesses have created a solution for noncontact body temperature sensing and facial identification to deliver smooth service. Many hotels in Guangzhou have used this technique, which not only decreases the danger of infection but also saves time and reduces hotel congestion (Lau, 2020).

Still, there is a lingering fear that threatens to outlive the virus itself. Social distancing has taken place of a life-saving necessity that has scarred our psyches and bodies.

KEYWORDS

- **physical distancing**
- **tourists**
- **behavior**
- **attitudinal change**
- **tourism industry**

REFERENCES

Alexis, P. R-Tourism: Introducing the Potential Impact of Robotics and Service Automation in Tourism. *Ovidius Univ. Ann. Ser. Econ. Sci.* **2017,** *17* (1).

Bartik, A. W.; Bertrand, M.; Cullen, Z. B.; Glaeser, E. L.; Luca, M.; Stanton, C. T. *How Are Small Businesses Adjusting to COVID-19? Early Evidence from a Survey*; National Bureau of Economic Research, 2020.

Brizek, M. G.; Frash, R. E.; McLeod, B. M.; Patience, M. O. Independent Restaurant Operator Perspectives in the Wake of the COVID-19 Pandemic. *Int. J. Hospital. Manage.* **2021,** *93*, 102766. https://doi.org/10.1016/j.ijhm.2020.102766

Durante, R.; Guiso, L.; Gulino, G. Asocial Capital: Civic Culture and Social Distancing During COVID-19. *J. Public Econ.* **2021,** *194*, 104342. https://doi.org/10.1016/j.jpubeco.2020.104342

Eze, V.; Okonkwo, U. U.; Chinyere, C. A. E.; Ukwueze, C.; Ezeugwu, F. O. COVID-19 and the Entertainment Industries in Nigeria. *Rupkatha J. Interdiscip. Stud. Human.* **2021,** *13* (1).

Greenstone, M.; Nigam, V. Does Social Distancing Matter? University of Chicago, Becker Friedman Institute for Economics Working Paper, 2020–26, 2020.

Ivanov, S. The Impact of Automation on Tourism and Hospitality Jobs. *Inf. Technol. Tour.* **2020,** *22* (2), 205–215. https://doi.org/10.1007/s40558-020-00175-1

Lau, A. New Technologies Used in COVID-19 for Business Survival: Insights from the Hotel Sector in China. *Inf. Technol. Tour.* **2020,** *22* (4), 497–504. https://doi.org/10.1007/s40558-020-00193-z

Mogaji, E. Impact of COVID-19 on Transportation in Lagos, Nigeria. *Transp. Res. Interdiscip. Perspect.* **2020,** *6*, 100154. https://doi.org/10.1016/j.trip.2020.100154

Moon, S. Effects of COVID-19 on the Entertainment Industry. *IDOSR J. Exp. Sci.* **2020,** *5* (1), 8–12.

Naisbitt, J.; Philips, D. *High Tech High Touch Technology and Our Accelerated Search for Meaning*, 2001.

Nhamo, G.; Dube, K.; Chikodzi, D. Implications of COVID-19 on Gaming, Leisure and Entertainment Industry. In *Counting the Cost of COVID-19 on the Global Tourism Industry*; Springer, 2020; pp 273–295.

Obi, D. Lagos Commuters Lose 75% of Weekly Working Hours to Traffic, 2018. Accessed April, 4, 2020.

Yang, Y.; Liu, H.; Chen, X. COVID-19 and Restaurant Demand: Early Effects of the Pandemic and Stay-at-Home Orders. *Int. J. Contemp. Hospital. Manage.* **2020,** *32* (12), 3809–3834. https://doi.org/10.1108/IJCHM-06-2020-0504

ONLINE SOURCES

https://www.indiatvnews.com/news/india/restaurants-sop-released-checklist-623460

https://www.cppr.in/centre-for-urban-studies/effects-of-covid-19-on-transportation-what-the-experts-say

https://www.businesstoday.in/opinion/columns/coronavirus-lockdown-impact-consumer-behaviour-post-covid-19-shopping-malls/story/406126.html

https://transfin.in/impact-of-coronavirus-pandemic-on-restaurant-food-industry-in-india

https://timesofindia.indiatimes.com/readersblog/creative-thoughts/impact-of-covid-19-on-the-usage-of-public-transport-21917/

https://theconversation.com/social-distancing-how-we-overcome-fear-of-one-another-to-embrace-a-new-normal-139089

https://retail.economictimes.indiatimes.com/news/industry/covid-19-impact-at-malls-the-last-to-recover-will-be-cinema-and-eateries-post-lockdown/75275855

https://www.nytimes.com/2020/04/21/world/americas/coronavirus-social-impact.html

https://www.thejakartapost.com/news/2020/05/06/hygiene-social-distancing-new-priorities-in-post-pandemic-tourism.html

https://cnr.ncsu.edu/news/2020/07/impact-of-covid-19-on-tourism/

https://www.cppr.in/wp-content/uploads/2020/09/Effects-of-COVID19-on-Transportation_What-Experts-Say.pdf

CHAPTER 19

FRUIT WINE PRODUCERS' PERCEPTIONS OF WINE TOURISM IN MEGHALAYA POST-COVID-19

ASHOK KUMAR[1,2]

[1]*Department of Tourism & Hotel Management, North-Eastern Hill University, Shillong, Meghalaya, India*

[2]*Institute of Hotel Management, Shillong, Meghalaya, India*

ABSTRACT

Wine tourism is an amalgamation of wine industry with tourism industry. There has been a tremendous interest shown by industrialists and academicians from the past two decades. Globally, cellar door sales and visitation to the wineries and wine tours are the major source of promoting the wine regions and wine tourism. Wine tourism is perceived as a marketing activity and an endorsement instrument by the small wine producers, to promote their local indigenous produce made out of indigenous fruits grown in rural settings. In 2019, recently, a "severe acute respiratory coronavirus-2 (SARS-CoV-2) illness," which was found in the city of China, Wuhan. It has contributed to a highly infectious sickness commonly known as coronavirus disease or COVID-19.

The virus was initially discovered in December 2019 and spread over other countries of the world in less than 3 months. This had enforced the World Health Organization (WHO) to announce a pandemic all over the world on March 11, 2020 as a precautionary measure for the safety of citizens. The current scenario of pandemic has stopped the movement of individuals, which had serious implications on businesses of most of the countries and

Post-COVID Tourism and Hospitality Dynamics: Recovery, Revival, and Re-Start.
Umendra Narayan Shukla & Sharad Kumar Kulshreshtha (Eds.)

has become a major threat to all activities related to tourism. Similarly, the Indian industry and all the stakeholders are also affected by the pandemic.

The main reason of the research is to observe if there are any disparities in perceptions of wine tourism among Meghalaya wine producers in Covid-19. A survey of 24 wineries was performed to look into the perceptions of wine tourism among wineries based on their production levels and cellar door sales. According to the findings, many winemakers have a good attitude toward wine tourism. They perceive the wine tourism as beneficial and full of opportunities for wine producers. They have trust that wine tourism can be used to gain the attentions of many private and governmental organizations.

19.1 INTRODUCTION

The ongoing pandemic Coronavirus, the "severe acute respiratory coronavirus-2 (SARS-CoV-2) syndrome," which has lately lead to a highly infectious disease called coronavirus disease or COVID-19. This virus originated in a city of China called Wuhan in December 2019. It was then spread to most of the counties of the world within 3 months. The World Health Organization (WHO) had declare a pandemic all over the world on March 11, 2020. The pandemic had a large impact on India's tourism industry both international and domestic tourism. Consequently, few more cases were confirmed among the students returned from China and other foreign countries. In light of the situation, and for containing the pandemic, India's Prime Minister declared a complete lockdown for the national interest on March 25, 2020.

19.1.1 HOW TOURISM INDUSTRY GOT AFFECTED BY COVID-19 PANDEMIC

The Corona Virus pandemic is the most serious threat faced by any country, with international travel expected to be reduced to 25% by 2020. Due to the pandemic, destination marketing services have undergone a significant shift, and restart of luxury travel will take some time to recover its place in every person's life. Due to the restrictions and awareness for social distancing and tendency of individuals to minimize the virus risk during luxury vacations, autonomous and self-driven housings where good amenities and separate quarters are provided like boutique hotels and homestays, and domestic markets will be more attractive till coming 1 year. This is because smaller properties provide better social distancing and reduced risk to virus due to less crowd in these places. The burgeoning tourism industry, as well as research and study on tourism during COVID-19, argues for a coordinated response to the epidemic as a catalyst for modifications (Siagalla, 2020). There was a fall in the trip budget for MICE Travel, and most organizations continued to adopt the Work from Home approach and allowed conference meetings/ calls instead of physical meetings to lower the risk of infection and cut costs. The leisure and luxury activities have seen an instant effect of Covid-19. The effect is seen on hotel and tourism industry across all topographical segments (consumers segmented based on region), including outbound, inbound and domestic travel, as well as nearly all vertical segments, including vacational, heritage, adventure activities, luxury, MICE (Meetings, Incentives, and Conferences & Exhibitions), Special Interest Tourism (SIT), and cruises. Industry needs to work not only to restore, but also replenish and restructure the new normal and financial order (McKinsey, 2020), and researchers need to use COVID-19 as a setting to duplicate present tourism dimensions and predictions (Gössling et al., 2020).

Most countries have regarded tourism as a major catalyst for the enhancement of economic growth and job creation for several decades, and it is commonly acknowledged that enormous investments will continue to happen into its expansion year after year around the world. According to the FICCI study (2020), India has 3961 centrally protected sites (monuments that are conserved and protected by Government of India), either as UNESCO world heritage sites or under the Archaeological Survey of India (ASI). The Qutub Minar in New Delhi and Agra Fort and Taj Mahal in Agra generate the maximum profits from tourism. In Agra alone, the lives of around 1.75 lakh daily wage earners have been impacted, including rickshaw drivers, tour guides, construction workers, andphotographers whose earnings are on daily basis. There is an additional difficulty for tour guides, which possesses government permits because they are not permitted to work in other occupations. Due to the pandemic scenario, majority of bookings for vacation travel in India (for the states of Goa, Rajasthan, and Kerela) have also been annulled, affecting domestic tourism. Because there was an anticipation that the count of COVID-19 cases going to increase high so as a precautionary measure, India's Prime Minister declared a four phases of lockdown order (LDO) to fully restrict the spread of virus.

The restrictions imposed by the government on air travel both domestically and internationally in 2020 had a substantial impact on the tourism industry in India. The tourism industry is required not only to mend, but also restructure, reproduce, and reorganize the new normal and global financial and commercial order (McKinsey, 2020), and researchers are required not merely repeat existing information in the measurement and forecasting of COVID-19 impacts on tourism industry as a background setting (Gössling et al., 2020; Hall et al., 2020). How this catastrophe may motivate industry to modify, how firms can leverage this catastrophe into revolutionary innovative approaches, and how effort can be done to consent, educate, and reorganize a new norm will be the largest obstacles, rather than possibilities. Crises have sparked change in the past, but none has resulted in a substantial transition (Hall et al., 2020).

Due to the global pandemic, the attitudes, potential, and behaviors toward change are deeply influenced by the awareness, empathy, and ability to understand and respond to the pandemic of those who are involved in Covid-19 project. Therefore, the perception of such people needs to be considered by the researchers doing studies. Gretzel et al. (2020) advocated for transformational e-tourism research, which has the potential to impact the future course of tourism through higher and more flexible institutional logics, theoretical perspectives, value systems, and technological concepts. However, tourism is a useful source for community advancement, but it is not the only source. Consequently, in all tourist sites, a social equilibrium in context to equality and justice, and also economic balance is required (Romagosa, 2020).

19.1.2 WINE MANUFACTURERS/ENTREPRENEURS

TABLE 19.1 Wine Manufacturers/Entrepreneurs of East Khasi Hills District of Meghalaya.

1.	ACE's Archie's premium wines, Pokhteh, Shillong
2.	Bari's Wine, Jaiaw, Shillong
3.	Breezing wine, Nongthymmai, Shillong
4.	B. Kharshandi,Kyiem, Shillong
5.	Daa-Mee Variety wine, mawroh, Golf link, Shillong
6.	Daly's Wine, Jaiaw, Shillong
7.	Damad food processing, Pohkseh, Shillong
8.	Donbok Kharkongor, Smit
9.	Dr. Lode, Laitumkhrah, Shillong
10.	Forever young club, Jaiaw, Shillong
11.	Green Dale wine,Nongrim Hills, Shillong
12.	Mawphlang Distillery, Mawphlang
13.	Merry Dew, Upper Nongthymmai, Shillong
14.	Mummy's wine, Laitumkhrah, Shillong

TABLE 19.1 *(Continued)*

15.	Ninlun Chongloi, Bishnupur, Shillong
16.	Old red wines, Risa colony, Shillong
17.	Pinnacle wines, Pohkteh, Shillong
18.	Three grace wine, Laitumkhrah, Shillong
19.	Nongsiej Food Processing Unit, Shillong
20.	WISE, Shillong and jowai

Source: Author's Compilation

19.1.3 WINE TOURISM IN INDIA

The Government of India's, Ministry of Tourism (MOT) has taken many measures to encourage tourism in the country. A variety of themes based on tourism have been introduced by government to improve the number of tourists. Wine tourism is the newest form of tourism, yet it has raised a lot of eyebrows because India has always been considered as a country focused on culture, traditions, and legacy. The government, on the other hand, has recognized the importance of wine tourism in India and pushed it as a special interest tourist after much deliberation. Although still relatively young, wine tourism is gaining up with other forms of tourism and will soon be a serious competitor.

India has made significant advances in the enhancement of technologies for making wines from variety of fruits like grapes, pear, mango, plum, apple, cashew, peach, banana, pomegranate, berry, strawberry, and kinnow (Joshi and Attri, 2005). Wine is a term used to describe the process of fermentation of by-products of grapes (*Vitis Vinifera*), but it can also apply to any un-distilled alcoholic fruit product obtained after fermentation. Other fruits produce a very little amount of wine when compared to grapes. In India, the wine industry is still in its early stages (Joshi and Attri, 2005). Fruit wines are alcoholic beverages that have a distinct flavor of a specific fruit, such as apple, pear, plum, apricot, cherry, or mango.

From a tourist standpoint, the term "wine route" refers to a path or road connecting the numerous wine-producing regions. Natural attractions such as hills, mountains, heritage sites, wineries, tasting rooms for wine tasting, ethnic eateries, and spiritual places adjacent areas, as well as local culture and traditions and museums, highlight this route. We can say that, tourists that take the wine tour and pass through that route can enjoy a variety of experiences such as trips to vineyards, wine-tasting sessions, gaining awareness about the winemaking procedure, gastronomy, wine and food synchrony, and local cuisine and culture. They can also sample regional cuisine and purchase things that are unique to the area while taking in the environment (Brunori and Rossi, 2000).

India has a long history of winemaking, but its winemaking industry is still in its early stages. In India, wine tourism is gaining a lot of traction. Many wineries in India have set up tasting facilities so that wine enthusiasts can be entertained by visiting and discovering the country's wine regions. Wine tourism is growing at fast pace and is competing well with other tourism themes giving them a run for their money. India is establishing itself as a world-class wine producer.

19.1.4 WINES OF NORTH EAST INDIA DURING COVID 19

The states, Arunachal Pradesh, Nagaland, Manipur, Sikkim, Assam, Mizoram, Tripura, and Meghalaya, are the eight states that make up North East India. Northeast India's soil, climate, and topography are noted for their diversity. Many flowers, fruits, vegetables, especially orchids, bamboo, spices, and medicinal and fragrant plants grow abundantly in this region. Northeast India has the greatest array of commercial fruits, including citrus, bananas, and jackfruit. Other tropical and subtropical fruits from the genera *Garcinia, Artocarpus, Annona, Phyllanthus, Persia, Averrhoa, Tamarindus, Aegle, and Passiflora,* among others, have been reported from the area.

The region's indigenous and minor horticulture crops, on the other hand, are underutilized. Through processing and value addition, such crops have the potential to reduce poverty, nutritional insecurity, and food insecurity (Deka, et al., 2012). The majority of these horticulture crops grown in the area are high in multiple vitamins, minerals, and other bioactive compounds that are useful in the pharmaceutical, aroma therapy, and food processing industries. The physiological and climatic conditions of Northeast India makes it favorable for cultivation of variety of citrus species and their hybrids in wild or semiwild form without any commercial cultivation and much care. Due to this, Northeast India is also known as the citrus depository of our country. The widely cultivated species in the area is Khasi mandarin *(Citrus Reticulata)*. Other species grown in the region are Sweet orange *(Citrus Sinensis), Citrus indica Tanaka (Indian wild orange),* Citrus assamensis, *C.medica,* C. ichangensis Swingle (Ichang Papeda), etc. Jackfruit is a native fruit that is grown abundantly in Tripura, Nagaland, Assam, and Meghalaya and needs attention (Deka et al., 2012).

TABLE 19.2 Fruit Wines of Northeast India.

Common/Local Name	Scientific Name	Family	Distribution
Wild kiwi Soh Kiwi (Khasi)	Actinidia strigosa	Actinidaceae	Sikkim
Wild kiwi Soh Kiwi (Khasi)	A. callosa	Actinidaceae	Arunachal Pradesh
Leteku	Baccaurea sapida (Roxb.)	Euphorbiaceae	Assam, Sikkim,Tripura, Meghalaya, A.P.
Carambola (Star fruit), Soh Pyrshong (Khasi)	Averrhoea carambola L	Oxalidaceae	Meghalaya, Assam
Indian crab apple, Soh-pho (Khasi)	Docynia indica	Rosaceae	Sikkim D,Khasi hill (Meghalaya),
Aonla	Emblica officinalis	Euphorbiaceae	All NE States
Soh-shang (Khasi), E. pyrifolia	Elaeagnus latifolia Linn.	Elaeagnaceae	lower Assam Meghalaya
Thekera tenga (Assamese	Garcinia lanceaefolia	Clusiaceae	Meghalaya, Mizoram, Nagaland, Assam
Box Berry Soh-phie (Khasi)	Myrica esculenta M. Nagi	Myricaceae	All region of north eastern hill
Soh-phie (Khasi)	Myrica fraquhariana	Myricaceae	Naga hills, Sibsagar (Dikho valley Assam), Khasi and Jaintia hill in Meghalaya
Passion fruit Soh-rub (Khasi)	Passiflora edulis P. edulis var. Flavicarpa	Passifloraceae	Meghalaya, Sikkim,Mizoram, Nagaland, Manipur,
Himalayan Pear Soh-shur (Khasi)	Pyrus pashia	Rosaceae	Khasi & Jaintia hills in Meghalaya
Blue Berry Soh-iong (Khasi)	Prunus nepalensis	Rosaceae	Khasi and Jaintia hills in Meghalaya
Otenga (Assamese)	Dillenia indica	Dilleniaceae	Assam,Meghalaya
Percea Pumsi (Sikkim	Machilus edulis King. Syn.	Lauraceae	A.P., Sikkim and other north eastern Region

Source: Authors Compilation

The state of Arunachal Pradesh produces a large quantity of pears, kiwis, and plums but due to lack of market and so less demand, most of them are wasted. Mizoram is well known for its export quality Grape wines. During the global pandemic of Covid-19 many small businesses suffered then a woman winemaker, Tage Rita, from Ziro Valley of Arunachal Pradesh, focussed on innovative technologies and advancements in wine making and helped the struggling farmers in the region.

> *"Lockdown was very tough for me. When works in the factory stopped, my focus and attention were on the gardens. I was looking for ideas to survive after the pandemic and so came the idea of improving my kiwi wine and add plum and pears into the list to offer something new to the wine connoisseurs after the pandemic,".*

During the pandemic, Rita, has introduced a new variety of wines under the brand name *Naara Aaba*. Naara Aaba is a special wine produced from organically harvesting kiwis in the region. Rita revived the local farming community by launching the wine Naara Aaba prepared from exotic fruits grown in the area in abundance and preserving the nutritional value of fruits. Earlier these fruits (kiwi, plums, and pears) got decay due to lack of marketing and storage facilities. As soon as travel bans were lifted by the government and the state was open for tourists, wine specialists and enthusiasts started looking for such special variety of wines, all made from organically grown kiwis, plums, and pears. Naara Aaba is a brand of special wines manufactured by company Lambu-Subu Food and Beverages Limited situated at Hong village located in Ziro Valley of Lower Subansiri district. "The products will not only offer something new to tourists but also help the poor farmers get the price of the fruits they have traditionally grown in their backyards," she said. Earlier, Rita was an agricultural engineer who resigned from engineering work with Arunchal Pradesh government. She switched to wine making industry and became the first organic kiwi wine maker. Purpa Tsering is another winemaker from Dirang Valley, similar to Rita, is helping local farmers in the region by purchasing, in large quantities, most of the locally cultivated fruits in the hills.

The Arunachal Pradesh Chief Minister Pema Khandu praised the hard work and efforts of Rita and said that value addition is provided to locally grown fruits of the state, the wine brand will help farmers to increase their earnings and at the same time offered new product to tourists. Similarly, Egam Basar, who is the head of Arunachal Pradesh Horticulture Research and Development Institute, said that wine policy laid down by the state government of Meghalaya in 2016 has encouraged the local business persons to adopt and produce the wine made up of kiwis and other locally available fruits (https://www.deccanherald.com/national/east-and-northeast/kiwi-pears-and-plum-wine-bring-new-hope-for-covid-19-battered-tourism-and-farmers-in-arunachal-pradesh-907918.html).

19.1.5 WINE MARKET IN MEGHALAYA

The niche market (where a specific product is focused) of fruit wines in Meghalaya can prosper at best, when the abundant supply of fruits is organized and handled properly. Meghalaya contains a variety of fruits, including pineapple, strawberry, banana, sohiong, passion fruit, and that might make good wines if they were given the right attention and economic development. All of this is only conceivable when the administration will take favorable efforts to allow the fermenting potential of its plentiful fresh fruits to be fully realized. The manufacture of indigenous beverages could be a useful source of income for ethnic communities as well as a source of cash for the government. Cottage and small-scale industries could benefit from increased production if the wine is excellent.

For centuries, alcoholic beverages produced by fermentation have been widely enjoyed by people in practically every country (Fellows, 2000). These fermented alcoholic beverages are typically made from locally accessible ingredients using traditional procedures (Rose, 1997), and the art of traditional methods of wine-making is thought to be passed down through generations through traditional and cultural values. The processing is improved by trial and error throughout the generations (Food and Agricultural Organization, 2012).

19.1.6 WINES OF MEGHALAYA DURING COVID-19

During Covid-19, the government has legalized the fruit wine production. Fruit winemakers in Meghalaya, can now utilize the rich agro-biodiversity and an abundance of fruits such as *grapefruit, mulberry, Mandarin orange, banana, litchi, pineapple, Jackfruit, strawberry, Plums, Pears, Sohiong, and Soh Phoh* for wine production. Amateur fruit winemakers have the potential to turn Meghalaya into India's fruit wine capital. During the ongoing Corona virus Pandemic, the state of Meghalaya declared plans to give licenses for the production and sale of handmade fruit wines in order to promote them in both international and domestic markets. The draught rules are available on the Meghalaya State Portal website, and suggestions are encouraged. The steps by government and formulation of plans and procedures toward wine industry has taken place after 17 years and a global pandemic which has brought fruit winemakers in Meghalaya one step closer to success.

TABLE 19.3 Native Wines Found in Meghalaya.

S. no.	Wine name (local names)	Botanical names
1	Orange (*Citrus sinensis* L.)	Orange *(Citrus sinensis)*
2	Cherry wine	Cherry *(Prunus cerasus L.)*
3	Cashew Nut Apple	Cashew Nut Apple *(Anacardium occidentale)*
4	Ginger wine	Ginger *(Zingiber Officinale)*
5	Mulberry Wine	Mulberry *(Morus)*
6	Strawberry Wine	Strawberry *(Fragaria)*
7	Passion fruit Wine	Passion fruit *(Passiflora)*
8	Blackberry Wine	Blackberry *(Rubus)*
9	Plum Wine	Plum *(Prunus Domestica)*
10	Banana Wine	Banana *(Musa)*
11	Jackfruit Wine	Jackfruit *(Artocarpus Heterophyllus)*
12	Peach Wine	Peach *(Prunus Persica)*
13	Pear Wine	Pear *(Pyrus)*
14	Pine Apple (Soh trun)	Pineapple *(Ananas Comosus)*
15	Kiwi Wine *(Soh Kiwi)*	Kiwi Fruit (Actinidia strigosa)
16	Black Cherry (Soh iong)	Black Cherry *(Prunus Nepalensis)*
17	Amla *(Soh Mylleng)*	Amla *(Emblica officinalis)*
18	Huyna (*Soh Mon*)	Huyna *(Meyna laxiflora)*.

Source: Compiled by the author.

They could make fruit wine for themselves and friends under the excise laws before Covid 19, but they could not sell it. Commercial wine production, particularly low-alcohol fruit wines, was prohibited in a state where grapes, pineapples, strawberries, passion fruits, and a local black cherry-like fruit known as Sohiong, as well as the indigenous pear variety Soh Phoh, abound. The reason for this was that Meghalaya was up till now subject to the preindependence, preconstitution era when Assam Excise Act of 1945 had prohibited the manufacturing of wine or any other alcoholic beverage. Thankfully, after an 18-year battle, the government has decided to legalize the manufacturing of fruit wines. The manufacture and sale of homemade products were introduced by the ruling Meghalaya Democratic Alliance Government.

After the cabinet meeting, Chief Minister, Mr. Conrad Sangma said that the Meghalaya Government has amended the excise rules. Due to the relaxation in wine-related regulations, individuals can prepare handmade wines and sell them in the market. He believes that the state government's strategic action will go a long way toward assisting farmers who grow a variety of fruits in the state to sell their harvest locally, which may then be sold to local wineries. Since 1992, a retired banker from Meghalaya who is 73 years of age, Brian K Daly, has been producing fruit wine. "In the hills of Meghalaya, wine can be made from every possible fruit," said Daly, who specializes in three types: pineapple, and two native abundantly grown fruits in Meghalaya: Soh Mon (plum) (Meyna laxiflora) and Sohiong (black cherry). Dr. Arnest Marbaniang, 40, has owned and operated a modest business. "During the lockdown, I realized how popular fruit wines were. Buyers would request for home delivery but due to restrictions it could not be done," said Marbaniang.

Voice for Local Businesses

Similar to other states, Meghalaya, seeking for methods to rebuild their devastated economies in the wake of COVID-19, formed an economic and financial committee led by Chief Minister Mr. Conrad Sangma. However, the state has imposed a tax of 25% on booze to raise income, the urge to commercialize fruit wine happen at that point of time when there is a strong push to market and advertise a self-sufficient state economy and be "loud for local," as Prime Minister Narendra Modi put it. Fruit winemaking is self-sustaining, according to Mr. Pravin Bakshi, Commissioner of Excise, Meghalaya, and would protect the outflow of local underprivileged people from rural areas to urban agglomerations or beyond the state. Mr. Michael Syiem, President of the Meghalaya Wine Makers Association, claims that the cottage industry (a small-scale manufacturing business operated out of home, for example, cotton weaving, carpet making, metal handicrafts, etc.), if exploited fully, can become a good source of job opportunities to many and a true utilization of the agricultural capability of the state.

19.1.7 REGULATION FOR MAINTAINING QUALITY OF FRUIT WINEMAKING

The making of fruit wines can take upto 1 year or more to mature before they can be sold in market. Mr. Marbaniang, a doctor by profession, has taught the skill of winemaking to a few family members and acquaintances. These wines are currently sold to friends and acquaintances under the labels "Khasi Wines" and "Nes-Ki Wines." Fruit wine, he claims, encourages healthy drinking habits because it contains less alcohol than hard beverages. "Alcohol level ranges from 8 to 15%, depending on the type of fruit and yeast strain," he stated. Home-made wines should "not exceed 14% percent ABV (alcohol by volume)," as per to the government's draft regulations.

A license, according to Daly, will go a long way toward standardizing the product. The local population responds enthusiastically to the Meghalaya wine festivals. Buyers from outside the state have shown interest as well. When the policy is finalized, the government should retain fruit winemaking as a small-scale or cottage industry that benefits the local community and farmers directly. Despite the unpleasantness brought on by Covid-19, Michael Syiem is upbeat and says, "At least we have something to cheer about in the midst of all the confusion, anxiety, and dread caused by the continuous pandemic."

19.2 REVIEW OF LITERATURE

19.2.1 WINE TOURISM AND WINE PRODUCERS

Wine tourism is a type of special interest tourism (SIT) that encompasses a variety of activities aimed at knowing the environment of wine such as tourist visits to vineyards, and wine regions (George, 2006). The routes of wine tours are specifically designed to wine producing regions and wine-related activities like tasting sessions. There have been so many definitions of Wine tourism given by eminent scholars. The definition of wine tourism, from the point of view of visitors of wine tour is extensively used. Wine tourism, according to Hall, is described as visits to vineyards, regions of wine production, theme-based wine festivals, and wine exhibitions where the key motivators for tourists are tasting sessions for wines made-up of grapes and having the experience of the features of a grape-wine region (Hall, 1996). The above definition of wine tourism is from the perspective of consumers. According to Getz (2000), multiple dimensions are considered while defining wine tourism. Firstly, from destination perspective wine tourism is a structured plan to create, enhance and advertise wine-related attractions, secondly, from a consumer's perspective it is their motivational factors and perception toward wine leading to desire a wine destination. Thirdly, from wine manufacturer's perception, it is a chance to spread awareness about wine and sell their wine directly to consumers. The wine tourism educates the tourists about the farming, brewing, and storage procedures along with the etiquettes and courtesies that go along with wine consumption. Wine tourism can be described as a marketing strategy for wine producers to increase awareness, build brand name, and direct selling of wine to consumers (Getz, 2000).

Among the different forms of tourism and touristic product, wine tourism or eno-tourism is a new kind of tourism product that is challenged by multiple serious growth and advancement issues, which have capability to negatively impact its sustainability and profits in long-run (O'Neill and Charters, 2000). A main problem is perception of wine manufacturers toward wine tourism. The tourists of wine tourism are greatly influenced by wine producers and wineries but some wine producers do not want to move in the direction of developing wine tourism (Getz and Brown, 2006a). The key focus of wine producers is on grape/fruit farming and fruit wine production whereas they consider wine tourism as less important. According to them, wine tourism is a kind of diversification from agricultural production to the value-added services and wine-related experiences (Macionis, 1998). The major challenge is that most wine producer's perception toward wine tourism as a secondary activity, however, wine tourists or cellar door sales are economic essential for small wineries. The key focus lies on wine producing and the final product, with less knowledge, awareness, and understanding of marketing, tourism and tourism needs, distribution channels and various service standards (Beames, 2003). For several years, however, wine and tourism are well-related to one another but the vital role of wine tourism as a basic reason of returns is recently, realized by industry experts and researchers (Dodd and Kolyesnikova, 2005).

19.2.2 PROS AND CONS OF WINE TOURISM

From the wine producer's point of view, Wine tourism has many pros and cons. The wine sales from visits to cellar doors plays a vital part in wine production. Cellar doors are buildings where underground storage of wines takes place along with tasting rooms and sale of wines. The cellar doors, nowadays, exists to have wine tourists an experience of wine manufacturing in ancient days when cool room technology was not developed and therefore underground storage of wines was performed. Due to minimum distribution costs in cellar door wine sales, allowing higher profit margins makes wines sales at the cellar door attractive for wine manufacturers (Charters and O'Neill, 2001). According to Fraser and Alonso (2006), wine tourism also increases the contact and communication between wine producers and wine consumers. This leads to the education of tourists about wine and wine producing regions, and have a wine experience during wine tasting sessions. It includes local cultures and wine escapes. The wine manufacturers tried to increase the short-term volume of wine sales, create brand loyalty and product loyalty in long-term by educating the tourists through variety of wine tourism activities like theme-based wine festivals, cellar door sales and tasting sessions, cellar visits, and vineyard tours (Hall and Mitchell, 2000). Aside from the financial increase created by cellar door sales, building brand names enhancing the awareness for product and thereby facilitating the repurchase of same brand variety is an implied justification offered for these operations (Barry, 2008). For the wineries that often do not have sufficient production of wines, in such cases wine tourism creates prospects for improved profits and brand loyalty by increasing brand awareness (Dodd, 1995). It also helps wine producers to successfully brand (brand is a name or logo or some other attribute which helps the consumer to differentiate one company's product with another company) the wine product and winery (Hall et al., 2000). Therefore, small wine producers are mostly dependent on wine sale at their location through cellar doors. The wine sales directly to wine tourists removes the transportation and retail costs therefore, leading to more profit margins (Fraser and Alonso, 2006).

Conclusively, wine tourism is important for all types of wineries-large or small as it can: improve volume of wine sales; educate and increase awareness among visitors about brand leading to brand loyalty; attract new market segments, for example, niche markets for wine product; higher profit margins from wine sales because of reduced distribution and retail cost due to direct selling of wine to tourists; improved contacts with wine business; create new and different types of networks and groups (e.g., with tourism industry) and taste new variety of wines (Dodd, 1995; Getz, 2000).

Despite many advantages, there are drawbacks associated with winery involvement in tourism—capital required for wine production, increased costs and time management, and incapability to considerably improve sales volume (Dodd, 1995). However, cellar door wine sales improved the sales volume, still wine tourism is perceived as distraction from wine business, by some wine manufacturers and therefore do not consider themselves as part of tourism industry (Beames, 2003). Many vineyards are important entities among local groups. Although, they don't consider wine visitors as an essential part of wine business, still they are included in the wine business to enhance regional wine tourism (Beames, 2003).

19.3 RESEARCH QUESTIONS

RQ 1. What are the perceptions of Fruit wine makers toward wine tourism during and post Covid-19?
RQ 2. What are the advantages and disadvantages of wine tourism?
RQ 3. What are the different components of wine tourism?

19.4 OBJECTIVES OF RESEARCH

- The study is done with the following objectives:
- To study the status of wine tourism during Covid-19
- To determine the perceptions of Fruit wine makers toward wine tourism during and post Covid-19 in Meghalaya.

19.5 RESEARCH METHODOLOGY

The current research was conducted at Shillong, Meghalaya's state capital. An exploratory research approach was used in this article, which was based on qualitative research approaches. In this regard, there were 25 participants belonging to wine manufacturer's group located at and around Shillong. The data is collected through phone interviews and questionnaire sent by email. The survey questionnaire consisted of 20 close-ended questions to obtain the information about perception of wine producers toward wine tourism. The parameters selected for determining attributes were benefits, expenses, and business prospects, during and post-Covid 19. The importance of each factor is assessed on a Likert scale. Likert scale is a rating scale widely used to represent people's perception toward a topic. It is a five-point rating scale varying from 1 representing strongly disagree to 5 representing strongly agree. Secondary data was gathered from reputable tourist research publications, books, and other sources, annual reports of Ministry of Tourism, national and local newspapers, Meghalaya Tourism, Government of India, and some useful websites.

19.6 RESULTS AND DISCUSSION

On the basis of the review of literature, it was figured out that wine tourism involves high costs (including both fixed and variable costs) which is a major hindrance to wine producer's participation in wine tourism. This factor outweighs the advantages associated with wine tourism. The costs incurred in wine tourism are: training, skill development and awareness of staff, arrangement of wine festivals, harvesting workshops, vineyard nature walks, and tasting sessions. The wine tourism cost is a major obstacle for small-size wineries whereas the effect of costs reduces as size of wineries increases. This modifies the perception of winery visits with the variation in size of winery. Table 19.4 states the responses of wine producer's perception of wine tourism in terms of cost. It can be visualized from Table 19.4, there is an additional cost of educating the staff about wine tours which is agreed by wine manufacturers. The fixed costs involves cellar equipment, refrigeration system, fermentation and storage, tasting rooms and variable costs includes packaging costs, bottling, labor cost, arranging tours/trips to vineyards and arranging wine festivals, tasting sessions, marketing. Investments needed for fruit orchards/wineries are also seen as a cost for wine producers. The importance of new technologies of marketing and advertisement of wines e.g., website, web pages are well-known in Meghalaya. This is done to improve the engagement with wine consumers, promote gastro-tourism, wine trails, wine festivals, and experience of wine tourism. Therefore, wine producers consider it as an expensive area. The answers of all the wine producers directs that "need of latest production technologies," "quality control and benchmarking," and "need new sales store" are not perceived as an expense for wine producers.

This is significant that advantages of wine tourism are realized by wine producers. This awareness of potential benefits from wine tourism increases the hopes of wine manufacturers from wine tourism and thereby increases the involvement of wine producers in wine tourism. We can visualize from Table 19.5, replies of all wine producers displays that "Showcasing indigenous wine making practices" helping local farmers", "enhancing sales", "brand loyalty" and "brand awareness" are major advantages of enotourism/wine tourism with respect to wine producers. Wine tourism is viewed as an important marketing strategy to advertise and promote the brand name thereby increasing brad loyalty in wine industry. The results indicate that Livelihood/ entrepreneurship development can decrease the rate of migration from rural settings to the urban areas.

TABLE 19.4 Cost Dimension of Wine Producers' Perceptions of Wine Tourism.

S. No	Statements	Total	
		Mean	**S D**
1	Need to increase volume of production to cover up the Production Cost	3.34	0.644
2	Need for new investment to Fruit Orchards/Wineries	3.80	1.291
3	Need for Information Technologies (Internet/Social Media)	3.56	1.032
4	Necessity to provide Skill Development and tourism awareness training to sales employee	3.52	1.323
5	Need of wholesale/retail stores	3.48	1.196
6	Need to introduce latest Production Technologies	2.96	1.556
7	Need to invest in Quality Control Assurance/Benchmarking	2.68	1.180

Source: Field Survey.

TABLE 19.5 Perception of Wine Producers Towards Benefits of Wine Tourism.

S. No	Statements	Total	
		Mean	**S D**
1	Showcasing Indigenous Wine making practices & Consumption Culture	3.15	1.324
2	Helping Farmers and encouraging them to grow more Fruits/ Vegetables	3.60	1.120
3	Brand awareness	3.64	1.036
4	Building Brand loyalty	3.92	1.288
5	Enhancing sales	3.94	0.800
6	Opportunity to collect data about customers, Feedbacks	3.10	1.234
7	Livelihood/ Entrepreneurship development	3.09	1.342
8	Decrease the rate of migration to Urban Areas	3.80	1.208

Source: Field Survey.

The third aspect to be considered is opportunities from wine tourism for wine producers. With the awareness and information about the opportunities provided by wine tourism can give the wine manufacturers a competitive advantage. Table 19.6 gives the information about the replies of wine producers about their perception toward opportunities and participation in wine tourism. Wine producers agree that wine tourism is a major attribute, which contributes in the development of wine industry/Cottage Industry, promote sustainable livelihood, and attracting the attention of State and Central government. The analysis of answers of wine producers shows that whole groups of different categories of wine producers thinks same on the "contribution to development of wine industry," "Support for development of wine producers," and "ensure competitive advantage for wine producers."

TABLE 19.6 Perception of Wine Producers Toward Opportunity from Wine Tourism.

S. no	Statements	Total	
		Mean	**S D**
1	Wine tourism contributesto the advancement of wine industry/ Cottage Industry	3.20	1.001
2	Wine tourism support the growth and progress of wine producers	4.40	0.577
3	Network with other forms of tourism	3.92	0.989
4	Wine tourism is a major source of attention from Meghalaya Government and Central Government directed at wine industry	3.74	1.175
5	Promote sustainable livelihood	3.16	1.104

Source: Field Survey

19.7 SUMMARY AND CONCLUSIONS

Wine tourism, also called enotourism, is a new type of tourism and is termed under Special Interest Tourism (SIT), which caters to the special wine-related interests of groups. The industrialists have shown interest in wine tourism and academicians have started studying about it from past few decades. Wine tourism is a type of food tourism with the objective of educating, increasing awareness, and unique experience of the environment of wine production and wine tasting. Wine tourism acts as a catalyst to local economy by job creation possibilities, increase in tourism offers, and small towns in interior gains new businesses. When tourists visit the vineyards and wineries during wine tours, the wine producers get an opportunity to launch a new brand, improve sales through direct selling, create brand loyalty that makes customers buy the same brand of wine again and building customer relationship. The key factor in determining the quality of wine tour experience is wine producers. Most of wine producers concentrates on wine production only and some remaining on tourism due to which they consider wine tours as additional costs. We studied, Fruit Wine producers' perceptions of Wine Tourism in Meghalaya during and Post Covid 19 has been determined. Due to Covid-19 the whole country of India was under lockdown which halted all the tourism activities. The fruit wines in Meghalaya are produced using indigenous exotic fruits and traditional old practices. The passionate fruit wine makers of Meghalaya with rich agro-biodiversity and abundance of fruits like *grape fruit, Mandarin orange, litchi, mulberry, strawberry, banana, pineapple, Jackfruit, Pears, Plums, Sohiong, and Soh Phoh* has got an opportunity to gain benefits from wine business during the ongoing Covid-19 pandemic because the Meghalaya State had legalized the production of fruit wine. The amateur fruit winemakers have the capability of developing Meghalaya into a fruit wine capital of India.

During the ongoing Coronavirus Pandemic, after a long time, Meghalaya government has taken steps to wine industry by introducing the Manufacture and Sale of Home-made fruit wines rules, 2020 that provided the licenses and making it legal to manufacture and sell homemade fruit wines. This was done to enhance the wine industry in domestic as well as international markets. It has enabled the individuals to sell their homemade wines into the market. The regulations and guidelines are available on the Meghalaya State Portal website, and is open for suggestions. This was a much-awaited step from the government, which took place after 17 years of long wait. The reason was that Meghalaya was still controlled by the Assam Excise Act of 1945 of preindependence era and pre-Constitution era, which completely restricted the production of wine or any alcohol. After a long fight of over 17 years, the Meghalaya government has finally decided to legalize the production and sale of fruit wines.

The present study has concluded that numerous wine producers of Meghalaya are having positive perception for wine tourism. They accept that there exists several benefits and potential prospects of wine tourism for wine producers which facilitate toward participation of them in wine tourism. Their response is that networking with wine tourism businesses and educational and awareness activities about wine tourism will be incorporated in their corporation. Many wine producers believe that participation in wine tourism can significantly help them in launching new brand, brand loyalty, and advertising and promotional activities. Wine producers considers wine tourism as beneficial for growth and advancement of the wine industry and also it can be a good source of attention from many governmental and private organizations.

KEYWORDS

- **wine tourism**
- **wine Producer**
- **tourism**
- **Covid-19**
- **pandemics**
- **India**

REFERENCES

Beames, G. The Rock, the Reef and the Grape: The Challenges of Developing Wine Tourism in Regional Australia. *J. Vacation Market.* **2003,** *9*, 205–212.

Brown, G.; Getz, D. Linking Wine Preferences to the Choice of Wine Tourism Destinations. *J. Travel Res.* **2005,** *43*, 266–276.

Brunori, G.; Rossi, A. Synergy and Coherence Through Collective Action: Some Insights from Wine Routes in Tuscany. *Sociologia Ruralis* **2000,** *40* (4), 409–423.

Charters, S.; O'Neill, M. Service Quality at the Cellar Door: A Comparison Between Regions. *Int. J. Wine Market.* **2001,** *13* (3), 7–17.

Dodd, T. H. Opportunities and Pitfalls of Tourism in a Developing Wine Industry. *Int. J. Wine Market.* **1995,** *7* (1), 5–12.

Dodd, T. H.; Kolyesnikova, N. Visitors to Texas Wineries and the Role of Gratitude in Wine Purchases. Texas Wine Marketing Research Institute. http://www.depts.ttu.edu/hs/TexasWine/docs/2005_VISITORS_TO_TEXAS_WINERIES_REPORT.pdf

FAO. Traditional Fermented Food and Beverage for Improved Livelihoods, 2012.

FAOSTAT. 2012. http://faostat.fao.org/site/339/default.aspx,

Fellows, P. *Food Processing Technology: Principles and Practice*, 2nd ed.; CRC Press LLC: Baca Raton: 2000.

Fraser, R.; Alonso, A. Do Tourism and Wine Always Fit Together? A Consideration of Business Motivations. In *Global Wine Tourism*; Carlsen, J., Charters, S., Eds.; CABI, 2006; pp. 19–26.

George, B. P. Wine Tourist Motivation and the Perceived Importance of Servicescape: A Study Conducted in Goa, India. *Tour. Rev.* **2006,** *61* (3), 15–19.

Getz, D. *Explore Wine Tourism: Management, Development & Destinations*; Cognizat Corporation, 2000.

Getz, D.; Brown, G. Benchmarking Wine Tourism Development: The Case of the Okanagan Valley, British Columbia, Canada. *Int. J. Wine Market.* 2006a, *18* (2), 78–97.

Getz, D.; Brown, G. Critical Success Factors for Wine Tourism Regions: A Demand Analysis. *Tour. Manage.* 2006b, 27, 146–158.

Gössling, S.; Scott, D.; Hall, C. Pandemics, Tourism and Global Change: A Rapid Assessment of COVID-19. *Journal of Sustainable Tourism* **2020,** 1–20. DOI: 10.1080/09669582.2020.1758708 10.

Gretzel, U.; Fuchs, M.; Baggio, R.; Hoepken, W.; Law, R.; Neidhardt, J. et al. e-Tourism Beyond COVID-19: A Call for Transformative Research. *Inf. Technol. Tour.* **2020,** 22 (2), 187–203. DOI: 10.1007/s40558–020–00181–3.

Hall, C. M. Wine Tourism in New Zealand. In *Proceedings of Tourism Down Under II: Towards a More Sustainable Tourism*; Kearsley, G., Ed. Centre for Tourism, University of Otago: Dunedin: Dunedin, 1996; pp 109–119.

Hall, C.; Scott, D.; Gössling, S. Pandemics, Transformations and Tourism: Be Careful What You Wish For. *Tour. Geog.* **2020,** 1–22. DOI: 10.1080/14616688.2020.1759131

Hall, C. M.; Johnson, G.; Cambourne, B. N.; Macionis, N.; Mitchell, R.; Sharples, L. Wine Tourism: an Introduction. In *Wine Tourism Around the World*; Hall, C. M., Sharples, E., Cambourne, B., Macionis, N., Eds.; Butterworth Heinemann: Oxford, 2000; pp 1–23.

Hall, M. C.; Mitchell, R. Wine Tourism in the Mediterranean: A Tool for Restructuring and Development. *Thunderbird Int. Busi. Rev.* **2000,** *42* (4), 445–465.

Joshi, V. K.; Devendra, A. Panorama of Research and Development of Wines in India. *J. Sci. Ind. Res.* **2005,** *64*, 9–18.

Macionis, N. Wineries and Tourism: Perfect Partners or Dangerous Liaisons. Wine Tourism—Perfect Partners. Proceedings of the First Australian Wine Tourism Conference, Margaret River, Western Australia, Bureau of Tourism Research, 1998; pp 35–50.

McKinsey & Company. Beyond Coronavirus: The Path to the Next Normal, 2020. https://www.mckinsey.com/_/media/McKinsey/Industries/Healthcare%20Systems%20and%20Service/Our%20Insights/Beyond%20coronavirus%20The%20path%20to%20the%20next%20normal/Beyon dcoronavirus-The-path-to-the-next-normal.ashx (accessed July 11, 2020).

Mohanty et al. *SA J. Tour. Herit.* July **2015,** *8* (2).

O'Neill, M.; Charters, S. Service Quality at the Cellar Door: Implications for Western Australia's Developing Wine Tourism Industry. *Manag. Serv. Qual.* **2000,** *10* (2), 112–122.

Romagosa, F. The COVID-19 Crisis: Opportunities for Sustainable and Proximity Tourism. *Tour. Geog.* **2020,** 1–5. DOI: 10.1080/14616688.2020.1763447

Rose, A. H. Alcoholic Beverage. In *Economic Microbiology I*; Rose, A. H., Ed.; Academic Press, UK, 1997.

Sevil, G.; Yüncü, H. R. İçki Tütün ve Tütün Ürünleri Sanayii, Özel İhtisas Komisyonu Raporu 2007–2013. *Tour. Rev.* *57* (4), 477–487. http://ekutup.dpt.gov.tr/imalatsa/tutun/oik677.pdf

Siagalla, M. Tourism and COVID-19: Impacts and Implications for Advancing and Resetting Industry and Research. *J. Busi Res.* **2020,** *117*, 312–321. DOI: 10.1016/j.jbusres.2020.06.015

WEB REFERENCES

https://www.deccanherald.com/national/east-and-northeast/kiwi-pears-and-plum-wine-bring-new-hope-for-covid-19-battered-tourism-and-farmers-in-arunachal-pradesh-907918.html (accessed on Jan 26, 2021).

https://www.deccanherald.com/national/east-and-northeast/kiwi-pears-and-plum-wine-bring-new-hope-for-covid-19-battered-tourism-and-farmers-in-arunachal-pradesh-907918.html (accessed on Jan 26, 2021).

https://indianexpress.com/article/north-east-india/meghalaya/how-the-pandemic-spells-hope-for-fruit-wine-makers-of-meghalaya-6419998/ (accessed on Jan 26, 2021).

https://www.telegraphindia.com/north-east/coronavirus-lockdown-licence-to-a-homemade-heady-brew-in-shillong/cid/1774335 (accessed on Jan 26, 2021).

https://www.indianwineacademy.com/item_3_859.aspx (accessed on Jan 26, 2021).

https://www.nrclitchi.org/uploads/research-papers/R-K-Patel/37.pdf (accessed on Feb 9, 2021).

https://en.wikipedia.org/wiki/Timeline_of_the_COVID-19_pandemic_in_India (accessed on Feb 12, 2021).

CHAPTER 20

RECONNECTING INDIAN TOURISM AMID COVID-19: THE ROLE OF GOVERNMENT FOR OUTBREAK CONTROL

DIVYA TALREJA[1], KULDEEP SINGH[2], and RAJESHRAMASAMY[3]

[1]*School of Studies, Tourism and Travel Management (SOSTTM), Jiwaji University, Gwalior, Madhya Pradesh, India*

[2]*School of Hospitality & Tourism, GD Goenka University, Gurugram, India*

[3]*Department of Tourism & Hospitality Management, Mizoram Central University, Aizawl, India*

ABSTRACT

In recent times, a novel virus named COVID-19 has affected the tourism sector severely. India's major revenue comes from the tourism sector, but due to this pandemic in 2020, there can be a devastating effect on tourism. In such a crisis, the role of the government is very significant during the recovery stage. Therefore, this study explores the role of the government for tourism recovery post-COVID-19 and to see how the outcome of tourism recovery is associated with the public evaluation of the measures taken by the Indian Government against the pandemic. Swift attention is needed for this event and the future of travel and tourism. So, this chapter also investigates whether the government's efficacy will increase if the government performs well in dealing with the pandemic.

Post-COVID Tourism and Hospitality Dynamics: Recovery, Revival, and Re-Start.
Umendra Narayan Shukla & Sharad Kumar Kulshreshtha (Eds.)

The sample size was 250 respondents, out of which 30 responses were missing incomplete and 220 were found fit for this chapter. To accumulate the responses, convenient sampling was adopted. The present COVID-19 circumstances in India were not favorable to opt for an offline survey and other sampling methods such as random probability. Therefore, data collection was conducted online during the pandemic period when most of the citizens of the country were facing lockdown. The results of this chapter were analyzed by SPSS 22.0. Percentage and frequency were calculated to analyze the responses. Further, the Pearson correlation test and multiple regressions were performed for data analysis. The negative emotion had the highest influence on the anticipated recovery period, whereas government performance (in dealing with the pandemic) did not influence the anticipated recovery period. Therefore, the role of the government as a stakeholder of tourism is very important for tourism recovery. At last, this chapter is concluded with some strategies and recommendations with future implications for the government.

There are some limitations in this chapter. First, data collection was finished within 2 weeks; as a result, the size of the sample was not large. Second, the nonprobabilistic method which is convenience sampling was applied and it becomes a challenge for the researcher to represent a sample. Third, this chapter was only restricted to India and is a specific focus on Indian respondents. There was also a time limitation in the research as impacts and level of virus outbreak differ from time to time. As people react to hazards differently, so generalization of implications should be cautious (Aliperti and Cruz, 2019). As the coronavirus pandemic is unfolding, a general view about the role of government is being submitted.

20.1 INTRODUCTION

Coronavirus is also known as COVID-19 which was termed by the WHO (World Health Organization). It already created very serious health problems all over the world. It started from Wuhan city in China at the end of 2019. Further, it spread rapidly all over the world and more than 7 lakhs people died due to COVID-19. According to the WHO, the COVID-19 was declared as a "public health emergency of international concern" (WHO, 2020). Numerous studies were conducted to explicate the coronavirus attributes which cause death rate, respiratory problem, and diffusion (Perlman, 2020). The Indian government initiated various strategies to control the spread of coronavirus by adopting lockdown, compulsory facemasks, encouraging the concept of work from home, social distancing, regular washing hands, and so on. The USA was the most affected country with more than 5.02 million followed by Brazil 2.97 million and India 2.09 million confirmed cases. At present, the total confirmed cases all over the world are more than 19 million; out of these data, more than 11 million recovered and 720,000 died. India was completely under lockdown and people were in their cities and states, which had a direct impact on the tourism businesses.

UNWTO (2020) had insisted with a strong statement "stay home today, travel tomorrow" to control the spread of COVID-19. It declines the movement of tourist arrivals by 57% in March 2020 now at present more than 95% were completed shutdown. It results reduced the travel movement of 67 million international arrivals and cut down 80 billion USD from international tourism (Arokiaraj, 2011). The tourism industry declines from 58% and now jumped to 78% (UNWTO, 2020). Indians have already lasted more than 40 million jobs (FICCI, 2020). The role of stakeholders is playing a vital position in determining an optimistic outlook of tourism and allied section revival in the future. Therefore, it is essential to understand the factors to determine the anticipated recovery period, the role of government activities, and self-efficacy (Lawrence et al., 2020). This study explores tourism recovery in relationship with the public assessment of the measures taken by the Indian government against the pandemic. This chapter also investigates whether the government's efficacy will increase if the government performs well in dealing with the pandemic.

20.1.1 COVID-19 IMPACT ON INDIAN TOURISM

Due to this pandemic, the tourism industry is the most affected when compared to other sectors. At present, the service providers like travel agents, hoteliers, and tourist guides will get affected. Besides, the Meetings, Incentives, Conventions, Exhibitions (MICE) sector will also face a crisis if things do not get normal shortly (Kumar, 2020). Subsequently, the tourism sector such as domestic, inbound, and outbound has a negative impact due to this pandemic. Tourism and the airline industry have forced the employees to leave their jobs (David et al., 2020). For a developing country like India, it becomes very difficult to deal with the government and other stakeholders. Therefore, proper strategies and planning are needed from the government for the betterment of the tourism sector.

New business models have a frame to support the tourism sector because of COVID-19 and also changed the customer's behavior and their choices

(Thames et al., 2020). To reduce the risk of COVID-19 dispersion and to increase international travel before a successful trial of the vaccine, methods like quarantine, artificial intelligence, and data mining are important. In the future, travel agents will focus more on special interest tourism, like nature-based tourism, responsible tourism, and ecotourism (Ozdemir and Yildiz, 2020). Also, Strielkowski (2020) stressed on the web application that demonstrates the patient's condition according to color codes. In the Indian context, more focus should be on domestic tourism to revive the tourism industry and attract tourists postpandemic period (Kumar, 2020). New technology can be updated in travel systems like sensor devices, CCTV cameras, thermal cameras, automatic functioning of doors, and uses of electronic cards (Olliver and Gupta, 2020; Ponkshe, 2020).

20.1.2 PRIMARY STAKEHOLDERS IN TOURISM

It is noted that there is a multifaceted network of interrelated stakeholders. It has a wide range of consumers to service providers, government, workers, shareholders, and the local population. In the tourism sector also, many stakeholders played a vital role like the government, private sector, tourist, local people, and NGOs. It is a complex task for the stakeholders as tourism is dynamic, so it is not easy for them to interact with each other (Stainton, 2020). The tourism sector is multifaceted and various stakeholders are interrelated as shown in Figure 20.1.

FIGURE 20.1 Visual depiction of the key stakeholders in the tourism industry.

Source: Reprinted with permission from Stainton (2020).

20.2 THEORETICAL REFLECTIONS

A review of the literature was conducted about the COVID-19 impacts on the tourism industry (Pearson and Milroff, 1993; Coombs, 1999; Henderson, 2007). Assessment of a preview of recent developments brings up a couple of questions and brings to the forefront the requirement for the vital activity to get ready for an eventual fate of strengthening dangers to human well-being and social prosperity in a world of increased travel (Jamal and Budke, 2020). Considering 2019, the travel industry information distributed by the World Tourism Organization (UNWTO), it is seen that the number of overall tourists expanded by 4% contrasted with the previous year and arrived at 1.5 billion (Meltem and Sercan, 2020). Tourism revenue was also nearly $2.9 trillion in statistics. Also, the travel industry is one of the most significant factors that increases the income of countries in the world (UNWTO, 2020a). Though it is an essential industry for the market, it features threats like the COVID-19 pandemic today (Meltem and Sercan, 2020). It is to be noted that tourism has conveyed enormously favorable conditions for this global pandemic (UNWTO, 2020b). In this unique circumstance, the travel sector is required as a trade generally influenced by COVID-19 within the outline of quarantine measures, from hospitality management to aircraft transportation (Meltem and Sercan, 2020). The reports distributed by the World Tourism Organization uphold this induction. Indeed, 20–30% misfortune would potentially acknowledge in worldwide tourists' insights for 2020. Hence, this loss will directly affect $300–400 billion regarding worldwide travel industry incomes (UNWTO, 2020c). OECD (2020) has stated that immediate measures taken to control the pandemic are of prime concern for the government which has negative consequences on the economy in the short run, but if the government performs well in controlling the pandemic, it will ultimately build public confidence in the long run as this was experienced during the crisis of 2008–2009 (David et al., 2019; Latha et al., 2020). It is to be noted that residents shape an optimistic viewpoint if the government executes well in managing the pandemic (Lawrence et al., 2020), and performance is not the only criterion that is used by the citizens to evaluate the government (Van de Walle and Bouckaert, 2003). As a result, this study explores whether the Indian government's efficacy will increase if the government performs well in dealing with the pandemic.

The pandemic uncovers the nature of pioneers and individuals a country has. Governments that are fruitful at controlling the spread of SARS-CoV-2 and limiting fatalities of COVID-19 will appreciate significantly increased public trust and social cohesion, while those that conceded incredible mediations to control its spread will see more prominent distrust and social stress, ensuing in the paralyzation of the public trust (Marlon, 2020). The performance of some individual government agencies builds a general attitude toward the government (other individual government agencies) and this says a lot about microperformance theory (Van de Walle and Bouckaert, 2003). This microperformance theory further extends by adding "Self-Efficacy" as a mediator (Lawrence et al., 2020).

"Self-efficacy is an overall idea alluding to the degree to which individuals accept they have the capability to manage with tasks or stressors" (Bandura, 2006). Self-efficacy influences how individuals feel, think, and act concerning risk-taking behaviors (Wong and Yang, 2020). It is an individual conviction on one's capacity to impact the events in his/her life, which is a psychological construct that drives the arrangement of positive anticipations about social/environmental events (Gavrilov-Jerkovic et al., 2014). An explicit self-efficacy relies upon the government's efforts in managing the pandemic (Lau et al., 2010) and tends to be the mediating factor (Lawrence et al., 2020). High government efficacy encourages trust in the public arena (Catterberg and Moreno, 2006), whereas social trust nurtures an optimistic view of the world (Uslaner, 2010). Individuals who doubt the adequacy of government performance and perceived high susceptibility are bound to feel stressed (Lau et al., 2010). Government viability in the tourism recuperation may prompt the expectation of a fast recuperation (Lawrence et al., 2020).

Emotional concern emerging from a pandemic can too be reduced through the performance of the government (Lau et al., 2010). Tourists, within the sight of an excessive amount of apparent physical risk, will, in general, abstain from traveling (Mair et al., 2016; Ravi et al., 2018; Sudhakar et al., 2017). Likewise, in the nearness of risk, the choice to adopt particular conduct is also connected to personality uniqueness of the visitors (Riad et al., 1999) and the point of reliability that they get through different modes of information (Spencer et al., 1992). To encourage the process of risk information, various crowds may entail diverse correspondence advances since they may have distinctive threat observations (Yosie and Herbst, 1998). Being presented with high personal risk conditions, people become overwhelmed by negative emotions (Aliperti and Cruz, 2019). General well-being mindfulness/instruction ought to, in this manner, give clear messages to rectify relevant perceptions. Cognitions including unproven convictions about methods of transmission, susceptibility, and perceived seriousness were related to some of the evasion behavior and emotional distress variables (Lau et al., 2010). Salem et al. (2021) in their study explored the outcome of government hold of hotels on the hotel's staff hotels (i.e., staff training, staff retention, and health support). Further, Mendolia et al. (2021) entailed that government-forced guidelines report for the majority of the cutback in the mobility observed throughout the pandemic period worldwide.

20.2.1 LESSONS FROM PAST PANDEMICS

In the early 2000s, a virus named severe acute respiratory syndrome (SARS) broke out in Guangdong Region, Southern China. This virus had also similar syndromes like COVID-19 and transmitted from human to human. Later, the WHO in 2003 announced tourist attractions like Hong Kong and Guangzhou as containment zones and subsequently declared several regions of China as endemic zones (Jamal and Budke, 2020).

Similarly, a different coronavirus middle-eastern respiratory syndrome (MERS) also spread in other regions of the world. It is widely assumed that all these viruses cause an animal-to-human transmission (Jamal and Budke, 2020). As compared to COVID-19, the MERS virus is less dangerous and spread at a slow rate from human to human, though the risk of death rate is high for the affected person.

According to Mair et al. (2016), many scholars from tourism learned from (SARS) and argued for proper planning and strategic management in such planning. Cooper (2005) investigated that previous pandemic viruses like SARS and Ebola turned out to be a source of misinformation and propaganda in the affected nations where these outbreaks broke out. Further, Novelli et al. (2018) in their study have taken examples from previous pandemics like HIV/AIDS, Ebola, and the Zika virus and evaluated that situations like climate change, social reaction, fear, and racism are also the major factors of disease outbreaks. According to McKercher and Chon (2004) and So¨nmez et al. (2019), international bodies like the WTTC and UNWTO should provide health and medical facilities to the countries that are suffering from these viruses and help the tourism sector.

20.3 EXPERIMENT METHODS AND MATERIALS

It is an exploratory study based on current issues related to COVID-19 in the Indian context. The researchers use the exploratory study as there were very few studies that tell the role of government and other stakeholders during COVID-19 (Arokiaraj, 2015). Data collection was conducted during the pandemic period when most of the citizens of the country facing lockdown. Therefore, an online survey was conducted, in which 220 total respondents from India were collected within 2 weeks (when the COVID-19 outbreak began to increase in India). The survey was conducted in the month of the last week of June 2020 and completed in the second week of July 2020.

Convenient sampling was applied in this study. The system recorded a total of 250 responses, out of which 30 responses were missing and incomplete. Therefore, a total of 220 responses were analyzed in the research, which was performed using SPSS 20. The questionnaire consists of seven demographic items and three open-ended questions related to the anticipated recovery period and the role of NGOs helping the tourism recovery. The multi-item measures of government performance, self-efficacy, and negative emotion were adapted from Fong et al.'s (2020) study on COVID-19 in China. Further, the 5-point Likert scale was applied to the variable "Government Performance," covering 1 = strongly disagree, 2 = disagree, 3 = neutral, 4 = agree, and 5 = strongly agree. A total of 13 items were taken in this variable. Later, three items of the variable self-efficacy were taken, covering 1 = very unconfident, 2 = not very confident, 3 = neutral, 4 = fairly confident, and 5 = very confident. Further, five items of the variable negative emotion were taken, covering 1 = strongly disagree, 2 = disagree, 3 = neutral, 4 = agree, and 5 = strongly agree. Lastly, the single-item measure of perceived government efficacy was also adapted from Fong et al. (2020).

20.4 RESULT AND DISCUSSIONS

Descriptive statistics were performed using percentage analysis on demographic variables. In addition, percentage analysis was also performed on the variables "anticipated recovery" and "role of NGOs helping the tourism recovery." Further, the Pearson correlation test was performed to find out the relationship between a dependent variable and independent variables. At last, multiple regressions were conducted to find out the influences of independent variables on the dependent variable.

The demographic profile of the respondents is given in Table 20.1. It is found that 50% of respondents are female, and 47% of respondents are aged between 25 and 44 years old. Around 71% of the respondents were unmarried. Most of the respondents are holding postgraduate degrees in their hands (34%) and most of them are students (34%) and government employees (27%). The average income level of the respondents is Rs. 30,000–60,000 per month by 38% and around half of the respondents are living in the urban area.

TABLE 20.1 Demographic profile of the respondents.

Demographic profile	Number of respondents	Percentage of respondents
Gender		
Male	104	47.3
Female	111	50.5
Others	5	2.3
Marital status		
Unmarried	157	71.4
Married	52	23.6
Divorced	6	2.7
Widow	5	2.3
Occupation		
Unemployed	8	3.6
Private job	38	17.3
Government	60	27.3
Self-employed	36	16.4
Student/Scholar	75	34.1
Others	3	1.4
Income group		
Below 10,000	69	31.4
10,001–30,000	17	7.7
30,000–60,000	84	38.2
60,000 and above	50	22.7
Age of respondents		
18–24	63	28.6
25–44	103	46.8
45–64	32	14.5
65–79	13	5.9
80 and above	9	4.1
Educational qualification		
Uneducated	12	5.5
Below matriculation	10	4.5
Matriculation	8	3.6
10 + 2	10	4.5
Graduation	53	24.1
PG	74	33.6
Ph.D./M.Phil.	44	20.0
Others	9	4.1
Type of area		
Rural	8	3.6
Semi-urban	61	27.7
Urban	113	51.4
Metropolitan	38	17.3
Total	220	100

Source: Authors' original work.

TABLE 20.2 Anticipated recovery period.

When do you estimate that the local tourism will recover and return back to normal?	Number of respondents	Percentage
Sep–Dec (2020)	19	8.6
Jan–Apr	59	26.8
May–Aug	32	14.5
Sept–Dec	96	43.6
Period after	14	6.4
Total	220	100

Source: Authors' original work.

TABLE 20.3 NGO helping the government.

Have you ever come across any NGO helping the government and create awareness about COVID-19 among locals?	Number of respondents	Percentage
"Yes"	81	36.8
"No, I do not"	139	63.2
Total	220	100

Source: Authors' original work.

In Table 20.4, it was found that the majority of the respondents (44.1%) think that NGO plays a major role in tourism recovery. In contrast, it was found that only (16.8%) think that it does not play major role, whereas (39.1%) feel that they are not sure about NGO role in tourism recovery.

TABLE 20.4 Role of NGO in tourism recovery.

Do you think NGOs/voluntary associations/self-help groups as tourism stakeholders are helping in the tourism recovery?	Number of respondents	Percentage
"Yes"	97	44.1
"No, I do not"	37	16.8
Total	86	39.1

Source: Authors' original work.

The Pearson correlation test was performed to find out the relationship between dependent variable and independent variables. The anticipated recovery period is selected as a dependent variable and independent variables are perceived government efficacy, government performance, self-efficacy, and negative emotion.

TABLE 20.5 Pearson correlation test for the anticipated recovery period.

Pearson correlations		Anticipated recovery period	Perceived government efficacy	Government performance	Self-efficacy	Negative emotion
Anticipated recovery period	Pearson correlation	1				
	Sig. (two-tailed)					
Perceived government efficacy	Pearson correlation	0.196*	1			
	Sig. (two-tailed)	0.004				
Government performance	Pearson correlation	0.060	0.758*	1		
	Sig. (two-tailed)	0.378	0.000			
Self-efficacy	Pearson correlation	0.081	−0.096	0.017	1	
	Sig. (two-tailed)	0.231	0.157	0.802		
Negative emotion	Pearson correlation	−0.518*	−0.130	−0.019	0.174*	1
	Sig. (two-tailed)	0.000	0.055	0.784	0.010	

**Correlation is significant at the 0.01 level (two-tailed).*

Source: Authors' original work.

Based on the above result, it is found that the perceived government efficacy had a positive relationship with the anticipated recovery period ($r = 0.196$), whereas the negative emotion had a negative relationship with the anticipated recovery period ($r = 0.518$). Further, it is found that government performance and self-efficacy had no relationship with the anticipated recovery period.

TABLE 20.6 Multiple regression test for the anticipated recovery period.

Model summary[b]	R	R^2	Adjusted R^2	Std. error of the estimate	Change statistics		df1	df2	Durbin-Watson	
					R^2 change	F change			Sig. F change	
	0.574[a]	0.330	0.317	0.941	0.329	26.387	4	215	0.000	0.192

[a]*Predictors*: Constant, negative emotion, government performance, self-efficacy, perceived government efficacy.

[b]*Dependent variable*: Anticipated recovery period.

Source: Authors' original work.

TABLE 20.7 Coefficients of anticipated recovery period.

Coefficients[a]	Unstandardized coefficients	Standardized coefficients		T value	P value	Collinearity statistics	
	β	Std. error	β			Tolerance	VIF
Constant	4.192	0.580		7.227	0.000		
Perceived government efficacy	0.257	0.086	21	2.999	0.003	0.404	2.474
Government performance	−0.255	0.145	−0.153	−1.761	0.080	0.413	2.421
Self-efficacy	0.423	0.121	0.200	3.490	0.001	0.950	1.052
Negative emotion	−0.785	0.086	−0.522	−9.092	0.000	0.947	1.056

[a]*Dependent variable*: Anticipated recovery period.

Source: Authors' original work.

Anticipated recovery period = 4.192 (0.264* perceived government efficacy) + (0.20* self-efficacy) + (−0.522* negative emotion).

Multiple regressions were conducted to find out the influences of independent variables on the dependent variable. The perceived government efficacy, government performance, self-efficacy, and negative emotion are selected as independent variables. The anticipated recovery period is selected as a dependent variable.

Based on the above result, it is found that the perceived government efficacy (β=0.264), self-efficacy (β=0.20), and negative emotion (β=−0.522) influence the anticipated recovery period. Further, it is identified that the performance of the government did not influence the anticipated recovery period. So overall, all the independent variables have jointly supported the model by 33%.

20.5 SUGGESTIONS

i. The government should promote India as a safe destination for tourists and should equip all the tourist places with sanitizers and disinfectants. Wearing masks and using hand sanitizers should be made compulsory.
ii. To come up with legal guidelines for maintaining sanitization and hygiene along with regular audits by government officials to earn back tourist confidence.
iii. The administration should work on public–private partnership model.
iv. Strict restrictions on travel can surely help the community to survive during this pandemic.

v. The government should focus more on new and innovative ways for contactless tourism.
vi. Tourism infrastructure should be converted into hospitals as the present health sector is not sufficient enough to cater to the needs.
vii. Gradual opening, initially it could be the movement of people within the locality or places, followed by interstate and international tourism.
viii. Providing people leverage on taxation and duties.
ix. All stakeholders must be brought under one roof.
x. Promote tourism as a part of tax saving; people spending money on travel within India should be allowed to take exemption in ITR.
xi. COVID-19 checkpoints should be there along with the facility of thermal scanning.
xii. COVID-19 clearance certificate should be a must for tourists to see that they are fit to travel.
xiii. The government should develop destinations near urban or metropolitan areas.
xiv. A proper sustainable approach must be taken to revive the tourism sector.
xv. Government should allow a limited number of visitors to a specific attraction at one particular time.
xvi. The government should encourage people to travel by road.
xvii. Bring a change in tourism activities and norms, make people especially locals, and tourists understand this new normal.
xviii. The government should focus on rural tourism (village, one of the important tourism products) and prepare communities to be ready to host tourists.
xix. The Ministry of Tourism, Government of India, can identify hotspots and promote lesser known destinations through virtual tours.
xx. The government should monitor the health status of tourists by checking the temperature, checking COVID-19 reports.
xxi. Need to develop innovative solutions to track and monitor tourists' movement and activities through effective application of ICT as in this age of industry 4.0 (Artificial Intelligence, BlockChain, Big Data, etc.).

20.6 CONCLUSIONS

It was found that the performance of the government was the lowest and it did not influence the anticipated recovery period. Therefore, there should be an improvement in the government's role to control this pandemic. In contrast, it was found that negative emotions have the highest influence on the anticipated recovery period. This shows that people feel fear, afraid while traveling in COVID-19, and are not ready to plan to travel shortly. In addition, it was found that government performance and self-efficacy had no relationship with the anticipated recovery period.

Though there is no surety when the local tourism industry will recover and return to normal, it was found that most of the people estimated the period of September–December 2020 for the recovery of tourism. During the pandemic period, the government of affected areas must focus on anticipatory measures against pandemic and exercised fewer attempts on involvement that outlined the people's viewpoint of tourism recovery. People from the country shaped a constructive outlook on the performance of the government in controlling the pandemic.

There are some limitations to this chapter. First, data collection was finished within 2 weeks; as a result, the size of the sample was not large. Second, the nonprobabilistic method which is convenience sampling was applied and it becomes a challenge for the researcher to represent a sample. Third, this chapter was only restricted to India and is a specific focus on Indian respondents. There was also a time limitation in the research as impacts and level of virus outbreak differ from time to time. It is to be noted that there is a scope of future research taking samples from other countries by applying probability sampling. Fourth, due to single-item measures, reliability can be hazarded and must be avoided in further study. Lastly, this research did not integrate perceived pandemic issues, leaving scope for future study.

KEYWORDS

- **COVID-19**
- **tourism recovery**
- **stakeholders**
- **Indian government**
- **self-efficacy**

REFERENCES

Altınay Ozdemir, M.; Yıldız, S. How Covid-19 Outbreak Affects Tourists' Travel Intentions? A Case Study in Turkey. *Int. Soc. Mental. Res. Think. J.* (Issn:2630–631X) **2020,** *6* (32), 1101–1113.

Arokiaraj, D. The Green Market: The Way to Save the World. *Busi. Strat.* **2011,** 41–44.

Aliperti, G.; Cruz, A. Investigating Tourists' Risk Information Processing. *Ann. Tour. Res.* **2019,** 79. https://doi.org/10.1016/j.annals.2019.102803.

Arokiaraj, D. *A Study on Environmental Responsibility of the Stakeholders of Auto Industry in Chennai*; Doctoral dissertation, 2015.

Bandura, A. Guide for Constructing Self-Efficacy Scales. In *Self-Efficacy Beliefs of Adolescents*; Pajares, F., Urdan, T., Eds., Vol. 5; Information Age Publishing, 2006; pp 307–337.

Banumathi, M.; Arokiaraj, D. Eco-labeling–The Need for Sustainable Marketing. In *National Conference in the era of Global Recovery-2011 (SGEGR***2011,** 2011; pp 511–515.

Cooper, M. Japanese Tourism and the SARS Epidemic of 2003. *J. Travel Tour. Market.* **2006,** *19* (2–3), 117–131. DOI: 10.1300/J073v19n02_10

Coombs, W. T. *Ongoing Crisis Communication: Planning, Managing, and Responding*; London: Sage Publications, 1999.

Catterberg, G.; Moreno, A. The Individual Bases of Political Trust: Trends in New and Established Democracies. *Int. J. Public Opin. Res.* **2006,** *18* (1), 31–48.

David, A. *Corporate and Individual Environmental Responsibility Towards Automobile*; Book Rivers, 2020.

David, A.; Ahmed, R. R.; Ganeshkumar, C.; Sankar, J. G. Consumer Purchasing Process of Organic Food Product: An Empirical Analysis. *Qual.-Access Success* **2020,** *21* (177).

David, A.; Nagarjuna, K.; Mohammed, M.; Sundar, J. Determinant Factors of Environmental Responsibility for the Passenger Car Users. *Int. J. Innov. Technol. Explor. Eng. (IJITEE)* **2019**. ISSN, 2278–3075.

Fong, L. H. N.; Law, R.; Ye, B. H.; Outlook of Tourism Recovery Amid an Epidemic: Importance of Outbreak Control by the Government. *Ann. Tour. Res.* **2020**. https://doi.org/10.1016/j.annals.2020.102951

FICCI. Travel & Tourism: Survive, Revive and Thrive in Times of COVID-19, 2020. https://www.grantthornton.in/globalassets/1.-member-firms/india/assets/pdfs/travel-and-tourism-in-times-of-covid-19.pdf

Gavrilov-Jerkovic, V.; Jovanovic, V.; Zuljevic, D.; Brdaric, D. When Less Is More: A Short Version of the Personal Optimism Scale and the Self-Efficacy Optimism Scale. *J. Happiness Stud.* **2014,** *15*, 455–474.

Henderson, J. *Tourism Crises: Causes, Consequences, and Management*; Butterworth-Heinemann: Oxford, 2007.

Jamal, T.; Budke, C. Tourism in a World with Pandemics: Local-Global Responsibility and Action. *J. Tour. Futures* **2020**. https://doi.org/10.1108/JTF-02–2020–0014

Kumar, U. The Economic and Psychological Impact of SARS-Cov2 on Travel and Tourism Industry: With Special Reference to India May 2020.

Lofredo, M. P. Social Cohesion, Trust, and Government Action Against Pandemics. *Eubios J. Int. Asian Bioethics* **2020,** *30* (4), 182–189.

Latha, C. J.; Sankriti, R.; David, A.; Srivel, R. IoT Based Water Purification Process Using Ultrasonic Aquatic Sound Waves. *Test Eng. Manage* **2020**. ISSN, 0193–4120.

Lau, J. T. F.; Griffiths, S.; Choi, K. C.; Tsui, H. Y. Avoidance Behaviors and Negative Psychological Responses in the General Population in the Initial Stage of the H1N1 Pandemic in Hong Kong. *BMC Infect. Dis.* **2010,** *10* (1), 139. https://doi.org/10.1186/1471–2334–10–139

Lau, J. T. F.; Griffiths, S.; Au, D. W. H.; Choi, K. C. Changes in Knowledge, Perceptions, Preventive Behaviours and Psychological Responses in the Precommunity Outbreak Phase of the H1N1 Epidemic. *Epidemiol. Infect.* **2011,** *139*, 80–90.

Mair, J.; Ritchie, B. W.; Walters, G. Towards a Research Agenda for Post-Disaster and Post-Crisis Recovery Strategies for Tourist Destinations: A Narrative Review. *Curr. Iss. Tour.* **2016,** *19* (1), 1–26. DOI: 10.1080/13683500.2014.932758.

McKercher, B.; Chon, K. The Over-Reaction to SARS and the Collapse of Asian Tourism. *Ann. Tour. Res.* **2004,** *31* (3), 716–719.

Mendolia, S.; Stavrunova, O.; Yerokhin, O. Determinants of the Community Mobility During the COVID-19 Epidemic: The Role of Government Regulations and Information. *J. Econ. Behav. Org.* **2021,** *184*, 199–231.

Novelli, M.; Gussing Burgess, L.; Jones, A.; Ritchie, B. W. "'No ebola…still doomed'—The Ebola-Induced Tourism Crisis. *Ann. Tour. Res.* **2018,** *70*, 76–87.

OECD. Tourism Policy Responses to Coronavirus (COVID-19), 2020. https://read.oecd-ilibrary.org/view/?ref=124_124984–7uf8nm95se&title=Covid-19_Tourism_Policy_Responses

Olliver, G.; Gupta, N. Rethinking India's Public Transport After the COVID-19 Lockdown Is Over, May 26, 2020. https://blogs.worldbank.org/endpovertyinsouthasia/rethinking-indias-public-transport-after- ovid-19-lockdown-over (accessed on July 18, 2020).

Perlman, S. Another Decade, Another Coronavirus. *N. Engl J Med.* **2020,** *382*, 760–762. http://dx.doi.org/10.1056/ NEJMe2001126

Pearson, C. M.; Milroff, J. From Crisis Prone to Crisis Prepared: A Framework for Crisis Management. *Acad. Manage. Exec.* **1993,** *7* (1), 49–59.

Ponkshe A. Steps to Restart Public Transit After Lockdown Ends, May 7, 2020. https://www.orfonline.org/expert-speak/steps-restart-public-transit-lockdown-ends-65814/ (accessed on July 18, 2020).

Ravi, S.; David, A.; Imaduddin, M. Controlling & Calibrating Vehicle-Related Issues Using RFID Technology. *Int. J. Mech. Product. Eng. Res. Dev.* **2018,** *8* (2), 1125–1132.

Salem, I. E.; Elbaz, A. M.; Elkhwesky, Z.; Ghazi, K. M. The COVID-19 Pandemic: The Mitigating Role of Government and Hotel Support of Hotel Employees in Egypt. *Tour. Manage.* **2021,** *85*, 104305.

Stainton, H. Stakeholders in Tourism: Who Are They and Why Do They. **2020,** 1–20.

Strielkowski, W. International Tourism and COVID-19: Recovery Strategies for Tourism Organizations, 2020. https://doi.org/10.20944/PREPRINTS202003.0445.V1

Sudhakar, B. D.; Kattepogu, N.; David, A. Marketing Assistance and Digital Branding-an Insight for Technology Up-Gradation for MSME's. *Int. J. Manage. Stud. Res.* **2017,** *5* (1), 2455–1562.

Swetha, G.; Anantha Eashwar, V. M.; Gopalakrishnan, S. Epidemics and Pandemics in India Throughout History: A Review Article. *Indian J. Public Health Res. Dev.* **2019,** *10* (8), 1570–1576. https://doi.org/10.5958/0976–5506.2019.02328

Spencer, J. W.; Seydliz, R.; Laska, S.; Triche, E. The Difference Influences of Newspaper and Television News Reports of a Natural Hazard on Response Behavior. *Commun. Res.* **1992,** *19*, 299–325.

Uslaner, E. M. Trust and the Economic Crisis of 2008. *Corporate Reputation Rev.* **2010,** *13* (2), 110–123.

UNWTO. COVID-19 Tourism Recovery Technical Assistance Package. https://webunwto.s3.eu-west-1.amazonaws.com/s3fs-public/2020–05/COVID-19-Tourism-Recovery-TA-Package_8%20May-2020.pdf

UNWTO. International Tourism Highlights 2019 Edition, 2020a. https://www.eunwto.org/doi/pdf/10.18111/9789284421152 (accessed April 11, 2020).

UNWTO. COVID-19: Putting People First, 2020b. https://www.unwto.org/tourism-COVID-19 (accessed April 5, 2020).

UNWTO. Impact Assessment of the Covid-19 Outbreak on International Tourism, 2020c. https://www.unwto.org/impact-assessment-of-the-covid-19-outbreak-on-international-tourism (accessed April 10, 2020).

Van de Walle, S.; Bouckaert, G. Public Service Performance and Trust in Government: The Problem of Causality. *Int. J. Public Admin.* **2003,** *26* (8–9), 891–913.

World Health Organization, India Novel Coronavirus Situation Report-1, 2020. https://www.who.int/docs/default-source/wrindia/india-situationreport-1.pdf?sfvrsn=5ca2a672_0

Wong, J. C. S.; Yang, J. Z. Seeing Is Believing: Examining Self-Efficacy and Trait Hope as Moderators of Youths' Positive Risk-Taking Intention. *J. Risk Res.* **2020,** 1–14. https://doi.org/10.1080/13669877.2020.1750463

Yosie, T. F.; Herbst, T. D. *Using Stakeholder Processes in Environmental Decision Making: An Evaluation of Lessons Learned, Key Issues, and Future Challenges*; Ruder Finn: Washington, DC, 1998.

CHAPTER 21

INNOVATIONS AND DIGITAL SOLUTIONS FOR TOURISM AND HOSPITALITY IN COVID-19 PANDEMICS

AMARJIT SINGH

Department of Tourism Management, K.M. Government College Narwana, Chaudhary Ranbir Singh University, Jind, Haryana, India

ABSTRACT

Prior to pandemic much discussion was on sustainability now covid-19 is challenging the tourism. The global diseases corona provides evidence that traveler can be agent and carrier of virus. Globally, Hall (2020) transport is a distributor of pathogens. Ivanov et al. (2020) investigated transmission of infection during traveling can be controllable with the application of electronic tourism, automation in technology such as Robot services and self-service kiosks have created a shield among suppliers and customers and also customers themselves. Rosen (2020) stated Hotel utilized Robot services in cleanliness of rooms, in addition, Wolfe (2020) some restaurants have introduced temperature recognition technology for customers and employees and also focused upon contactless payments. Benckendorff et al. (2005) also state that most research dealing with information technology and the Internet would fall into the category of "technology as a tool of the tourism industry." Networking among suppliers and tourist indicate the innovative progress of travel trade. Werthner and Klein (1999) accepted that electronic tourism acts as a supportive product and empowers the working organizations of travel

Post-COVID Tourism and Hospitality Dynamics: Recovery, Revival, and Re-Start.
Umendra Narayan Shukla & Sharad Kumar Kulshreshtha (Eds.)

sector. This chapter discusses global infectious disease and focuses upon the digitalization of travel sector. COVID-19 outbreak recommendations state the positiveness in adoption of technologies by tourism sector. In nutshell, chapter outlines how travel demand is influenced by technology and investigates that pandemic stimulated the stakeholder to go for electronic tourism acceptance.

21.1 INTRODUCTION

Generally, pandemic comes with the travel restrictions and affect global economy. In the beginning of 21st centuries, the pandemics severe acute respiratory syndrome (SARS) in 2002, "Bird flu" and Swine flu in 2009, MERS in 2012, restrict tourism activities in respective regions where these diseases spread. These crisis impacts were noticed as behavioral change of traveler and they were sensitively prone to hygienic and safety aspect of affected regions. So-called Covid-19 is a worldwide influenza and emerged as global crisis the earth is badly facing this by travel restrictions, social distancing, curfews, surveillance, and monitoring of human being. Its impacts on tourism can be seen as cancelation of events and shutdown that leads to no demand. Once those attractions are facing mass number now have non, Coronavirus is changing the living pattern of society and less travel is on first priority result can be seen as restaurant, shopping Mall, market center, and recreational center are now emptied. At present speculation are of what kind of tourism would be after virus outbreak but they belief it will rebound and boom may be not down after the crisis. Blogs and newspaper articles are depicting popular tourism sites in "before" and "after" photographs (Holtham, 2020). As vaccine invention is going on and in future the tourism industry will revive with new dimension, broadly, technological innovation in travel trade is going to become common and the new concept of smart and sensor tourism will be in practices. Gössling et al. (2020) reconsidered transformation in global tourism system over the crisis. Even though e-trend is common among the stakeholders and likely in near future the digitalization of everything will be helpful in tourism revival. The most noteworthy changes would be of location-based system and digital along with sensor inclusion in traveling means a new reality. Benckendorff et al. (2019) stated involvement of technology in tourism have changed the structure and every subdomain of industry.

21.2 TRAVELER'S RISK BEHAVIOR TO PANDEMIC

Consumer behavior was searched in marketing later some author has reviewed social and psychological variable then elaborate this in tourism. Mill and Morrison (2002) generalized travel behavior as a continuous process and analyzed it in intercorrelated forms. As traveling decision correlates to motivational forces, the tourist's behavior also changes accordance. Lee and Chen (2011) stated pandemic restrictions influences the perceptions of health risk and consumption capacity of disposable income result in behavior change is not down. Because variables regional disparity, socioeconomic status, education, and psychological play key role in determination of behavior, Wen et al. (2005) analyzed the influence of SARS on Chinese consumer behavior and their demand of leisure travel further stated women and lower group were more affected. Tourists tend to visit the places for various motives but communicable disease affect their behavior and reverse relation to tourism demand is noticed. Travelers perceive the risk of traveling if the destination is disease affected in addition media report or travel advisory of self-government made him sensitive. Page (2009) also stated tourist avoid the visiting of diseases affected countries. In case of viral disease, the behavior relies upon risk perception of individual and affects their traveling decision. Liu et al. (2011) examined the effect of SARS on air travel in U.S., China, Hong Kong, and Taiwan fear of the risk was perceived differently between countries, thus reflecting different lifestyles. Earlier study says natural disaster SARS, bird flu 2008, swine flu, and MERS broken out as change in behavior of society. A sudden and accidental incident is sufficient to cause instability of a whole system, even to a large extent of changing the internality and continuity of a system (Gleick, 1987; Faulkner and Russell, 1997; Peat, 1991). Tourist's consumption pattern also could be seen as travel avoidance for short duration. Zhang Wen et al. (2005) investigated that people did not completely stop leisure travel, but also have shorten the length of stay and tend to be lower expending tourist. Other element also supports situations such as querying family and friends and (NFO, 2003) research shows that quite a lot of Americans intend to increase domestic leisure travel and cut outbound trips in addition preference of transportation means of has changed; for instance, anticipation of rural in tourism and short duration trip has increased. Rittichainuwat and Chakraborty (2009) explored repeat travelers are destination-aware tourists who are more knowledgeable of the range of risks available at the destination than first-time travelers as experienced tourists they have had opportunities to participate in a number of activities and may be able to differentiate between the real and perceived risks at the destination and thus are less likely or unlikely to be influenced by media coverage. Maddux and Rogers (1983) have shown fear appeal can change intention toward a behavior that reduces the risk, also called the protection motivation theory (Roger, 1975). It is also stated that a higher risk perception is negatively correlated with purchase intentions (Yeung and Morris, 2001). Madelon Hutjens (2014) stated Consumers tend to retain a feeling of immunity, thinking that they, and people in their close proximity, will have a lower risk of experiencing a certain event, compared to others. Rosselló et al. (2017) investigated that in the choice of destination their behavior relies upon spread of disease in concerned area. Particularly, if both visiting country and originate are suffering of viral pandemic then risk exist and tourist oppose to travel coincide it low risk destination are inevitable for them. Roehl and Fesenmaier (1992) studied tourist's perception of risk and found travel behavior correlates to the situation and proposed that traveler's risk perception is differ to destinations. However, when people in their proximity are affected, the perception of risk and subsequently fear will increase significantly, compared to consumers without personal experiences (Wahlberg and Sjoberg, 2000). So, travel behavior is affected by purchase intention, which is determined by the perception of risk toward a product (Madelyn Hutjens, 2014).

21.3 SMART TOURISM: ACCEPTANCE OF INNOVATION

Today crisis in tourism may be prolonged and the demand-driven product requires innovative ideas. Innovation is a process that turns a problem-solving idea into practice (Ottenbacher and Gnoth, 2005: 206) in addition change in habit means invitation of smart living.

Gretzel (2011) suggested the evolution of information technologies in tourism leads toward digital dimensions in the forms of artificial intelligence. Likely, technology can manage and significantly continue the business if being used properly. Xiang and Fesenmaier (2017) Advanced application such as cloud computing, sensors and GPS widespread use, virtual and augmented reality and the full adoption of social media and mobile technologies have pushed the emergence of smartness in tourism. Gajdošík (2018) explored technology uses at destination where tourists utilize the discounted services through wearable gadgets and smart cards. Tech savvy tourist perceives risk while visiting and go for intelligent automation at the destination. Loureiro et al. (2020) emphasized the need to develop a particular framework for virtual reality (VR) and AR technology for analyzing tourists' behavior and their intention to use these technologies. In Smart Tourism Experience, Femenia-Serra, and Neuhofer (2018) explored it as multilayered type that is achievable with the stakeholder interactions and innovative spirit result in data-driven experience is constructed in a real-time. Wang et al. (2016) also stated most important items preferred by tourists are tourist attraction homepage, smart vehicle-scheduling, personal-itinerary design, free Wifi, smart cards (bands), intelligent-guide system, crowd handling, mobile payment, tourist-flow monitoring, and online information. Along these, the inclusion of digitization, artificial intelligence, and location-based system also can reduces the risk perception of tourist and may be helpful in upraise of travel demand. At present most of the destination does not cover above mentioned infrastructure but smart tech user has pressurized the Tourism Organization go for innovation in this area. Table summarizes the technologies that are being presented for the solution of various virus outbreaks. Examples of focus area are also given with the application of innovations.

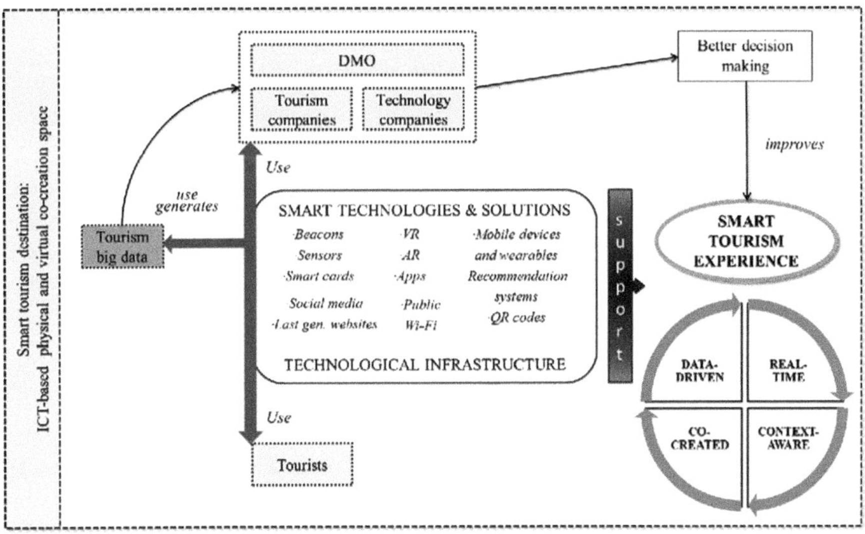

FIGURE 21.1 The smart tourism experience: dimensions and creation in the smart destination context.

Source: Reprinted with permission from Femenia-Serra and Neuhofer (2018)

TABLE 21.1 Digital Solution and Innovation in Various Virus Outbreaks.

Various virus outbreaks	Focus areas: tourism, hospitality, aviation, and destination	Digital solutions & innovations
SARS (2003)	Tourism and Hospitality	Audin (2004). Catastrophe theory stated recovery pattern is the cause of natural and social phenomena. Online travel portals Expedia, Priceline, Orbitz, and Travelocity recorded considerable growth in the booking revenue in virus outbreak.
SWINE FLU (2009)	Destination	Market of Vietnam, Philippines, Malaysia, Taiwan, and India diverted towards domestic tourism in crisis.
ZIKA (2015)	Cruises	This year Celebrity Cruises adopted Omni channel approach at Miami through digital experience along with mobile application that opened new horizons of personalization as well as strengthened its brand.
MERS (2012)	Semi luxury product	South Korean adopted online shopping behavior during epidemic result in willingness to innovation such as preparedness and contact tracing. (Jung and Sung, 2017)

21.4 DIGITAL DRIVEN EXPERIENCE; PERSONALIZATION OF TOURISM FOR ENCOUNTER OF COVID-19

Tourists are habitual of mobile technology whenever participate in event and many companies extended this terminology by offering wearable gadgets, e.g., smart watches, Google glass, and bendable iPad screen. Presently, in crisis the travel company have time to go for Pilot study, for fruitful result personalized service can be assess, earlier results of Virgin Atlantic (Laura Entis, 2014) stated the passengers arrived at Heathrow Airport welcomed by staff with the wearing of Smart Watch and Google Glass. Although the scheme was only to know the experience of traveler but may be result oriented in case of physical distance. Consumer wants fingertip services and also support to such type of solution. Meanwhile, United is first airline to offer an app for Google Assistant the Fitbit Ionic. This will allow customers to receive flight status information and their boarding pass on their smart watch (Lewis Lazare, 2017) and also such types of devices is helpful in knowing more about the travelers as well as personalize the service. Travelport Mobile Agent a mobile application allows travel suppliers to work on their Apollo, Galileo, and Worldspan terminal, these offer services without office meeting and full mobility to travel agent at anytime and anywhere. Akihiko Miura

(2019) ANA uses hearable technology to simplify communication the attendants are equipped with Samsung Gear S3 and S2 smart watches. Mostly airlines are utilizing automated passport control kiosk with new facial recognition technology (Katherine Khashimova Long, 2020) Sea-Tac innovation are incorporation of LED lights with both Bluetooth Low Energy iBeacons and visual light communication capability by smartphone app passenger can expect submeter location accuracy. Along this, VR and robot technology for challenged person at airport (Dave Wilson, 2018) leading the way in innovation and also can apply for all passengers. Jim Peters (2017) SITA extend HoloLens in air transport industry to manage the complex operation, offered mixed reality of augmented and VR. For Smart Screening, Doha airport deploys a wearable intelligent helmet that is effectively portable and enables contactless temperature measurement (Engr. Badr, 2020), latest addition of smart accessory is Spryng by Sri Lankan airlines (2019) offered a wireless active pneumatic compression wrap by which transit passengers can use it for muscle relief before to board a flight, the airlines also going to introduce this service onboard. Therefore, Smart tourist' can be described as an exigent and well-informed person (Mascaró 2012).

21.5 LOCATION-BASED TOURISM SYSTEM

Tourist looks controlled and safety base environment, hence use of location-based technology may be an innovativeness in viral pandemic. Location-based system relies upon smartphone services these made travel personalized with technological app. Nowadays e-trends is common feature of society and viral pandemic nature motivate them to make demand of advanced app having functional qualities. So multiagent travel app requirement increased with the time, in recent years we have seen the emergence of new social network platforms where users can easily and are willing to update that kind of information, for example, TripAdvisor (Castillo et al., 2008). Mobile app recommendations of nearby location of restaurants, transportation, oil gas station, and other daily routine things made it compulsory instrument of trip. Tour recommender system works in combination with the smartphone's positioning system to collect tour information that more closely meets the user's needs, providing a higher quality location-based services (Chen and Tsai, 2017). In the pandemic period, nation issued advisory to residents for downloading the COVID—19 contact tracing app in their mobile and by this one can search himself that he is in safe location. As technology is growing with the time these advisory app can be developed in new mode and recommend healthy guidelines, which can improve the idea of secure traveling. Mobile system equipped with advanced app assist Travel Company to offers service where tourist stands. Indian Railways introduces QR Code to ensure contactless ticket checking now passengers will be able to avail this facility as a result there won't be any contact between passengers and railway employees (Yadav, 2020). Three factors were found to give users the sense that the system was easy to use, allowing them to operate the system quickly: (a) the quality (i.e., accuracy and timeliness) of the information produced by the system, (b) the stability and clarity of the system architecture, and (c) the users' perception of the convenience brought by the system (Chen and Tsai, 2017). In current scenario, e-Tourism system is common among user where they review on site about the services. Demand of ontology-based domain can be seen in future as user will consider security on priority and may prefer SigTur type system. SigTur stores in each node of the ontology the preference of the user (and the confidence of the system on that preference) the user performs actions on the recommended activities (e.g., adds an activity to the current travel plan) the system updates the preferences on the concepts associated to that activity and spreads this information through their parents in the ontology (Borràs et al., 2014). The rapid development of technology has allowed tourism recommender systems to increase their functionality with new artificial intelligence technology such as multiagent technology, optimization algorithms, cluster analysis, and so on (Chen, 2017).

21.6 ADOPTION OF INTELLIGENT AUTOMATION

Adoption of automation in tourism industry means inclusion of artificial intelligent system, Ivanov et al. (2019) stated the characteristic as product or service is delivered by an autonomous agent without the human involvement. Influenza outbreak come as changes in people's behavioral intensity and to reduce the risk it pushes them toward inclusion of artificial intelligence in every sphere. Self-service culture is going to be common and virtual agents such as Chatbots assist the customer during service and create digital relationship among suppliers and users. Although facial recognition system is common in some area, for example, boarding pass at door entry, access to their hotel room if this technique is utilizing in authorize payment it works efficiently without interaction to Kiosk. MasterCard have implemented the facial recognition systems by which customer can pay through camera of vendor. Now principal suppliers have to install this in every segment of establishment and untouched experience will remain core component of their product. Del Rio et al. (2016)

find out that airport uses face recognition technology to identify the passengers at boarding gates and passport control. Innovation of voice control technology may be a boom for travel trade the devices works on voice recognition pattern with internet connectivity and respond accordingly. Wynn hotel in Las Vegas has announced that it will introduce in all its rooms Amazon's Echo voice-controlled speaker equipped with the Alexa digital assistant while Aloft Hotels use Siri. Liu and Hung (2019) fast food restaurants install self-service kiosks to sell burgers, Wei et al. (2016) while airports and hotels for self-check-in. In restaurant adoption of mobile kitchen, an advanced system that contain features of touch-free chamber and air pressure system serve with safety aspect (Sarah McBride, 2020). A completely automated restaurant is also forthcoming in New York (Marks, 2016). Bittendorfer et al. (2019) Automation personalizes the services by flight information, reservation of car from car rentals or can control devices in hotel room. Traveler have to provide ID many time, blockchain technology presents a share digital database that quickly verify without waiting further with this guest can access their own account. Surfacing in hospitality, Japan Introduced a robot hotel (Martin, 2016) and robotic information agents (Pan et al., 2015) Robotic floor cleaners (Sung et al., 2008). Virus-killing Robot may be a new development in this segment and installing of these in super market or tourist places can make a destination standardized one having securable aspect. The Westin Houston Medical Centre is the first and only hotel in the U.S. to utilize Light Strike Germ-Zapping Robots to sanitize and disinfect its guest rooms and common areas to help combat the threat of coronavirus (Rosen, 2020). Engr. Badr (2020) According to Doha airport the robots are being deployed in vulnerable high-passenger flow areas to reduce the spread of pathogens. Robotic vending machine can sell noodles, snacks, wash paper, beverages, and paper towel at destination or on the way to travel sights. Japan National Tourism Organization is expected that the concept of R-tourism will offer business opportunities, new or better services, novel experiences, customization, and improvement of access to activities for different customer segments. Yang et al. (2020) stated that COVID-19 has forced the companies to use robots and automation to fight the pandemic), Wolfe (2020) included hospitality companies.

21.7 CONCLUSIONS

Post-COVID effect may change people life pattern and tourist will consider security on priority and perceive risk during participation in event. Current situation of avoidance of physical contact state digital solution of travel trade and tourist event can successful with artificial intelligence as use of labor is probable a threat. So, traveler behavior may change in the coming time, Webster and Ivanov (2019) generalized the perception towards technologies and stated its impact a great deal in specific uses of automation in travel, tourism, and hospitality. Day by new concepts is emerging in tourism sector and progress is also noticed over the years. Internet system of social sites such as twitter and facebook has transformed the structure of industry and VR is a new way of business and beyond the screen, it moved to voice recognition. Now time has come and people have to learn different way of interaction in a given environment. The new foundation of personalized interaction show travel sector focus is upon this and various devices are in operation. For long perspective, National Tourism Administration should emphasis on safety image of attraction by offering smart tourism experience, since various techno-based model in tourism are available, now tourist behavior is changing and their intensity of purchase require VR and augmented reality in this sector.

KEYWORDS

- **COVID-19**
- **digitalization**
- **virtual network**
- **automation**
- **travel demand**
- **smart tourism**

REFERENCES

Akihiko, M. All Nippon Airways Adopts "Hearable Technology" to Simplify Communication Between Flight Attendants, 2019. https://www.futuretravelexperience.com

Angulo, C.; Pfeiffer, S.; Tellez, R.; Alenyà, G. Evaluating the Use of Robots to Enlarge AAL Services. *J. Ambient Intell. Smart Environ.* **2015,** *7* (3), 301–313.

Aubin, D. Forms of Explanations in the Catastrophe Theory of René Thom: Topology, Morphogenesis, and Structuralism. In *Growing Explanations: Historical Perspective on the Sciences of Complexity*; Wise M.N., Ed.; Duke University Press; Durham, NC, 2004; pp 95–130.

Badr. HIA Adopts Robotics and Wearable Tech for Thermal Screening, 2020. https://www.futuretravelexperience.com/2020/05/hia-adopts-robotics-and-wearable-tech-for-thermal-screening/#

Benckedorff, P.; Moscardo, G.; Murphy, L. High Tech versus High Touch: Visitor Responses to the Use of Technology in Tourist Attractions. *Tour. Recreat. Res.* **2005,** *30(3),* 37–47.

Benckendorff, P. J.; Xiang, Z.; Sheldon, P. J. *Tourism Information Technology*, 3rd ed.; CABI: Boston, 2019.

Berezina, K.; Ciftci, O.; Cobanoglu, C. Robots, Artificial Intelligence, and Service Automation in Restaurants, 2019. https://doi.org/ 10.1108/978-1-78756-687-320191010.

Bittendorfer, T.; Bunt, J.; Grunder, L.; Riedel, D.; Magnus, B.; Salzlecher, T. Technology in Tourism: How Voice Assistants Influence the Hospitality Industry. In *ISCONTOUR 2019 Tourism Research Perspectives*; Maurer, C., Siller, H., Eds.; Proceedings of the International Student Conference in Tourism Research, Norderstedt, 2019; pp 328–338.

Borràs, J.; Moreno, A.; Valls, A. Intelligent Tourism Recommender Systems: A Survey. *Expert Syst. App.* **2014,** *41* (16), 7370–7389.

Castillo, L.; Armengol, E.; Onaindía, E.; Sebastiá, L.; González-Boticario, J.; Rodríguez, A. Samap: An User-Oriented Adaptive System for Planning Tourist Visits. *Expert Syst. App.* **2008,** *34*, 1318–1332.

Chen, C. C.; Tsai, J. L. Determinants of Behavioral Intention to Use the Personalized Location-Based Mobile Tourism Application: An Empirical Study by Integrating TAM with ISSM. Future Gen. Comput. Syst. **2017**. http://dx.doi.org/10.1016/j.

Dave, W. Sea-Tac Airport Creating "An Entirely New Passenger Experience", 2018. https://www.futuretravelexperience.com

Del Rio, J. S.; Moctezuma, D.; Conde, C.; de Diego, I. M.; Cabello, E. Automated Border Control E-Gates and Facial Recognition Systems. *Comput. Secur.* **2016,** *62*, 49–72.

Entis, L. Virgin Atlantic Is Using Google Glass to Greet Customers, 2014. https://www.entrepreneur.com/author/laura-entis.rss

Faulkner, B.; Russell, R. Chaos and Complexity in Tourism: In Search of a New Perspective. *Pacific Tour. Rev.* **1997,** *1* (2), 91–106.

Femenia-Serra, F.; Neuhofer, B. Smart Tourism Experiences: Conceptualization, Key Dimensions and Research Agenda. *J. Reg. Res.* **2018,** *42*, 129–150.

Gajdošík, T. Smart Tourism: Concepts and Insights from Central Europe. *Czech J. Tour.* **2018,** *7* (1), 25–44. DOI: 10.1515/cjot-2018–0002.

Gleick, J. *Chaos: Making a New Science*; Heinemann: London, 1987.

Gössling, S.; Scott, D.; Michael Hall, C. Pandemics, Tourism and Global Change: A Rapid Assessment of COVID-19. *J. Sustain. Tour.* **2020**. DOI: 10.1080/09669582.2020.1758708

Gretzel, U. Intelligent Systems in Tourism: A Social Science Perspective. *Ann Tour Res.* **2011,** *38* (3), 757–779.

Hall, C. M. Biological Invasion, Biosecurity, Tourism and Globalisation. In *Handbook of Globalization and Tourism*; Timothy, D., Ed., 2020.

Holtham, A. Before and After: How Coronavirus Has Emptied Tourist Attractions Around the World. Cond_e Nast Traveller, March 30, 2020. www.cntravellerme.com

Hutjens, M. *The Influence of Fear on the Buying Behaviour of Consumers in Case of an Animal Disease Outbreak*; Wageningen University: Wageningen, The Netherlands, 2014.

Ivanov, S.; Webster, C. Perceived Appropriateness and Intention to Use Service Robots in Tourism. In *Information and Communication Technologies in Tourism 2019*; Pesonen, J., Neidhardt, J., Eds.; Cham: Springer, 2019a.

Ivanov, S.; Webster, C.; Stoilova, E.; Slobodskoy, D. Biosecurity, Automation Technologies and Economic Resilience of Travel, Tourism and Hospitality Companies. SocArXiv 2hx6f, Center for Open Science, 2020.

Jim, P. SITA Explores Emerging World Of Mixed Reality for Air Transport. Industry, 2017. https://www.sita.aero/about-us/contact-us/get-in-touch

Jung, E.; Sung, H. The Influence of the Middle East Respiratory Syndrome Outbreak on Online and Offline Markets for Retail Sales. *Sustainability* **2017,** *9* (3), 411–412. https://doi.org/10.3390/su9030411

Katherine, K. L. Port of Seattle Moves Toward Implementing Facial Recognition. *The Seattle Times (TNS)*, December 12, 2020. https://www.governing.com

Lee, C.-C.; Chen, C.-J. The Reaction of Elderly Asian Tourists to Avian Influenza and SARS. *Tourism Management* **2011,** *32* (6), 1421–1422.

Lewis, L. United Airlines Is First Carrier to Offer Dedicated Apps for Google Assistant. Fitbit, 2017. https://www.bizjournals.com

Liu, C.; Hung, K. Understanding Self-Service Technology in Hotels in China: Technology Affordances and Constraints. In *Information and Communication Technologies in Tourism 2019*; Pesonen, J., Neidhardt, J., Eds.; Springer: Cham, 2019; pp 225–236.

Liu, J.; Moss, S. E.; Zhang, J. The Life Cycle of a Pandemic Crisis: SARS Impact on Air Travel. *J. Int. Busi. Res. 2011. 10* (2), 63–78.

Loureiro, S.M.C.; Guerreiro, J.; Ali, F. 20 Years of Research on Virtual Reality and Augmented Reality in Tourism Context: A Text-Mining Approach. *Tour. Manage.* **2020,** *77*, 104028.

Maddux, J. E.; Rogers, R. W. Protection Motivation and Self-Efficacy: A Revised Theory of Fear Appeals and Attitude Change. *J. Exp. Soc. Psychol* **1983,** *19* (5), 469–479.

Marks, G. A Fully Automated Restaurant Just Opened in New York City (114 Years After the Last Automated Restaurant Opened). *Washington Post*, December 19, 2016. https://www.washingtonpost.com

Mascaró, T. Smart@Tourism. Internet y la innovación aplicadas al nuevoparadigma delas Smart cities y el Turismo, April 20, 2012. http://www.slideshare.net

Mill, R.; Morrison, A. *The Tourist System*, 4th ed.; Kendall/Hunt: Dubuque, IA, 2002.

NFO. Plog Research Impact Study CTC Market Research, 2003. http://www.canadatourism.com.

Ottenbacher, M.; Gnoth, J. How to Develop Successful Hospitality Innovation. *Cornell Hotel Restaurant Admin. Quart.* **2005,** *46* (2), 205–222.

Page, S. J. Current Issue in Tourism: The Evolution of Travel Medicine Research: A New Research Agenda for Tourism? *Tour. Manage.* **2009,** *30*, 149–57.

Pan, Y.; Okada, H.; Uchiyama, T.; Suzuki, K. On the Reaction to Robot's Speech in a Hotel Public Space. *Int. J. Soc. Robot.* **2015,** *7* (5), 911–920.

Peat, F. D. *The Philosopher's Stone: Chaos, Synchronicity and the Hidden Order of the World*; Bantam: New York, 1991.

Rittichainuwat, B. N.; Chakraborty, G. Perceived Travel Risks Regarding Terrorism and Disease: The Case of Thailand. *Tour. Manage.* **2009,** *30* (3), 410–418. DOI: 10.1016/j.tourman.2008.08.001.

Roehl, W.; Fesenmaier, D. Risk Perceptions and Pleasure Travel: An Exploratory Analysis. *J. Travel Res.* **1992,** *30* (4), 17–26.

Rogers, R. W. A Protection Motivation Theory of Fear Appeals and Attitude Change. *J. Psychol.* **1975,** *91* (1), 93–114.

Rosen, S. You Can Stay at a Hotel That Uses Virus-Killing Robots to Clean Its Rooms, March 28, 2020. https://thepointsguy.com/news/virus-killing-robot-hotel/

Rosselló, J.; Santana-Gallego, M.; Awan, W. Infectious Disease Risk and International Tourism Demand. *Health Policy Plan.* **2017,** *32* (4), 538–548.

McBride, S. Robot Food Start Ups Have a New Pitch: No Humans Touch Your Lunch, 2020. https://www.bloombergquint.com

SriLankan Airlines. Introduces Wearable Technology for Transiting Passengers, 2019. https://www.futuretravelexperience.com/2019/08/srilankan-airlines-wearable-technology transiting-passengers/#

Sung, J.-Y.; Grinter, R. E.; Christensen, H. I.; Guo, L. Housewives or Technophiles? Understanding Domestic Robot Owners. *Proceedings of the International Conference on Human Robot Interaction* 2008.

Wahlberg, A. A.; Sjoberg, L. Risk Perception and the Media. *J. Risk Res.* **2000,** *3* (1), 31–50.

Wang, X.; Li, X. R.; Zhen, F.; Zhang, J. How Smart Is Your Tourist Attraction? Measuring Tourist Preferences of Smart Tourism Attractions via a FCEM-AHP and IPA Approach. *Tour. Manage.* **2016,** *54*, 309–320.

Wen, Z.; Huimin, G.; Kavanaugh, R. R. The Impacts of SARS on the Consumer Behaviour of Chinese Domestic Tourists. *Curr. Iss. Tour.* **2005,** *8* (1), 22–38. DOI: 10.1080/ 13683500 508668203.

Werthner, H.; Alzua-Sorzabal, A.; Cantoni, L.; Dickinger, A.; Gretzel, U.; Jannach, D. et al. Future Research Issues in IT and Tourism. *Inf. Technol. Tour.* **2015,** *15* (1), 1–15.

Werthner, H.; Klein, S. *Information Technology and Tourism: A Challenging Relationship*; Springer-Verlag: Wien, 1999.

Wolfe, A. Cali Group Expands Use of Facial Recognition Tech. *Hospitality Technology*, March 24, 2020. https://hospitalitytech.com/cali-group-expands-use-facial-recognition-tech

Xiang, Z. From Digitization to the Age of Acceleration: On Information Technology and Tourism. *Tour. Manage. Perspect.* **2018,** *25*, 147–150.

Xiang, Z.; Fesenmaier, D. R. Big Data Analytics, Tourism Design and Smart Tourism. In *Analytics in Smart Tourism Design, Concepts and Methods*; Xiang, Z., Fesenmaier, D. R., Eds.; Springer International Publishing Switzerland: Cham, 2017; pp 299–307. DOI: 10.1007/978-3-319-44263-1.

Yadav, V. K. Indian Railways Introduces QR Code for Tickets. IRCTC Website to be Revamped, 2020. https://www.indianrailway.com

Yang, G.-Z.; Nelson, B. J.; Murphy, R. R.; Choset, H.; Christensen, H.; Collins, S. H. et al. Combating COVID-19—The Role of Robotics in Managing Public Health and Infectious Diseases. *Sci. Robot.* **2020,** *5* (40), eabb5589. DOI: 10.1126/scirobotics.abb5589

Yeung, R. M.; Morris, J. Food Safety Risk: Consumer Perception and Purchase Behaviour. *Br. Food J.* **2000,** *103* (3), 170–187.

Zhang, W.; Gu, H.; Raphael, R. K. The Impacts of SARS on the Consumer Behaviour of Chinese Domestic Tourists. *Curr. Iss. Tour.* **2005,** *8* (1), 22–38. DOI: 10.1080/13683500508668203

CHAPTER 22

CONTACTLESS SERVICE IN POST-COVID-19: AN INSIGHT FROM THE HOSPITALITY INDUSTRY

PATITA PABAN MOHANTY[1] and ARSLAN MUNIR[2]

[1]*SOA University, Khandagiri, Bhubaneswar, Odisha, India*

[2]*Kansas State University, Manhattan, Kansas, USA*

ABSTRACT

The recent pandemic of COVID-19 has created havoc and catastrophic circumstances in every nook and corner of the universe. This pandemic has engulfed "man, machine, and material" and paralyzed them for an indefinite time. Furthermore, the pandemic has resulted in an uncertainty in the operation and management of industries relying on "man, machine, and material." Though the past reflection of the hospitality industry has been termed as the "sunshine" industry, the COVID-19 pandemic has turned it entirely into a "state of cloudiness." This people-intensive still embrace its patrons with folded hands inspired by the true essence and value of hospitality, but with a new design and protocol triggered by various technological innovations. To mitigate the spread of this deadly disease, "social distancing" has been the new norm of the hospitality business and its existence heavily depends on the mantra of "contactless" and "touchless." There is no exception to the hospitality industry, where contactless and touchless operations have been the new standard operating procedures (SOPs). Although these SOPs for the hospitality industry may seem to be "gloomy" and "spiritless," but for the sake of the bread and butter of employees and the safety of guests and employees, these are essential and bare needs of this hour. It is therefore

Post-COVID Tourism and Hospitality Dynamics: Recovery, Revival, and Re-Start.
Umendra Narayan Shukla & Sharad Kumar Kulshreshtha (Eds.)

imperative that hotels embrace a proactive approach to this paradigm change for sustainability in the post-COVID-19 period. This may focus on assimilating various technological and digital approaches, such as artificial intelligence, big data analytics, the Internet of things, 5G network, augmented reality, virtual reality, and robotics for fulfilling the guest service delivery and experience through contactless hospitality.

22.1 INTRODUCTION

The coronavirus pandemic has crippled the tourism industry, forcing travelers to return home and creating havoc simultaneously. After World War II, the COVID-19 epidemic has the most negative impact on global healthcare and economic structures, and has influenced every dimension of human life. In other words, the global pandemic COVID-19 is being treated as the biggest crisis since the World War II. The World Travel and Tourism Council (2020) warned that the COVID-19 pandemic could slash 50 million jobs in the travel, tourism, and hospitality industry internationally, with Asia, predicted to be the hardest hit. The report of the World Economic Forum reveals that the industry could take up to ten months for restoration and survival World Economic Forum (2020). Similarly, the report of the WTTC also reveals that presently the pandemic has become a curse to the tourism industry that was estimated to engulf the sector by up to 25% in 2020, and reduce the travel market by up to 25% in 2020 BBC News (2020).

The global economy was shuttered almost overnight due to the COVID-19 pandemic UNWTO (2020). The pandemic has resulted in an immense challenge for the hospitality industry. Various techniques to flatten the COVID-19 curve, such as lockdowns, social distancing, stay-at-home orders, limitations on travel and mobility, have resulted in the temporary closing of many hotel companies and dramatically reduced the hotel turnover Bartik et al. (2020). The Indian hospitality industry is one of the largest service industries in the world, and more than 2000 hotels and 1000 restaurants are registered under the Federation of Hotel and Restaurant Associations of India (FHRAI). The tourism industry is among the hardest hit of all economic sectors due to the COVID-19 as foreign tourist influx has tremendously decreased. The foreign tourist arrivals are the most lucrative source of foreign exchange earnings as well as revenue of the hospitality industry. The lockdown due to COVID-19 has resulted in a temporary halt of hotel reservations as well as cancellation of prior bookings, thus creating a huge loss of revenue. Europe, Italy, France, and Spain are among the hardest-hit countries due to the COVID-19 crisis, and their hospitality and tourism industry has suffered a great disruption. In the future, this effect on the European tourism industry can also be seen in the Indian hospitality industry as a significant number of travelers from Europe visit India.

Indeed, COVID-19 has accelerated the advent of emerging technology in hospitality operations and applications. In a time when certain individuals are cautious of traveling, technical advances that may have been intended specifically for launching as innovations or additional services have unexpectedly become obligations. Technological innovations offer the hospitality industry a chance to restore itself, after the massive blow of COVID-19. All sectors across the globe are trying to maintain customer trust and confidence and keep the economy running; the hospitality industry is no exception. As emerging norms of daily life, such as mask-wearing, hand sanitizing, and social distancing are embraced, touch-free, or contactless devices are seen as ways to provide people services while maintaining social distancing. Consequently, hoteliers must think out of the box for meeting consumer demands during COVID-19, and to revive their trust and loyalty. Because the COVID-19 issue is expected to have a massive impact on hotels throughout the world, academics in this critical hospitality area should seek to enhance theory and knowledge to assist hotels in becoming more robust and operate successfully postpandemic.

The present chapter identifies and investigates the impact of technological innovations facilitating contactless services in the hospitality industry as well as identifies various areas where the technological innovations can be incorporated post-COVID-19. The study will also provide future directions of hotels' operations post-COVID-19 and also highlight various emerging research areas that can facilitate contactless services in the hospitality industry.

22.2 METHODOLOGY

The present study summarizes a literature review relevant to the impacts of COVID-19 in the hospitality industry by using the "COVID-19" keyword along with associated keywords, such as lockdown, "hospitality," "new normal," "touchless," "contactless," "service quality," "social distancing," "technological innovation," "artificial intelligence (AI)," and "Internet of things" from global database sources like PubMed, Publon, Elsevier, Google scholar, and Bio med central. Apart from the above, various international and national peer-reviewed indexed journals have been referred for the current study.

22.3 LITERATURE REVIEW

Due to the COVID-19 global health crisis, the hospitality industry is "already facing collapse" and is "in a battle for survival" (Guevara, 2020). The recent pandemic has decreased the revenue generation of hotels due to slow-paced tourism and travel activities (Hoisington, 2020). The COVID-19 pandemic has resulted in drastic blows to hoteliers internationally as activities across the globe continue to be suspended and hotel occupancy rates plunge. The revenue per available room (RevPAR) is anticipated to remain in decline in the U.S., Europe, and Asia as leisure and business travel is disrupted or discontinued due to the threat of COVID-19 (Courtney, 2020). The hotel industry is vulnerable to threats posed by unexpected catastrophes such as epidemics, natural disasters, and terrorist attacks (Hung et al. 2018; Chan and Lam, 2013). Various forms of disasters have diverse business ramifications and encourage hoteliers to adopt steps to overcome different crisis-induced challenges. For example, Hong Kong hoteliers improved security by installing the closed-circuit television systems and increasing safety training for hotel workers after the 9/11 attacks (Chan and Lam, 2013). Similarly, the hotels in Korea developed a full range of hygiene equipment and offered health awareness training programs to employees post-SARS-3 epidemic.

Currently, the COVID-19 pandemic has had a major negative effect on many sectors, with hospitality and tourism being the hardest hit sectors, as individuals are forced to avoid physical contact and other types of social interaction that are also integral to allowing hoteliers to provide the style of service that is traditionally associated with this industry. Hence it is acknowledged that after the COVID-19 pandemic, the sustainability of hotels depends on the capacity of hotels to be innovative and adaptive if they need to be profitable. As a "high-touch" business, the hospitality and tourism industry and its enablers, such as aviation, sports, and transport, have come to a standstill due to social distancing and travel bans (Jamal and Budke, 2020). Although it is too early to predict exactly what tourism and travel will look like post-COVID-19, the World Health Organization (2020) recommends that social distancing is an effective way of mitigating the transmission of the virus, and thus social distancing practices could continue for years post-COVID-19. It is therefore crucial that hotels implement a proactive approach to this paradigm shift for survival in the post-COVID-19 period. This may include the implementation of different types of digital solutions complemented by 5G cellular networks, AI, Internet of things (IoT), and even robots to help carry out physical distance protocols.

The COVID-19 lockdowns may have revolutionized the way hospitality business would be conducted in the future, provided the current hygiene and social distance rules and regulations together with more reluctant and confused clients. Although it is too early to predict how and when the hospitality sector will recover from the COVID-19 pandemic, it is apparent that technological innovations will play a major role in this recovery. Such technical advances, if developed correctly, can also accomplish some of the objectives of providing services to the customers in a more responsible way. The community spread of this disease is fatal and needs to be mitigated by maintaining a healthy physical distance among the individuals residing in the community (Fong et al., 2020). These social distancing techniques have had a huge impact on industries that rely on high levels of meaningful human contact, such as hospitality and tourism, both of which are struggling (Hoque et al., 2020). History has demonstrated that a crisis can spur technological innovation and development (Colombo et al., 2016). Now, the most important issue would be how to address this situation, the solution of which would rely heavily on the digitalized transformation of services through contactless mediums. The potential conglomeration of the Internet of human things and Web 4.0 and 5.0 with AI and robotics could trigger this turnaround in the hospitality industry during and post-COVID-19. This techno-centric approach has the potential to convert the entire hospitality ecosystem toward the path of sustainability, and can also engender a high volume of trust and confidence among the customers.

Robotics, AI, IoT, and related technologies can play a pivotal role in the recovery of the hospitality industry during and post COVID-19 as people would be more safety-oriented, when stepping out, and safety, hygiene, and sanitation would be the principal factors for customers while choosing the hotels and hotel services (Hall et al., 2020). Therefore, in the future, hospitality sectors may go for the extra mile by investing heavily in technological innovations to provide physical distancing between customers and employees.

22.4 WHY CONTACTLESS SERVICES IN HOSPITALITY INDUSTRY?

The COVID-19 pandemic is extremely infectious through individual-to-individual spreading (Chan et al., 2020) and therefore World Health Organisation (2020) has advised the entire community residents to minimize the contact and increase physical distance. Hence, in order to achieve the minimum contact, the introduction of AI, IoT, and robotics technology into the hospitality industry can become a necessity. In particular, many hotels are implementing "unmanned" systems in order to offer touchless hotel services through robotics applications. It has been speculated that the severity of the pandemic will ultimately replace the human touch services by providing basic hospitality services through different types of robots. In addition, according to Wen et al. (2020), due to the persistent risk associated with this pandemic, during and after the pandemic, the hospitality and tourism industry will think vigorously about the upkeep of the health of their employees. According to an IBM analysis, 85% of all service interactions around the world will be done without any human involvement in the future (IBM, 2017).

It is predicted that the traditional hospitality services rendered by the human touch will be soon replaced by technology-oriented machines to provide touchless and/or contactless services. Touchless services refer to technology-based services that can be requested by guests by means of gestures and/or speech instead of touching a control panel or interface. Similarly, contactless services refer to technology-based services that can be requested by guests without any physical contact with a machine. An example of a contactless service is a smart point-of-sale (POS) terminal that uses radio signals to connect to a smart card thus obviating any physical contact. In the coming days, the introduction, implementation, and application of contactless services in many aspects of hotel operation will establish a new "servicescape" and will be a newly growing topic of research. Sustainable hotel competitiveness can be ultimately enhanced by the service innovation induced by contactless technology. In order to surmise the incorporation of contactless services in hotels post-pandemic, it is essential to assess the hotels' capabilities to adopt the technological setup and also the latest developments (Kuo et al., 2017).

Hotel settings that get daily social and human interaction, as indicated by Park et al. (2019) are more vulnerable to contamination by touch and can become carriers of infectious illnesses, such as COVID-19 Chan et al. (2020). Post-COVID-19, the perception of safety and hygiene will go beyond the regular healthcare aspects. The next generation customers after realizing the fatal condition of corona disease will seek the highest degree of healthcare facilities and hotel-centric amenities. Many of them, on their personal level, will be willing to pay the extra price for a healthy diet and sleep hygiene to remain protected from diseases. The above implications can be solved by the contactless services offered by the hotel.

Given the increasing importance of maintaining hygiene in hotels, it is essential to conduct user awareness programs on cleanliness and sanitization to prevent the pandemic spread. People across the globe have been mentally influenced by the fear factor and the resulting pensive environment, and this will continue for some time in the future, even after COVID-19 vaccinations. Hence, in this catastrophic time, contactless service will be the only solace for guests and hoteliers. COVID-19 would also contribute to a restructuring of the service delivery, customer engagement, and delivery of services, considering mental, and behavioral conditions caused by the pandemic.

22.5 CONTACTLESS PLATFORMS AND THEIR APPLICATIONS

In order to facilitate the contactless hospitality services, it is necessary for hoteliers to be aware of the potential advantages of AI, robotics, augmented reality, virtual reality, and IoT in hotel service delivery (Zabin, 2019). A growing body of research has focused on implementations of AI and robotics at the personal and hotel organizational level (Li et al., 2019; Tussyadiah, 2020). Hence, considering social distancing is the preventive strategy to monitor and safeguard against this deadly pandemic, adopting AI and robotics can be a real savior for the guests and front-line employees. Therefore, contactless services leveraging AI, robotics, IoT, and 5G/6G cellular networks can be the emerging form of technological innovation for the survival of the hospitality industry postpandemic.

The developments in AI, robotics, IoT, and 5G/6G cellular communications have become progressively powerful in the hospitality and tourism industries to provide various personalized services to hotel patrons (Yu, 2020). Contactless services performed post-COVID-19 will replace many of the human expertise and labor through service automation by technology/machines. Moreover, keeping in mind the fear factor of a deadly pandemic, customers are more inclined toward the robotic services in every aspect of hotel operation offered to them. Additionally, during COVID-19, the service industries have come to recognize their ability to deliver an automated service to boost efficiency, service capability, offer reliable quality of service, enhance competition, minimize costs and improve financial performance.

The hospitality industry is human interaction and labor-intensive industry, whereas social distancing is the preventive measure for the COVID-19, which implies reducing human interaction, and thus hospitality industry is forced to handle this dilemma. Fortunately, the use of AI, robotics, IoT, and wireless communication provides a way to manage this dilemma. The understanding of the COVID-19 pandemic reveals that AI, robotics, and IoT technologies could be used quickly and successfully to clean, disinfect, distribute food and medicines or provide people with pandemic-related knowledge (Yang et al., 2020), since robots are not affected by the virus. Presently, the hospitality and tourism industry, like any other industry, has also been severely influenced by information and communication technologies (Benckendorff et al., 2019). In a short span of time, high-end hospitality business operations have enabled themselves to assimilate a multiple ranges of service and functional robots with a variety of techno capacities to perform the basic functional jobs in hotels (Miller and Miller, 2017; Russell and Norvig, 2016).

During the pandemic, hotel organizations are trying their best to provide contactless services through AI, robotics, IoT, and other platforms to continue the service delivery. Studies have discussed the use of AI and robotics in hotels from a multitude of perspectives. For example, the study of Tung and Law (2017), is directed toward research on customer experience derived from human-machine interaction. The research has further gone to explore the development and implementation of AI and robotics, for the continuation of service delivery in lieu of human touch (Huang and Rust, 2020). Moreover, specific varieties of AI (mechanical, thought, and feeling) could open up diverse challenges in the hospitality sector for its implementation and day-to-day operation. In consideration of COVID-19 and related health emergencies, AI may be adopted in smart hotels to perform varied types of hygiene and sanitation activities. For instance, cleaning robots may identify easily the high-touch areas minimizing the chances of infection of the particular environment. In particular, in order to revive the hotel industry during and post-COVID-19, particular attention needs to be paid toward hygiene, cleanliness, and related activities. Research indicates that when making purchase decisions in hotels and dining willingness in restaurants, customers are driven by hygiene and cleanliness conditions of the property (Choi, 2019). Hence postpandemic, health and hygiene rendered by the innovation-driven technology would be an added advantage for the upkeep of the hotel premises, and would help restore the customers' trust and loyalty.COVID-19 has altered the hotel cleaning practices, and some of them have begun supplying their guests with "social distancing seclusion" service (Glusac, 2020).

Figure 22.1 depicts some of the contactless hospitality services that can be enabled by various technologies. Guest-facing systems are the primary means of interaction between guests and hospitality service providers (Kansakar et al., 2019). Guest-facing systems (shown in Figure 22.1) include POS terminals, hand-held devices, thin-client terminals, and Kiosk machines. These guest-facing systems can be made contactless by using AI, IoT, and

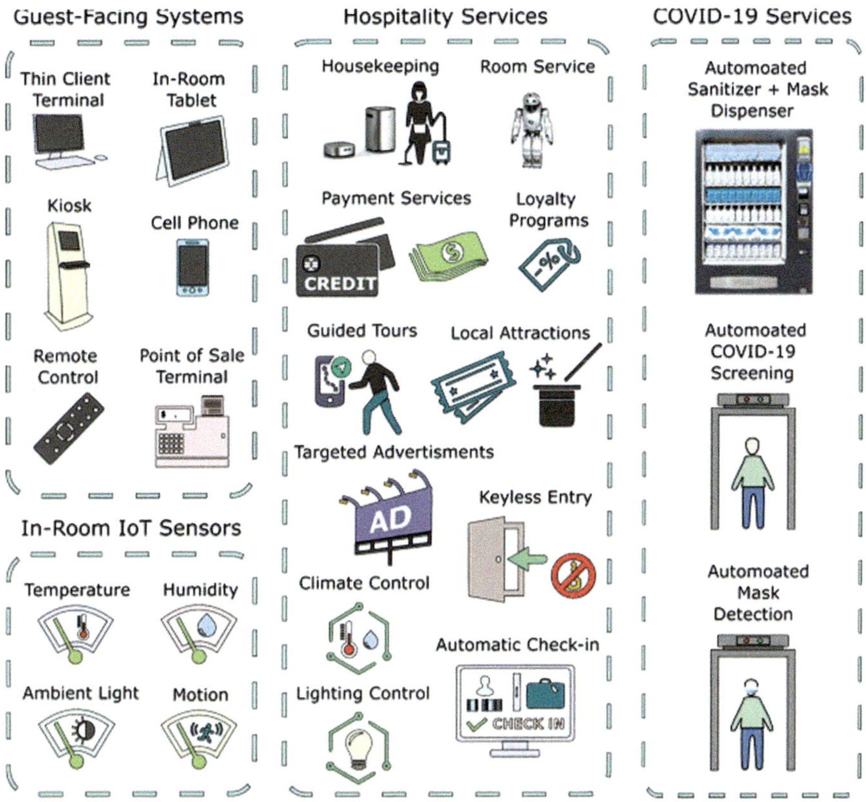

FIGURE 22.1 Contactless hospitality services during and postpandemic.

Source: Adapted from Kansakar et al. (2019).

wireless communications. For example, contactless payment systems use radio-frequency identification or near-field communication for making secure communications. Examples of contactless payment systems include Samsung Pay, Apple Pay, and Google Pay. Contactless guest-facing systems can provide guests automatic check-in and check-out services, and keyless entry services. Furthermore, IoT-based systems can be used to provide guests automatic control of in-room functions, such as automatic adjustment of in-room temperature based on a guest's body temperature, which can be measured by wearable sensors, and adjustment of in-room lighting based on a guest's sleep cycle. Similarly, mobile apps provided by hospitality service providers based on IoT, AI, augmented reality, and virtual reality can provide guests self-guided and interactive tours of property as well as nearby attractions. As shown in Figure 22.1, using AI, computer vision, and robotics, hotels can provide COVID-19-related services to guests, such as automated COVID-19 screening, automated mask detection, and automated sanitizer and face mask dispenser.

Robotics, in particular, is very crucial for providing contactless services in the hospitality industry. Henn-na Hotel, situated in Nagasaki, Japan, has become the world's first hotel to be entirely staffed by robots (REVFINE, 2021). In the Henn-na Hotel, the robots are deployed throughout the hotel and provide information, front desk services, storage services, and check-in and check-out services. Hilton has developed a robot that interacts with guests and responds to their question using speech recognition (REVFINE, 2021). An advantage of using robots for information and guest services is their ability to interact with guests in a variety of languages, which is often not possible with human-based services. Robots can also be used for providing butler and bellman services and can assist the guests with their luggage and other chores. Robots in the hospitality industry, in particular, during and post-COVID-19 not only provide contactless services and help mitigate the spread of the disease, but the interaction with robots itself is fascinating to many guests, and help to increase the hotel revenue and appeal.

Post-COVID-19, the customer's demand would be soaring high for contactless services compared with the human touch service considering the severity of the deadly disease. Hence, service personalization will be essential through contactless services powered by IoT, AI, deep learning, and big data analytics (Huang and Rust, 2018; Mariani, 2019). Service delivery performed by the employees will definitely be impacted by the contactless service as there would be a huge gap between the two. Therefore, the hotel organization will seek to harness the competitive advantage by personalizing the services to patrons through big data analytics. Thus, big data analytics and business intelligence will facilitate the provision of personalized contactless services (Mariani, 2019; Mariani et al., 2018; Mariani and Perez Vega, 2020).

For hotel professionals, AI has emerged to be exceptionally vital. For example, by giving hotel guests access to their rooms/venues using facial recognition software, the convenience and efficiency of the check-in and check-out processes are enhanced. The adoption of 5G, AI, IoT, and big data analytics in the hospitality industry provides a tremendous potential for real-time and personalized contactless services for hotel guests. Countries around the world are using AI for face recognition and body temperature scanning to reliably recognize individuals with the virus, and minimize the spread of the disease. High-end hotels can also provide COVID-19-related services, such as automated COVID-19 screening, automated mask detection, automated sanitizer, and mask dispenser as shown in Figure 22.1. These COVID-19 services at hotel premises can enhance the confidence and trust of guests and can enhance the hotel revenue as guests will feel comfortable that hotel management has taken necessary steps to ensure the guests' safety and health. The pandemic is leading many hotel sectors to embrace "contactless" as an alternative. Hotels are using robotics applications to maintain social distance and reduce the number of employees who are working physically. In many European and Southeast Asian hotels, robots are also being used for roles that hotel employees cannot perform during the pandemic time. Robots are used for the frontline activities such as in lobby areas and travel desks to protect hotel guests, and hotel staff. Similarly, robots can ease the work of room maids and stewards by laying the table in restaurants, preparing the coffee, tea, and sandwiches in pantries and coffee shops, as well as providing room service for the guests, and dispensing facemasks and hand sanitizers. An example is Henn-na Hotel that has become the world's first hotel to be entirely staffed by robots.

The robots can not only help protect the health of customers and prevent the spread of diseases but can also increase the customer experience and loyalty. In February 2020, for example, in Beijing, the food delivery giant MeituanDianping launched robots that helped carry food from kitchens to delivery workers and customers waiting for takeout orders (Toh and Wang, 2020). Additionally, heavy cleaning in hotels and kitchen food production processes are carried out by robots (Meisenzahl, 2020). Similarly, robots have played a pivotal role by loading and unloading various kitchen materials from stores to other locations. In case of quarantine guests, the hotel may use robots to deliver food, linen, and other things without compromising the health and safety of employees (Guzzo, 2020; Kent, 2020).

The humanoid service robots with various range of capacities, such as cooking apprentices, porters, room servers, room maids, hosts, waiters, captains, sommeliers, and bellboys, are performing a gamut of services from the lobby to floor to the kitchen to hotel rooms with great zeal and gesture (Drexler and Lapré, 2019). For instance, customers may wish to get physically distant service after the current pandemic, which may not be feasible with human employees offering service. Robotics can, therefore, become increasingly relevant in terms of providing hospitality and tourism facilities in physically remote places (Seyitoğlu and Ivanov, 2020). Furthermore, 5G networks can play an important role in delivering contactless services via different wirelessly connected machines and at remote locations in the property. For example, the combination of 5G and AI in the hotel industry, will allow faster check-in and payment through facial recognition, leading to dramatic improvements in service efficiency and safety. The hotels' meetings, incentives, conferences, and exhibitions segment can also adopt the contactless service by the high-end technology Wi-Fi-6 in the areas like conference, lobby, meeting, and convention with faster speed and larger coverage.

Post-COVID-19 remote working will emerge as a new paradigm for working. This paradigm shift will also influence the traditionally people-centric hospitality industry. Already technological innovations have automated many aspects of the hospitality industry. Without the presence of hotel managers, the cloud-based property management system can control and perform the hotel operations effectively and efficiently, which can minimize the chances of infectious diseases in hotel premises and can provide more safety for the customers and employees. These technological innovations are expected to further automate and provide even more contactless services post COVID-19. For example, through a mobile-installed concierge app, the guest will be able to order the food and beverage and other services without personal contact.

The pandemic has challenged the education sector by inflicting several changes, leaving a significant influence on students, teachers, and educational institutions. The impact of COVID-19 has impacted students' education due to the closure of all institutes impeding the learning of students. Many educational institutions have adopted learning systems to provide instruction and learning to students online (Dhawan, 2020). To make e-learning more effective and efficient amid such disasters, technology, and its utilization is of utmost importance. During this crucial juncture, making hospitality education and learning alive, an online version of pedagogy has been devised ushered by the latest technology. Similarly, virtual platforms, eBooks, and online sources are provided to students to enhance better and flexible education (Jena, 2020). Due to the advent of technology, the hospitality education has transferred from classroom teaching to the platforms like Youtube, Instagram, Facebook, and Twitter, powered by the latest inventions, such as 5G, AI, big data analytics, and IoT. However, for many researchers, the COVID-19 crisis is an opportunity for technological innovations to help recover many industries, such as airlines, beauty, and personal care, sports, movie theatres, restaurants, and hotels, that have been impacted the most by COVID-19 (Hall et al., 2020).

22.6 CHALLENGES IN POST-COVID-19

Keeping in view the pandemic, this present chapter has explored various technological innovations in the form of AI, robotics, big data analytics, 5G, and that can provide a framework for contactless services in the hospitality industry considering the guests' safety and security. There may be various challenges faced by the hoteliers and the hotel management systems to carry out the routine functions once the pandemic is over. It would be a challenge for big data scientists to analyze the vast hotels' data for identifying and tracking the epidemic among the guests, in a secure and privacy-preserving manner. Nevertheless, the practical implications to AI adoption in various types of hotels and among relevant customer market segments are absolutely fascinating. AI applications can be regarded as helping or complicating service interactions (Mariani and Perez Vega, 2020) and the success of these innovations is highly dependent on the particular customer's comfort with technology (Wen et al., 2020). As the present pandemic situation is full of fear and apprehension, it will be a daunting task for AI to stabilize the guests' emotions and psychological stress. Furthermore, even though AI and robotics will be able to provide a personalized service to the guests, it would be a challenging task to build a personal relationship with the hotel guest.

Considering the post-COVID-19 operation and future of the hotel service, the contactless service will be the new normal for hotel staff as well as hotel guests. However, it requires large capital investment for technological innovations and their return on investment will determine a hotel's willingness or reluctance toward robotic and AI-based operations. As AI is gradually replacing human workers in hotels, more employees will eventually lose their jobs, and a mismatch of "servicescape" will be generated between machine-induced services and human touch services, which may be sometimes not accepted by the hotel guests (Huang and Rust, 2018). When incorporating the contactless service in hotels, it would be a challenging task for hoteliers to restructure and reengineer the hotel product offerings and market these offers effectively. Due to technological innovations and demand, there may be huge variations in the supply chain, service chain, and production chain, which may cause some problems.

The incorporation of contactless services in the hospitality industry leveraging technological innovations also needs to consider security and privacy issues. Since training AI models rely on data, it is often challenging to acquire the hotels' data, and further train the model on this data in a secure and privacy-preserving manner. As the data is susceptible to cyberattacks, it is imperative that data security principles are adhered to while providing the data for training AI models for providing contactless services in the hospitality industry. Thus, security and privacy issues need to be considered while providing machine-based contactless services in the hospitality industry.

22.7 FEASIBILITY & FUTURE DIRECTIONS

For hoteliers all over the world, the threat of COVID-19 has posed an unparalleled challenge. In addition to the technical resources available, the use of technology to mitigate the danger is driven by comprehensive research into customer expectations and attitudes. The hospitality industry will focus on many issues and challenges post-COVID-19 when the new normal situation arises because traditionally hotels would strive to achieve the guest satisfaction level through human-rendered personalized services as compared to machine-rendered services. However, postpandemic, much more deliberation, and speculation would be focused on different kinds of processes and methodologies for carrying out the daily operations. Hotel guests will demand even tougher hygiene practices which will make contactless service a necessity. Furthermore, the importance of digital marketing will increase further post-COVID-19. The hospitality industry has invested in digital marketing during and post-COVID-19 to retain and enhance its revenue. Additionally, the hospitality industry has followed the Darwinian evolutionary theory concept of "survival of the fittest" during COVID-19 as the hotels and restaurants that have not adapted to the changing requirements due to the pandemic have gone out of business. On the other hand, hotels that have embraced technological innovations, such as contactless service, in their daily operations have sustained during the pandemic.

From the perspective of the hotel sector, the business is expected to rebound by the first cycle of 2022. The pandemic might be a catalyst triggering the hotel industry to be more tech-oriented. When this pandemic is over, it would be a great opportunity for researchers and practitioners to find the future directions for safe hotel operations by taking up researches from many dimensions. Post-COVID-19, the higher degree of sanitation and cleanliness would be a new way of life for customers, hence future research should investigate whether the younger generation customers show their willingness to pay a premium price for the redefined/upgraded level of sanitation and hygiene. Furthermore, hotel-centric healthcare aspects will be another new emerging area for research. Hotels will invest a great deal of money in providing hotel-centric health care facilities. Furthermore, to safeguard the health and safety of guests, a synergic relation will be established between hotel cleaning staff and AI-driven cleaning robots to provide enhanced hygiene and cleanliness in the hotel premises.

Considering the increased role AI and robotics can play in contactless services, further research needs to be conducted to identify the role of AI and its feeling for guest care and empathy during the services. There will be an increasing demand from travelers post-COVID-19 on hotel guest service, hence the forthcoming research would investigate the why and how AI is significant for hotel service delivery. Another new emerging research area for future research direction is how hotels will be able to analyzc real-time data to forecast future emergencies. The future of the hospitality industry postpandemic will greatly depend on the perception of hotel guests from different states and countries toward the satisfaction and experience of the contactless services digitalized by AI, robotics, big data analytics, 5G, and IoT. Apart from the above, crisis management is another significant subject necessary for further study, and it is important to investigate how hotels can build infectious disease control contingency plans.

22.8 CONCLUSIONS

The global uncertainty and chaos associated with COVID-19 may have a lasting effect on the hospitality industry. In order to reinstate the confidence of customers and to help the hospitality industry to rebuild and stabilize, there is a need for adhering to and implementing the healthcare guidelines provided by the health authorities. The time has come for the hospitality industry to identify the changing needs of customers and their satisfaction. The pandemic has put hidden hotel attributes like guest wellness, contactless service, and green concept at the forefront. Hotels' approach and embracing the contactless service will further open new avenues of skilled and professional personnel required for future hotel sustenance and survival. The current research agenda and framework in this article are not comprehensive but reveal many aspects in which the hotel industry and the service paradigm are expected to be influenced by COVID-19. This study future aims to explore the use of technological innovations, such as AI, robotics, big data analytics, 5G, and IoT in hotels to provide enhanced sanitation, automated COVID-19 screening, and personalized contactless services to hotel guests (Mariani, 2019; Mariani et al., 2018; Mariani and Perez Vega, 2020).

This study will pave the way for future studies aimed at exploring how technology-based solutions could be introduced in hotels to minimize future service disruptions and failures, and protect guest loyalty, and increase the number of repeat customers. Considering that the COVID-19 pandemic has heavily impacted the hospitality industry, this study will usher researchers and practitioners to have a deeper insight into the implications derived from this crisis, and develop newer safe and contactless service approaches to improve the contemporary service practices in the hospitality industry.

KEYWORDS

- **contactless**
- **COVID-19**
- **hospitality industry**
- **service quality**
- **artificial intelligence**
- **robotics**
- **Internet of Things**

REFERENCES

Bartik, A. W.; Bertrand, M.; Cullen, Z. B.; Glaeser, E. L.; Luca, M.; Stanton, C. T. *How are Small Businesses Adjusting to COVID-19? Early Evidence from a Survey (No. w26989)*. National Bureau of Economic Research, 2020.

BBC News. Virus 'Could Cost Millions of Tourism Jobs', 2020. https://www.bbc.com/news/business-51852505 (accessed on March 27, 2020).

Benckendorff, P. J.; Xiang, Z.; Sheldon, P. J. *Tourism Information Technology*, 3rd ed.; CABI: Wallingford.

Chan, E. S.; Lam, D. Hotel Safety and Security Systems: Bridging the Gap Between Managers and Guests. *Int. J. Hospi. Manag.* **2013,** *32*, 202–216.

Chan, J. F. W.; Yuan, S.; Kok, K. H.; To, K. K. W.; Chu, H.; Yang, J.; Xing, F.; Liu, J.; Yip, C. C. Y.; Poon, R. W. S.; Tsoi, H. W. A Familial Cluster of Pneumonia Associated with the 2019 Novel Coronavirus Indicating Person-to-Person Transmission: A Study of a Family Cluster. *Lancet.* **2020,** *395* (10223), 514–523.

Choi, J. Is Cleanliness Really a Reason for Consumers to Revisit a Hotel? *J. Environ Health.* **2019,** *82* (5), 16–22.

Colombo, M. G.; Piva, E.; Quas, A.; Rossi-Lamastra, C. How High-Tech Entrepreneurial Ventures Cope with the Global Crisis: Changes in Product Innovation and Internationalization Strategies. *Ind. Innov*. **2016,** *23* (7), 647–671.

Courtney, E. J. COVID-19 Will Cause a Significant Decline in Global RevPAR, Cash Flow, for Rated Lodging Companies, 2020. https://www.spglobal.com/ratings/en/research/articles/200311-covid-19-will-cause-a-significant-decline-in-global-revpar-cash-flow-for-rated-lodging-companies-11382396#ID493

Dhawan, S. Online Learning: A Panacea in the Time of COVID-19 Crisis. *J. Edu Tech Syst.* **2020,** *49* (1), 5–22.

Drexler, N.; Lapré, V. B. For Better or for Worse: Shaping the Hospitality Industry Through Robotics and Artificial Intelligence. *Res. Hospi. Manag.* **2019,** *9* (2), 117-120.

Fong, M. W.; Gao, H.; Wong, J. Y.; Xiao, J.; Shiu, E.; Ryu, S.; Cowling, B. J. Non-Pharmaceutical Measures for Pandemic Influenza in Non-Healthcare Settings—Social Distancing Measures. *Emerg. Infect. Dis*. **2020,** *26* (5), 976–984.

Glusac, E. Bring in the Robot Cleaners: Travel Industry Innovations for the Pandemic, March 28, 2020. https://www.nytimes.com/2020/03/28/travel/coronavirus-hotels-private-jets-virtualspas.html (accessed on March 29, 2020).

Guevara, G. Open Letter from WTTC to Governments, 2020. https://www.wttc.org/about/media-centre/press-releases/press-releases/2020/open-letter-from-wttc-to-governments

Guzzo, E. Coronavirus Pandemic: A Call to Action for the Robotics Community, March 20, 2020. https://spectrum.ieee.org/automaton/robotics/medical-robots/coronavirus-pandemiccall-to-action-robotics-community (accessed on March 28, 2020).

Hall, C. M.; Scott, D.; Gössling, S. Pandemics, Transformations and Tourism: Be Careful What You Wish For. *Tour. Geog.* **2020,** 1–22.

Hoisington, A. 5 Insights About How the COVID-19 Pandemic Will Affect Hotels, 2020. https://www.hotelmanagement.net/own/roundup-5-insights-about-how-covid-19-pandemic-will-affect-hotels

Hoque, A.; Shikha, F. A.; Hasanat, M. W.; Arif, I.; Hamid, A. B. A. The Effect of Coronavirus (COVID-19) in the Tourism Industry in China. *Asian J. Multidisc. Stud.* **2020,** *3* (1), 52–58.

Huang, M.; Rust, R. T. Artificial Intelligence in Service. *J. Serv Res*. **2018,** *21* (2), 155–172.

Huang, M. H.; Rust, R. T. Engaged to a Robot? The Role of AI in Service. *J. Serv. Res*. **2020,** *1*, 30–41. https://doi.org/10.1177/1094670520902266

Hung, K. K.; Mark, C. K.; Yeung, M. P.; Chan, E. Y.; Graham, C. A. The Role of the Hotel Industry in the Response to Emerging Epidemics: A Case Study of SARS in 2003 and H1N1 Swine Flu in 2009 in Hong Kong. *Global Health.* **2018,** *14* (1), 117.

IBM. 10 Reasons Why AI-Powered, Automated Customer Service is the Future, 2017. https://www.ibm.com/blogs/watson/2017/10/10-reasons-ai-powered-automated-customer-service-future/ (accessed on Nov 16, 2020).

Jamal, T.; Budke, C. Tourism in a World with Pandemics: Local-Global Responsibility and Action. *J. Tour. Futures*, 2020. https://doi.org/10.1108/JTF-02-2020-0014

Jena, P.Impact of Covid-19 on Higher Education in India. *Int. J. Adv. Educ. Res.* **2020,** *5* (3), 77–81.

Kansakar, P.; Munir, A.; Shabani, N. Technology in the Hospitality Industry: Prospects and Challenges. *IEEE Consum. Electr. Magaz.* **2019,** *6* (3), 74–82.

Kent, C. How Are Robots Contributing to the Fight Against Coronavirus? February 5, 2020. https://www.medicaldevice-network.com/features/coronavirus-robotics/ (accessed on March 28, 2020).

Kuo, C. M.; Chen, L. C.; Tseng, C. Y. 2017. Investigating an Innovative Service with Hospitality Robots. *Int. J. Contemp. Hospital. Manage.* **2017,** *29* (5).

Li, J. J.; Bonn, M. A.; Ye, B. H. 2019. Hotel Employee's Artificial Intelligence and Robotics Awareness and Its Impact on Turnover Intention: The Moderating Roles of Perceived Organizational Support and Competitive Psychological Climate. *Tour. Manage.* **2019,** *73*, 172–181

Lu, L.; Cai, R.; Gursoy, D. Developing and Validating a Service Robot Integration Willingness Scale. *Int. J. Hospital. Manage.* **2019,** *80*, 36–51.

Mariani, M. Big Data and Analytics in Tourism and Hospitality: A Perspective Article. *Tour. Rev.* **2019,** *75* (1), 299–303.

Mariani, M. M.; Perez Vega, R. Beyond the Hype: Psychological Mechanisms Enabling the Acceptance, Adoption, and Engagement with Artificial Intelligence Technology in Marketing. *Psychol. Market.* 2020.

Mariani, M. M.; Baggio, R.; Fuchs, M. and Höpken, W. Business Intelligence and Big Data in Hospitality and Tourism: A Systematic Literature Review. *Int. J. Contemp. Hospital. Manage.* **2018,** *30* (10), 3514–3554.

Meisenzahl, M. How China, the US, and Europe Are Using Robots to Replace and Help Humans Fight Coronavirus by Delivering Groceries, Sanitizing Hospitals, and Monitoring Patients, 2020. https://www.businessinsider.com/ robots-fighting-coronavirus-in-china-us-and-europe-2020-3 (accessed on March 28, 2020).

Miller, M. R.; Miller, R. *Robots and Robotics: Principles, Systems, and Industrial Applications*; McGraw-Hill Education: New York, 2017.

Park, H.; Kline, S. F.; Kim, J.; Almanza, B.; Ma, J. Does Hotel Cleanliness Correlate with Surfaces Guests Contact? *Int. J. Contemp. Hospital. Manage.* 2019.

REVFINE. 8 Examples of Robots Being Used in the Hospitality Industry, 2021. https://www.revfine.com/robots-hospitality-industry/

Russell, S. J.; &Norvig, P. *Artificial intelligence: A modern approach.* Harlow: Pearson Education Limited.

Seyitoğlu, F.; Ivanov, S. Service robots as a tool for physical distancing in tourism. Current Issues in Tourism (in press).

Toh, M.; Wang, S. Drones. Disinfecting Robots. Supercomputers. The Coronavirus Outbreak Is a Test for China's Tech Industry, February 24, 2020. https://edition.cnn.com/2020/02/23/tech/china-tech-coronavirus-outbreak/ index.html (accessed on March 30, 2020).

Tung, V. W. S.; Law, R. The Potential for Tourism and Hospitality Experience Research in Human-Robot Interactions. *Int. J. Contemp. Hospital. Manage.* **2017**.

Tussyadiah, I. A Review of Research into Automation in Tourism: Launching the Annals of Tourism Research Curated Collection on Artificial Intelligence and Robotics in Tourism. *Ann. Tour. Res.* **2020,** *81*, 102883.

UNWTO. Tourism and Covid-19, 2020. https://www.unwto.org/tourism-covid-19-coronavirus (accessed on March 29, 2020).

Wen, J.; Kozak, M.; Yang, S.; Liu, F. COVID-19: Potential Effects on Chinese Citizens' Lifestyle and Travel. *Tourism Rev.* **2020**. DOI: 10.1108/TR-03-2020-0110

WHO. Q&A on Coronaviruses (COVID-19), 2020. https://www.who.int/newsroom/q-a-detail/q-a-coronaviruses

World Economic Forum. This Is How Coronavirus Could Affect the Travel and Tourism Industry, 2020. https://www.weforum.org/agenda/2020/03/world-travelcoronavirus-covid19-jobspandemic-tourism-aviation/ (accessed on March 20, 2020).

World Travels and Tourism Council. Impact of Pandemic on Travel and Tourism, 2020. https://www.wttc.org/priorities/

Yang, G. Z.; Nelson, B. J.; Murphy, R. R.; Choset, H.; Christensen, H.; Collins, S. H.; Dario, P.; Goldberg, K.; Ikuta, K.; Jacobstein, N.; Kragic, D.; Taylor, R. H.; McNutt, M. Combating COVID-19—The Role of Robotics in Managing Public Health and Infectious Diseases. *Sci. Robot.* **2020,** *5* (40), eabb5589. DOI: 10.1126/scirobotics.abb5589

Yu, C. E. Humanlike Robots as Employees in the Hotel Industry: Thematic Content Analysis of Online Reviews. *J. Hospital. Market. Manage.* **2020,** *29* (1), 22–38.

Zabin, J. Artificial Intelligence: Working Hand in Hand with Hotel Staff, 2019. https://hoteltechnologynews.com/2019/07/artificial-intelligence-working-hand-in-hand-with-hotel-staff/

CHAPTER 23

GLOBAL CRISES: POST-COVID SUSTAINABLE STRATEGIES FOR THE TOURISM AND HOSPITALITY INDUSTRY

OMVIR GAUTAM[1], NAMAN SHARMA[2], and POOJA AGRAWAL[1]

[1]*Department of Commerce and Management, Vishwakarma University, Pune, Maharashtra, India*

[2]*Department of Management, Indian Institute of Foreign Trade, Kolkata, West Bengal, India*

ABSTRACT

The COVID-19 has emerged as challenging virus in the globe. Global tourism and hospitality have witnessed losses as there is no vaccine and merely *Nonpharmaceutical Interventions* are considering as a most feasible curable approach to combat this current pandemic. The-tourism industry has downturn and responsible for losses of millions of jobs, business opportunities for micro, small medium and large tourism, and hospitality enterprises globally, these virus outbreaks are considering the worst pandemic in last five decades. All most countries-imposed travel restrictions on airlines, tourism, and hospitality event sector worldwide. The tourism industry already experienced pandemics such crisis most recently, that is, SARS, Middle East respiratory Syndrome, and Zika has shown similar impacts over tourism and hospitality industry. An inclusive outline of pandemics and its impact is studied from the present scenario and available literature. This present literature and factors going to assist to understand the effect of

Post-COVID Tourism and Hospitality Dynamics: Recovery, Revival, and Re-Start.
Umendra Narayan Shukla & Sharad Kumar Kulshreshtha (Eds.)

Coronavirus on tourism and hospitality sector. Some factors studied and also suggest some sustainable strategies in this tough pandemic situation which affect tourism and hospitality domain at world level. This chapter will focus post covid sustainable measures along with present challenges for tourism industry in order to be append by futuristic researchers, policy makers, and disaster management organizations to be prepared for such futuristic catastrophic endemics.

23.1 INTRODUCTION

The present virus of unexplained cause found in Wuhan, China, was emerged as disrupting toward global tourism and hospitality industry. The outbreak of this novel coronavirus (COVID-19) was first reported among 41 infected residents in inception of January 2020, China (Huang et al., 2020). The outrage of this virus was very rapid and unnoticed by various political leaders across different part of the world. The initial symptoms of this catastrophic event were issued by intelligence services; (Washington Post, 2020). The cause of this novel virus was unknown and many news confined the origination of this virus from Wuhan, China (CNN, 2020). It is the contagious disease caused by the most recently discovered coronavirus. It is now a pandemic affecting many countries globally either in economic growth or tourism sector (Miles and Shipway, 2020; Škare et al., 2021) There are many viruses found earlier across the world.

The tourism sector is often enforced to endure circumstances out of control, like natural calamities, earthquakes, endemics, global warning, and novel viruses like SARS-CoV (appeared and blowout in 2002–2003), Middle East respiratory Syndrome (MERS)-CoV (converted endemic in 2015), Zikavirus (well-known in 1952 in tropical Africa, Southeast Asia, and the Pacific Islands.), and COVID-19 (began burgeoning in December, 2019).

COVID-19 is more disastrous for hospitality and tourism industry across the globe than World War II (Hunt, 2020). Henceforth, struggling for sustainability Fotiadis et al. (2021). There are enduring and temporary layoffs foreseeable for the persistence of several global tourism and hospitality firms Shine (2020). The pandemic COVID-19 outbursts restrict tourist agility and hotels perceive a depression in tenancy tariffs. International, national and local travel has diminished to significant level. The number of infected patients globally doubled swiftly, since the infection rate speeded and turned to be a pandemic event (Johns Hopkins, 2020). The covid affected entire world significantly. Each nation started some measure to lift the lockdowns' purpose to make softer few short-range monetary losses instigated by COVID-19 crisis; therefore, world economy remains to decrease. The Governments and agencies are progressing their impetus as employment continues to down, and mutually consumption and demand declines for different sectors. This results in developed and developing nations come up with different forms of nonpharmaceutical interventions (NPI), comprising lockdown (quarantine, isolation) postponing or abandoning of major events (i.e., tradeshows, sports events, concerts, summer Olympics, political arguments, social gathering, conferences, seminars, and festive meets). These measures were only subjective measures as shortage availability to curb the disease and only medical interventions accessible to treat it. There were adverse effects over the economics of all the countries, the United Nations estimates global gross domestic product diminish to 3.2%. The developed nations will down to 5%, and their productivity drop to 0.7% (S&P Global Market Intelligence, 2020).

This article note suggests a new post-COVID-19 framework of hospitality and tourism sustainable strategies. The subsequent main outcomes from the Coronavirus (COVID-19) pandemic inform the framework.

1) Unavailability of sustainable standard operating procedures, even with a safety-net stacking, hospitality, and tourism people earn do not earn optimal income in order to fight against this pandemic.
2) Service industry affected most, as primarily it has been complete lockdown. But, the demand for essential services experienced a lot. The tourism and hospitality employees undergone into hibernation, while other sector professionals are highly engaged like medical filed, administration, and municipal corporations.
3) Intensive care and vaccine development felt highest growth and engagement (Humphrey and Nash, 2020); or transferrable skills enabling employee's mobility from dormant to swift services of the economy.
4) Development immunity is anticipated more often in future, as the employees will learn the resilient attitude toward precautionary measures (World Economic Forum, 2020).

This study gives new insights by putting forward a new pandemic framework of tourism and hospitality prospective challenges. Our study offers hospitality and tourism employer's enhanced levels of job prospective, ensure a higher and optimum utilization of existing workforce, the augmented skill sets more likely to be harness in endemic time for emergency contribution of employees Rastegar et al. (2021). The hands-on experience or value of this suggested framework lies in its contribution toward pandemic–proofing tourism and hospitality employers around the globe.

23.1.1 TOURISM, GLOBAL IMPACT, AND PANDEMICS

This pandemic thrashes all business across the globe like ski destination Ischgl town in Austria (Anderson et al., 2020). The tourism and hospitality found mostly affected business for the entire nations. One of the most impactful area under tourism is airline industry Budd et al. (2020), as it is directly affecting the national economies and also responsible for the economy loss of nations. The entire airline industry is complete lockdown and staggered 100% decline in bookings in this wretched situation Maneenop and Kotcharin (2020). The tourism systems elements such as international tours, travels, local tourism, day itineraries, cruises, public–private transport, cafés and restaurants, festive events, sports events and accommodation, hotel bookings, and international sports mega events like Olympics 2020 and ICC Cricket champions trophy World Cup has been affected and canceled long lasting. The impact is now unforeseen but in future, the losses to various nations will be in a significant economic crisis Baum and Hai (2020). The tourism sector is now one of the worst-hit by the outbreak of Corona Virus (COVID-19), affecting both demand and supply of travelers. This denotes one more shortcoming and risk with respect to a weaker global economy, geo-graphical, geopolitical, trade tensions, and irregular enactment between foremost outbound travel markets (UNWTO, 2020). This outbreak, that is, COVID-19 clear uncontrollable which adversely not affect the economies but also there are some pertinent factors and events at global level unpleasantly distress the tourism and hospitality sector and its outcome as a performance (Brouder et al. 2020).

The tourist across the different parts of the world were halted and due to this outbreak, they felt stressed and unsafe till now. The domestic and global tourism deceased swiftly over a span of just few months in 2020. In April 2021, many of international flights or tourism are at stake, hence no international flights no revenue generation to the hospitality allied businesses Lew et al. (2020). Currently, UNWTO predicts that international tourist appearances could deteriorate from 20 to 30% in 2020, it is also estimated that growth rate may remain from 3 to 4% estimate in initial of January 2020. Further, this will in result into a loss of US$ 30 to 50 billion in outlay by global visitors (international tourism receipts). Table 23.1 shows pandemic indicators like number of deaths and number of confirmed COVID-19 cases.

As tourism the tourism arrivals had considered for our study. The data has been extracted from the WHO and World Bank, May 2020. It is noteworthy that tourism has been exposed to a lot of challenges earlier also, but in March–April 2020, it was emerged as catastrophe. The major catastrophic events were like Hantavirus (1993), SARS outbreak (2003), ZIka virus (2005), global financial crisis in 2009, and MERS. All these crises have unfolded the tourism economy to slow down across the world. This present virus outbreak would affect the tourism and hospitality sector to maximum extent. It is also evident that the impact will be negative on the economies of the nations but will also recover from this pandemic as recovered in earlier crises also (Burkle, 2006).

TABLE 23.1 Matrix as Number of Confirmed, Deaths and Tourism Arrivals at Globally of COVID-19.

Nation	Number of Confirmed cases	Number of Deaths	Tourism arrivals
United States of America	1,432,265	8718	120,000
Russian Federation	290,678	2722	65,000
The United Kingdom	243,699	34,636	78,522
Brazil	233,142	15,633	14,562
Spain	23,135	2765	8922
Italy	225,435	31,908	21,054
Germany	174,697	7935	14,587
Turkey	149,435	414	4521
Iran	120,198	6988	3658
India	96,169	3029	7869

Source: Author compiled data from WHO and World Bank till 17 May, 2020.

Figure 23.1, demonstrates the data of top 10 nations in terms of highest COVID-19 confirmed cases with the Tourism footfalls and Number of deaths in Dec 2019–May 2021. As mentioned, this period was zero-hour period due to its mysterious nature and different tourism tycoons did not understand the reasons. So, series 1 denotes the relationship between number of confirm cases and number of deaths, covering statistics from December, 2019 to May 17, 2020. The series 2 displays the association among the confirmed cases and tourist arrivals.

The number of confirmed cases were highest in United Sates of America 1,432,265 followed by Russian federation 290,678, The United Kingdom 243,669, and so on as shown in Table 23.1.

The associations among pandemics and tourism are major issue to understand with global precautionary measures. Our study also establishes the relationship between pandemic and number of tourist footfalls and developed minimum rapid potential measures like lock down, NPI, and self/

imposed quarantine mechanism till May 2020. The exponential equation (trend line) mentioned for pandemics such as number of confirmed cases, no. of deaths, and tourism footfalls. It was predicted by researcher primary data for depicting only exponential function.

$$Y(c,t)=9.4327e^{0.002X} \ldots\ldots\ldots\ldots\ldots(a)$$

$$Y(c,d)=6.7565e^{0.0019x} \ldots\ldots\ldots\ldots\ldots (b)$$

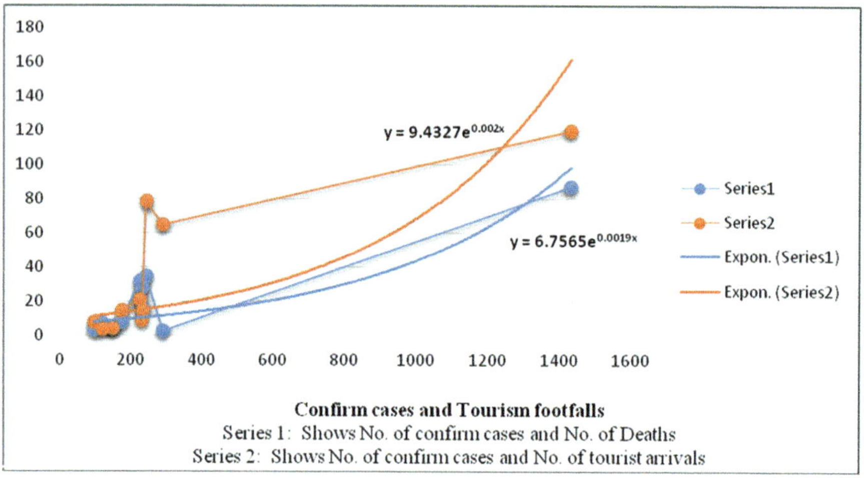

FIGURE 23.1 Impact of pandemics on world tourism (Dec 2019–May 2020).

Source: Researcher draw chart on basis of available data on WHO dashboard till May 2020.

Equation (a) and (b) Mayes (1994). We have tried to show the relationship among all required variables; equation (a) shows the confirmed cases with tourist rate and clearly, the relationship is exponential. Similarly, equation (b) reveals the relationship among confirmed cases with number of deaths. In both the equation, the relationship is exponential and depicts future concern for the hospitality and tourism industry.

23.2 GLOBAL TOURISM AND COVID-19

At international level, there is a huge loss of global tourists, which counts to loss of billions USD in export returns from global tourism. tourism jobs that are at risk, the policy paper said jobs in associated sectors, including food service, that provide employment for 144 million workers worldwide are also at risk. The jobs risk accounts nearly 120–144 million jobs cut or at risk in this pandemic situation (New Indian express, 25 Aug 2020). Currently, Covid-19 played a major role in downturn of the tourism. The world has witnessed major health crisis earlier also in last five decades, yet none had analogous impact over global economy as COVID-19 did. The international tourism has a loss of 0.85 billion to 1.1 billion global visitors in terms of counts (UNWTO, 2020). Epidemics like Ebola, SARS, and Zika measles were not asymptomatic transmissible like COVID-19 is, so employers adopting shedding mechanism against pandemic with certain known signs of virus enactment (Bai et al., 2020; Rothe et al., 2020). Consequently, people across the globe have initiated social distancing, covering nose, mouth, and wearing N 95 masks not to prevent virus spread through any of the measure like coughing, sneezing, or speaking closely. The hospitality and tourism industry do not have sufficient testing kits, unwittingly asymptomatic spread is functional (Li et al., 2020). Numerous studies explained the role of airline and impact of COVID-19. Mainly, the interest of researchers is to be reflective at the infancy stage of pandemic, since all different governmental research bodies and tourism researchers identified the cause of pandemic. Virus outbreaks existed many times in last several decades, also warned major threats and concerns for the tourism, society, and for the societal stakeholders (Gössling, 2002, 2016, 2020; Page and Yeoman, 2007; Scott and Gössling, 2015). The proposed reasons could be rapid population bombarding industrialization in food industry, international tourism leading to not disposal mechanism, global carriage networks as paths in spreading of virus or pathogens (Labonte et al., 2011; Gössling, 2020).

23.3 GLOBAL REVENUE AND REGIONAL TOURISM PERFORMANCE

The global revenue has down drastically in travel and tourism industry (Remeikis, 2020; Fotiadis et al., 2021). The income in certain nations in 2019 and probable effect of COVID-19 crisis in 2020 (in million U.S. dollars) are shown in Figure 23.2. The global revenues in millions U.S. dollars with respect to end of 2019 and start 3 months 2020. The United States revenue decreased from 156,266 USD to 111.600 USD noting a downfall of 28.6%, China recorded loss in revenues from 117,843 USD to 78,288 USD showing a downfall of 33.6%, Germany records losses from 31,821 USD to 19,468 USD with downtrend of 38.8% and Italy posited losses from 20,257 USD to 11,153 USD with fall of 49.3%. The subjected reasons could be as follows; Deterioration of tourist arrivals and receipts and export revenues, global tourism spending across globe slowdowns due to of COVID-19. As per Mobility Market Outlook on COVID-19, profits for domain of travel and tourism from Italy confirmed utmost affected by COVID-19, further, it declining from 20.3 billion U.S.D to 11.2 billion U.S.D as shown in Figure 23.2.

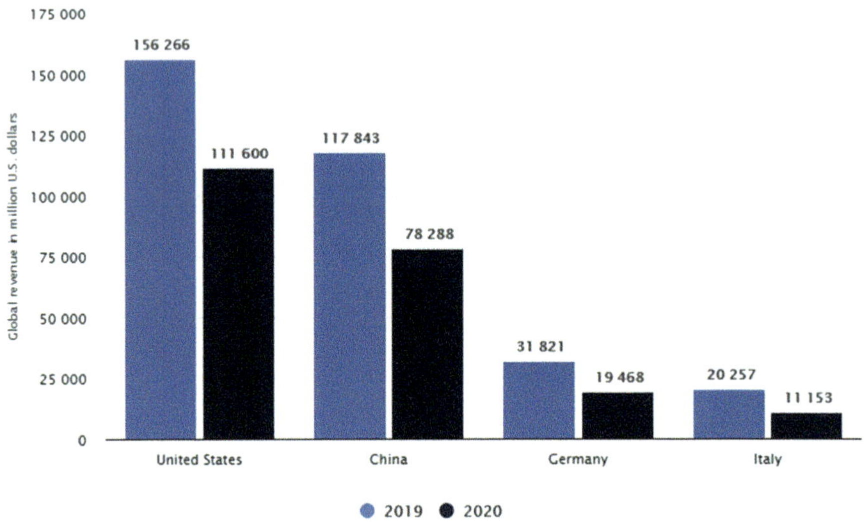

FIGURE 23.2 Revenue for the travel and tourism industry.

Source: Compiled by author from WHO and Statista 2020 and Mobility Market Outlook.

23.4 UNEMPLOYMENT (JOB LOSSES) AND MONTHLY LIFTING OF RESTRICTIONS AT GLOBAL LEVEL

The current pandemic crises forecasts lay-offs in travel and tourism domain across globally in 2020 (Lock, 2020). The outcome of the coronavirus pandemic is perceived in terms of impact over global travel and tourism. The repercussion of COVID-19 epidemic streaked about 81 million jobs in 2020 (ILO, 2020). The global nation records a biggest loss from Coronavirus is the Asia Pacific region, losing almost 63.4 million jobs, whereas Europe is estimated to second worst-hit with an anticipated down employment rate to 13 million (World economic Outlook, Oct 2020, IATA, 2020; UNWTO, 2020). While a decline of between 58 and 78% is predicted for 2020, this would imply a loss of between USD 910 billion and USD 1.2 trillion in tourism export revenue (UNWTO, May 2020). Tourism receipts are unlikely to recover to precrisis levels until 2023. According to current IMF research on tourism in a postpandemic world, global visitor arrivals decreased by >65% in the first half of this year, with a near-halt since April, compared to 8% during the global financial crisis and 17% during the 2003 SARS outbreak.

The upliftment of restrictions initiated across various nations with precautionary measures as shown in Figure 23.3. As per reports of UNWTO (2020) anticipated a 20–30% deterioration in 2020 global arrivals accounted for losses of tourism receipts of 300–450 USD billion. Henceforth, much more is anticipated by WTTC (2020), expecting a loss of up to 2.1 USD trillion in 2020 (Gössling, 2020; IATA, 2020). The monthly change has and monthly projections in terms of losses in USD has been estimated. It is clear that till May end, 2020 all estimates scenario 1. Scenario 2 and scenario 3 has been decreased with 56%, 70%, and 78%, respectively. Gradual opening of borders will start from June onwards and shown the recovery in terms of lifting travel restrictions tourism. Moreover, international tourism 2020 Circumstances have been witnessed. Projections for year 2020 have been reduced numerous times since outburst of virus, hence ambiguity endures to govern. As per contemporary situation is concerned, the tourist footfalls drops from 58 to 78% in 2020. Tourist footfalls relies upon the virus spread, duration of travel constraints, and international borders shutdowns. The subsequent settings for 2020 are founded upon three potential announcements (dates) for opening of global borders.

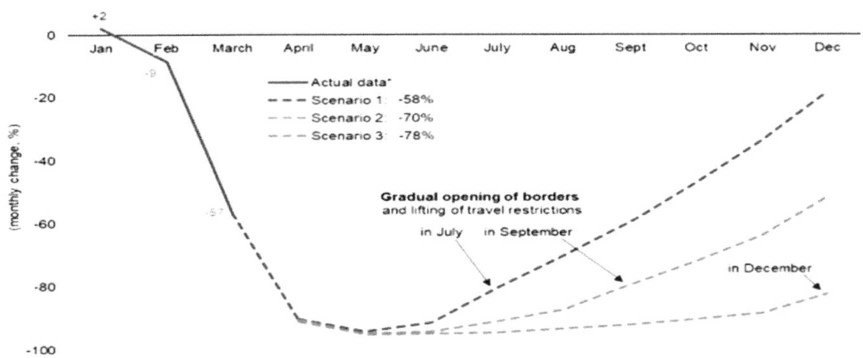

FIGURE 23.3 Monthly variation in tourist arrivals based on ease down of national borders and lifting travel restrictions.

Source: Researcher compiled data available with UNWTO, May 2020.

Scenario 1: (−58%) lifting of global borders to ease travel constraints in initial part of July this year

Scenario 2: (−70%) lifting of global borders to ease travel limitations in early of September

Scenario 3: (−78%) lifting of global borders to ease travel restrictions in early December

Travel experts perceive recovery in 2021 but it need to be well-directed through various measures. It was also estimated that domestic demand will recover faster than global demand.

As per reports of UNWTO Panel of Experts survey is concerned the symbols of regaining majorly by the last quarter of 2020, however mostly will be recovered in 2021. Similarly, earlier pandemic crisis, holiday/vacation travel is probable to improve earlier, mainly travel for spending vacations, to meet friends and kinsfolks, then commercial travel.

The projection about retrieval of global tourist and travel is affirmative in Africa and the Middle East with most experts anticipating recovery in 2020. Economy experts or professionals in the Americas are the minimum enthusiastic and minimum anticipated recovery in 2020, while in Europe and Asia the viewpoint is noteworthy, with half of the experts expecting to see recovery within this year. With estimated repossession hopes, leads to enhance the role of domestic travel. Survey from IATA among latest air tourists accompanied in April 2020 revealed that 58% are curbing their travel plans to domestic destinations only and even this journey is restricted.

Domestic Revenue Passenger Kilometers (RPKs) could recover by 2022 at global level. The reports also show that International RPKs are anticipated to normalcy by 2024 (Toursimanalytics, 2020). Mutual recognition was main deliverable of the COVID-19 Aviation Recovery Task Force (CART) of the International Civil Aviation Organization (ICAO, 2020).

The agency CART plays a pivotal and crucial role for boosting the international tourism. It must find a mechanism on the different measures required to curb coronavirus as airline industry revamps. The CART must design a framework between governmental bodies with different strategies including social distancing norms, NPI and quarantine facility with thermal detectors at different international airports, opening of various global borders with covered approach of measures and precautions which need to be implemented globally in a proper way.

23.5 IMPLICATIONS OF THE STUDY

For futuristic researchers, policy makers, and tourism educators, there is dearth to make certain amendments in methods, teaching capabilities, and sustainable strategies during and after crisis in future to relaxed the magnitude of the pandemic or epidemic with better proactive approach. In hospitality and Tourism, despite all these challenges, researchers', or educators' emotions (i.e., thwarting and fear) must not affect students' attitudes and performances.

As another implications, it is important to inspect the tourism and hospitality sector for drafting draft certain sustainable strategies to curb these challenges globally. Researchers or policy framers need to study and include pivotal stakeholders such as employees, businesses, clients, scholars, and talent to prepare a framework in order to survive before, during, and after a crisis sustainably?

Some research questions (RQ) developed as from study for future researchers as;

RQ 1: Is assessment of international tourism as a framework of recovery and sustained transformation to achieve robust destinations?

RQ 2: Do development framework will definitely curtail the loss in hospitality and tourism sector?

Moreover, apart from these two research questions, there is also dearth to understand the recovery phase after any pandemic. More focus on international tourism and hospitality with proper safety and health considerations must be given. It involves restructuring of global value chains as vector in pandemics (Gössling, 2002).

23.6 CONCLUSION AND POST COVID SUSTAINABLE STRATEGIES FOR TOURISM AND HOSPITALITY SECTOR

The pandemic Coronavirus offers a catalytic stimulus to the tourism sector, tourism researchers, and policymakers about the special effects of global slowdown. People had developed will power and also understand the post ill effects of Covid-19. The operational level assessment outraged the effect of outbursts on everyone's life. The unemployment rate risen exponentially in many nations (US Bureau of Labor Statistics, 2020). As final words of this article consist of post covid sustainable strategies for tourism and hospitality sector are catered. There is a dearth to formulate more sustainable strategies in order to get rid of catastrophic effects of this pandemic. The Triple Billion targets are an ambitious initiative to improve the health of billions of people by 2023. They are the foundation of WHO's Thirteenth General Programme of Work 13 acting as both a measurement and a policy strategy. As post sustainable strategy for Tourism and Hospitality Industry the following points are explained. First strategy could be defined as education, environmental and social justice, and racial healing must be prioritized in the industry (Stankov et al., 2020; Nepal, 2020). Second strategy is to focus more upon technological advancement, as it plays a significant role in growing flexibility in the tourism business. Natural disasters hasten technological change Hall et al. (2020). People received a lot of help from tech savyy during COVID-19. It can be extended in Big Data analytics forecasting. replacing people with robots, mobile applications tracking based services.

Thirdly strategy could be utilizing Sendai Framework's for tourism and hospitality resilience with Sustainable Development Goals. Fourth strategy would thrust over the sustained Corporate Social Responsibility goals unified by some integrated agency. Like Melinda & Gates foundations could associate with IMF or United Nations for unified strategy in order to contribute significantly in hours of economic crisis Mattera et al. (2021). The contest is now to mutually absorb from this international slowdown to hasten the revolution of sustainable tourism. One important factor for virus outbursts are living beings inquisitive and galloped with disease as habit in long term sustenance (Hall e al., 2013; Barlow et al., 2016; Lade et al., 2020; Hunt, 2020).

The current pandemic could centrally to a precarious review of the global tourism numbers, for consistent reasons of risks sustained in international travel and hospitality and its influence to climate change (World Economic Forum., 2019). The Tourism "success" predominantly defined as the segment of all tourism organizations – WHO, UNWTO, ICAO, and UNWTO as development as tourist counts. This outlook seen as in context of the global economic slowdown (Hall, 2009) and as the demerits of over domain like tourism, climate change, and pandemic COVID-19 has been ascertained for the conclusive remarks with respect to tourism and hospitality domain.

KEYWORDS

- **COVID-19**
- **global crisis**
- **hospitality**
- **pandemic**
- **outbreaks**
- **sustainable tourism**
- **strategies**

REFERENCES

Bai, Y.; Yao, L.; Wei, T.; Tian, F.; Jin, D.-Y.; Chen, L.; Wang, M. Presumed Asymptomatic Carrier Transmission of COVID-19. *JAMA* **2020,** *323* (14), 1406. https://doi.org/10.1001/jama.2020.2565

Barlow, J.; Lennox, G. D.; Ferreira, J.; Berenguer, E.; Lees, A. C.; Nally, R. M.; Thomson, J. R.; Ferraz, S. F. d B.; Louzada, J.; Oliveira, V. H. F.; Parry, L.; Ribeiro de Castro Solar, R.; Vieira, I. C. G.; Arag~ao, L. E. O. C.; Begotti, R. A.; Braga, R. F.; Cardoso, T. M.; de Oliveira, Jr, R. C.; Souza Jr, C. M.; Gardner, T. A. Anthropogenic Disturbance in Tropical Forests Can Double Biodiversity Loss from Deforestation. *Nature* **2016,** *535* (7610), 144–147. https://doi.org/10.1038/nature18326.

Baum, T.; Hai, N. T. T. Hospitality, Tourism, Human Rights and the Impact of COVID-19. *Int. J. Contemp. Hospital. Manage.* **2020.**

Brouder, P.; Teoh, S.; Salazar, N. B.; Mostafanezhad, M.; Pung, J. M.; Lapointe, D.; et al. Reflections and Discussions: Tourism Matters in the New Normal Post COVID-19. *Tour. Geog.* **2020,** *22* (3), 735–746.

Budd, L.; Ison, S.; Adrienne, N. European Airline Response to the COVID-19 Pandemic–Contraction, Consolidation and Future Considerations for Airline Business and Management. *Res. Transp. Busi. Manage.* **2020,** *37*, 100578.

Burkle, F. M. Jr, Globalization and Disasters: Issues of Public Health, State Capacity and Political Action. *J. Int. Affairs* **2006,** *59* (2), 231–265.

CNN. What Will Travel Look Like After Coronavirus? 2020. https://www.cnn.com/travel/article/coronavirus-travel-industry-changes/index.html (accessed on May 17, 2020).

Databank. **2020,** 17 May. https://data.worldbank.org/ (accessed on May 18, 2020).

Fotiadis, A.; Polyzos, S.; Huan, T. C. T. The Good, the Bad and the Ugly on COVID-19 Tourism Recovery. *Ann. Tour. Res.* **2021,** *87*, 103117.

Frempong, R. B.; Stadelmann, D.; Wild, F. *Coronavirus-Lockdowns, Secondary Effects and Sustainable Exit-Strategies for Sub-Saharan Africa* (No. 2020-09). CREMA Working Paper, 2020.

Gössling, S.; Scott, D.; Hall, C. M. Pandemics, Tourism and Global Change: A Rapid Assessment of COVID-19. *J. Sustain. Tour.* **2020,** 1–20.

Hall, C. M.; Scott, D.; Gössling, S. Pandemics, Transformations and Tourism: Be Careful What You Wish For. *Tour. Geog.* **2020,** *22* (3), 577–598.

Hall, C. M.; Gössling, S., Eds. *Sustainable Culinary Systems: Local Foods, Innovation, and Tourism & Hospitality*; Routledge, 2013.

http://bwhotelier.businessworld.in/article/International-Tourist-Numbers-could-fall-60-to-80-per-cent-in-2020-UNWTO-Reports/08-05-2020-191606/ (accessed on May 8, 2020).

http://tourismanalytics.com/news-articles.html (accessed on May 17, 2012).

https://covid19.who.int/?gclid=CjwKCAjwwYP2BRBGEiwAkoBpAh_xzMJFBW5x2T-WFj68u9e43OtdyTytjX4rCFDANhntyjqjf7agm4hoCExoQAvD_BwE (accessed on May 15, 2020).

https://documents1.worldbank.org/curated/en/280131589922657376/pdf/MENA-Crisis-Tracker-June-27-2021.pdf (accessed on June 27, 2021).

https://webunwto.s3.eu-west-1.amazonaws.com/s3fs-public/2020-04/COVID19_NewDS_.pdf (accessed on May 16, 2020).

https://www.icao.int/Newsroom/Pages/ICAO-Council-adopts-new-COVID.aspx

https://www.ilo.org/asia/media-centre/news/WCMS_763819/lang--en/index.htm (accessed on March 2021).

https://www.newindianexpress.com/world/2020/aug/25/covid-19-impact-global-tourism-lost-usd-320-billion-in-five-months-says-un-2187991.html (accessed on Dec 23, 2020).

https://www.spglobal.com/en/research-insights/articles/covid-19-daily-update-may-15-2020 (accessed on May 15, 2020).

https://www.statista.com/statistics/1102915/covid-19-forecasted-real-gdp-growth-g20-countries/ (accessed on May 18, 2020).

Huang, C.; Wang, Y.; Li, X.; Ren, L.; Zhao, J.; Hu, Y.; Zhang, L.; Fan, G.; Xu, J.; Gu, X.; Cheng, Z.; Yu, T.; Xia, J.; Wei, Y.; Wu, W.; Xie, X.; Yin, W.; Li, H.; Liu, M. et al. Clinical Features of Patients Infected with 2019 Novel Coronavirus in Wuhan. *Lancet* **2020,** *395* (10223), 497–506. https://doi.org/10.1016/S0140-6736 (20)30183-5; https://edition.cnn.com/2020/03/31/health/what-is-coronavirus-covid-19-wellness/index.html (accessed on May 4, 2020).

Humphrey, T.; Nash, D. UQ Fast Tracks Nurses in Response to COVID-19 Pandemic, 2020. https://www.uq.edu.au/news/article/2020/04/uq-fast-tracks-nurses-response-covid-19-pandemic (accessed on April 16, 2020).

Hunt, R. Marriott CEO: Coronavirus Is Hurting the Hotel Chain Worse Than 9/11 or World War II, 2020. https://www.nasdaq.com/articles/marriott-ceo%3A-coronavirus-is-hurting-the-hotel-chain-worse-than-9-11-or-world-war-ii-2020 (accessed on March 25, 2020).

Labonte, R.; Mohindra, K.; Schrecker, T. The Growing Impact of Globalization for Health and Public Health Practice. *Ann. Rev. Public Health* **2011,** *32* (1), 263–283. https://doi.org/10.1146/annurev-publhealth-031210-101225

Lade, S. J.; Steffen, W.; de Vries, W.; Carpenter, S. R.; Donges, J. F.; Gerten, D.; Hoff, H.; Newbold, T.; Richardson, K.; Li, R.; Pei, S.; Chen, B.; Song, Y.; Zhang, T.; Yang, W.; Shaman, J. Substantial Undocumented Infection Facilitates the Rapid Dissemination of Novel Coronavirus (SARS-CoV2). *Science* **2020.** DOI: 10.1126/science.abb3221

Lew, A. A.; Cheer, J. M.; Haywood, M.; Brouder, P.; Salazar, N. B. Visions of Travel and Tourism After the Global COVID-19 Transformation of 2020, 2020.

Maneenop, S.; Kotcharin, S. The Impacts of COVID-19 on the Global Airline Industry: An Event Study Approach. *J. Air Transp. Manage.* **2020,** *89*, 101920.

Mattera, M.; Gonzalez, F. S.; Ruiz-Morales, C. A.; Gava, L. Facing a Global Crisis-How Sustainable Business Models Helped Firms Overcome COVID. *Corp. Govern.: Int. J. Busi. Soc.* **2021**.

Mayes, R. L. Discovering Relationships: Logarithmic and Exponential Functions. *School Sci. Math.* **1994,** *94* (7), 367–370.

Miles, L.; Shipway, R. Exploring the COVID-19 Pandemic as a Catalyst for Stimulating Future Research Agendas for Managing Crises and Disasters at International Sport Events. *Event Manage.* **2020.** https://doi.org/10.3727/152599519X15506259856688.

Nepal, S. K. Adventure Travel and Tourism After COVID-19–Business as Usual or Opportunity to Reset? *Tour. Geog.* **2020,** *22* (3), 646–650.

Rastegar, R.; Higgins-Desbiolles, F.; Ruhanen, L. COVID-19 and a Justice Framework to Guide Tourism Recovery. *Ann. Tour. Res.* **2021,** 103161.

Remeikis, A. Australian Government Won't "Jump to a Solution" to Help Casual Workers in Coronavirus Crisis. *The Guardian Website*, 2020. https://www.theguardian.com/world/2020/mar/10/australian-government-warned-not-to-drag-its-feet-on-help-for-casual-workers-in-coronavirus-crisis (accessed on March 25, 2020)

Rothe, C.; Schunk, M.; Sothmann, P.; Bretzel, G.; Froeschl, G.; Wallrauch, C.; Zimmer, T.; Thiel, V.; Janke, C.;Guggemos, W.; Seilmaier, M.; Drosten, C.; Vollmar, P.; Zwirglmaier, K.; Zange, S.; W€olfel, R.; Hoelscher, M. Transmission of 2019-nCoV Infection from an Asymptomatic Contact in Germany. *N. Engl. J. Med.* **2020,** *382* (10), 970–971. https://doi.org/10.1056/NEJMc2001468

Sharma, G. D.; Thomas, A.; Paul, J. Reviving Tourism Industry Post-COVID-19: A Resilience-Based Framework. *Tour. Manage. Perspect.* **2021,** *37*, 100786.

Shine, R. Coronavirus Pandemic Leaves Businesses Struggling to Survive the COVID-19 Shutdown, 2020. https://www.abc.net.au/news/2020-04-04/coronavirus-shutdown-leaves-businesses-struggling-to-survive/12116334 (accessed on May 16, 2020).

Škare, M.; Soriano, D. R.; Porada-Rochoń, M. Impact of COVID-19 on the Travel and Tourism Industry. *Technol. Forecast. Soc. Change* **2021,** *163*, 120469.

Stankov, U.; Filimonau, V.; Vujičić, M. D. A Mindful Shift: An Opportunity for Mindfulness-Driven Tourism in a Post-Pandemic World. *Tour. Geog.* **2020,** *22* (3), 703–712.

UNWTO. UNWTO World Tourism Barometer May 2020—Special Focus on the Impact of COVID-19, May 2020. https://www.oecd.org/coronavirus/policy-responses/tourism-policy-responses-to-the-coronavirus-covid-19-6466aa20/ (accessed on April 24, 2021).

World Economic Forum. Outbreak Readiness and Business Impact Protecting Lives and Livelihoods Across the Global Economy, 2019. http://www3.weforum.org/docs/WEFHGHI_Outbreak_Readiness_Business_Impact.pdf (accessed on April 24, 2020).

Wright, N.; Fagan, L.; Lapitan, J. M.; Kayano, R.; Abrahams, J.; Huda, Q.; Murray, V. Health Emergency and Disaster Risk Management: Five Years into Implementation of the Sendai Framework. *Int. J. Disaster Risk Sci.* **2020,** *11* (2), 206–217.

CHAPTER 24

FOOD ON WHEEL: ANALYZING FACTORS AFFECTING CUSTOMERS' SATISFACTION TOWARD FOOD DELIVERY APPS IN NEW NORMAL

ALBATTAT AHMAD[1], AHMAD SAYUTI YAHYA[2], NOOR FARAHIYAH AZMAN[2], and FATIMAH ZAHRA RAMLY[2]

[1]*Graduate School of Management, Post Graduate Center, Management and Science University, Malaysia*

[2]*School of Hospitality and Creative Arts, Management and Science University, Malaysia*

ABSTRACT

In a recent situation where Covid-19 pandemic has erupted globally, shopping approach using Mobile Foods Ordering Apps (MFOAs) has significantly increased in numbers to cater to the changing consumption habit of residents and expedited the transition of traditional catering businesses to online services to survive the epidemic and maintain sustainable development. While internet applications have transformed the way people purchase for food, it is important to understand how customers see these services. The goal of this study is to examine factors affecting customer satisfaction with meal delivery applications among MSU students using the 7Ps of marketing mix. The researcher created a questionnaire to gather primary data. A structural questionnaire was administered to 264 students using a quantitative and descriptive manner. The study's findings were examined

Post-COVID Tourism and Hospitality Dynamics: Recovery, Revival, and Re-Start.
Umendra Narayan Shukla & Sharad Kumar Kulshreshtha (Eds.)

using the multiple regression test. The theoretical framework focuses on the seven Ps of marketing mix, which include product, physical proof, people, promotion, pricing, place, and process, which resulted in a positive and statistically significant relationship between customer satisfaction with meal delivery applications. The results of multiple regression indicated that product, advertising, and procedure all contributed significantly to customer satisfaction when utilizing a food delivery app. Apart from that, location, physical evidence, people, and pricing are determined to have no substantial effect on customer satisfaction when it comes to using meal delivery applications. Process and product are the two aspects that have the greatest impact on meal delivery apps.

24.1 INTRODUCTION

Recently, mobile food ordering apps (MFOAs) have been generally regarded as creative channels in the restaurant industry for reaching customers and providing them with high-quality services (Al Alwan, 2019). Food Delivery apps are well-known global online food delivery services that operate in over 20 countries. There are numerous meal delivery app companies emerging these days, like Grabfood, Dahmakan, and Foodpanda. These food delivery apps continue to see increased demand, particularly among young people who are too busy with job and education which limit them to cook or even purchase food for themselves. In a recent situation where Covid-19 pandemic has erupted globally, shopping approach using MFOAs has significantly increased in numbers to cater to the changing consumption habit of residents and expedited the transition of traditional catering businesses to online service to survive the epidemic and maintain sustainable development (Zhao and Bacao, 2020). Given the situation, it is expected that the revenue from this platform will grow globally at 12.49% from 2021 to 2025 (Statista, 2021). While internet applications have transformed the way people purchase for food, it is important to understand how customers see these services. Though academics are becoming more interested in studying different aspects of these applications, our thorough literature analysis indicates that existing information on consumer behavior in relation to 7Ps of marketing mix in general and MFOAs in particular still needs to be expanded. Thus, this study examined how the 7Ps marketing mix influences customer satisfaction using meal delivery applications.

24.2 LITERATURE REVIEW

24.2.1 PRODUCT

The term "product/service" refers to the goods or services that a business or individual provides. The marketing mix defines product as "what is being given to market for sale that may fulfill a demand." The product could be a tangible item, a service, a location, an organization, or an idea (Fakhimi Azar et al., 2011). They established that an organization that provides a high-quality service satisfies a variety of client needs. According to Ray (2009), customer satisfaction occurs as a result of the client purchasing and utilizing the product or service. When a product or service meets the expectations of the consumer, the customer is satisfied, and vice versa (Abdullah et al., (2016) The purpose of this study was to establish a link between the components of the marketing mix and consumer satisfaction in Malaysia's tourism business. The research findings indicate that there is a favorable correlation between all seven pillars of the marketing mix and consumer happiness. Additionally, the findings imply that the tourism business in Malaysia should consider the marketing mix while developing their marketing strategy.

24.2.2 PRICE

David (2008) defined price as the amount of money charged for a product or service, or the value exchanged by purchasers for the benefits of owning or consuming the product or service. The study's findings indicate that the marketing mix is a critical component of satisfying and retaining bank clients. Price was the most influential factor that customers identified with. According to an analysis of the drivers of customer satisfaction with supermarkets in Thai Nguyen, Vietnam, price is the most influential factor, followed by product (Reynold and Beautty, 2009).

24.2.3 PROMOTION

Webster (2010) defined promotion as acts taken to increase product or service acceptability and sales through advertising, publicity, or discounts. Nowadays, promotion encompasses the coordination of promotional efforts such as media advertising, direct mail, personal selling, sales promotion, public relations, packaging, retail displays, website design, and people to provide a consistent, customer-focused message (Ferrell and Hartline, 2008). Advertising is the primary facet of promotion, as it is the most visible component of an integrated marketing communication campaign (Ferrell and Hartline, 2008). The purpose of promotion is to spread the word about the benefits of a product or service to entice more customers to buy (Lamb et al., 2009) in order to garner attention, remind, and attempt to nudge the customer toward the end action of client loyalty (Cummins, 1998). Promotion is used to keep users informed about the product and to increase demand (Jager, 2007). To get optimal success, each product or service requires a unique mix of promotional activities.

24.2.4 PLACE

A delivery network is a collection of independent businesses that collaborate to get a product or service to consumers (Armstrong and Kotler, 2006). According to Kevin et al. (2009), an efficient and well-managed distribution network benefits customers. Ferdons and Towfique's (2008) study established a significant link between distribution (location) and customer happiness. Ataman et al. (2008) found that location (distribution) is critical in the development of new brands. According to Wairachu (2000), businesses should ensure accessibility to their products and services by establishing convenience stores. The ease with which items and services are accessible provides clients with flexibility and a positive perception of the purchase. This perception has a cumulative influence on a business's sales, which in turn affects its market share. Lu and Lukoma (2011) discovered that customers are satisfied with the supermarket's location, staff courtesy, and reliability when they researched buyer satisfaction levels in supermarkets. According to them, supplementary services rank last in terms of client satisfaction. As a result of these findings, the impact of these traits on customer satisfaction cannot be overlooked.

24.2.5 PEOPLE

These are human actors who contribute to the delivery of services and thereby influence the buyer's perspective. Appropriate staffing and personnel are critical components of providing a high-quality service to clients. Cengiz and Yayla (2007) conducted a study in which they examined the relationship between marketing mix, perceived value, quality, customer satisfaction, and loyalty, as well as positive word of mouth. Akroush (2011) reclassified the 7ps services marketing mix. The research demonstrated that the ingredients of the services marketing mix are 5ps in Jordan, not 7ps. There is compelling empirical data to support the classic four pillars of the marketing mix and the additional three pillars of the marketing mix for services. Due to the detachability of services, numerous shareholders participate in the buying and selling process. These shareholders include clients, other customers, and the organization's staff (Zeithaml et al., 2006). Individuals cannot be divorced from the overall service delivery process. In tourist experiences where first-class treatment is necessary since service directly affects overall consumer happiness, staff appearance, knowledge, and attitude should be enhanced to increase the tangibility of the product-service combination (Rodriguez, 20 13) Additionally, tourism is a labor-intensive industry, and the tourism experience is contingent on the tourist's interactions with indigenous populations and the presence of highly competent workers in such locations (UNWTO, 2007).

24.2.6 PROCESS

To generate and deliver product elements to customers, an effective process must be implemented. A process is defined as the sequence and sequence of events that occur during the performance of a service. Mucai et al. (2013) found that physical evidence does not have a major impact on customer satisfaction in non-star hotels in the municipality of Meru, but process marketing mix does. The study indicated that an organization's strategy should be founded on feasible fundamental ideals of discipline, i.e., the firm should avoid attempting to be all things to all people. The business must specialize in something so it can deliver an exclusive sense of fulfillment to a certain market. Mohammad (2015) discovered that customer happiness is not a one-time event. Because it is ongoing, management must be involved. Consumer satisfaction was measured using the 7ps in this study of retail bank clients in north-eastern Nigeria. The research discovered that while product, process, and physical evidence all had a substantial link with consumer satisfaction, price, promotion location, and people did not. This study demonstrated that process has the most impact on customer satisfaction, whereas pricing has the least. Customer satisfaction is a critical metric for determining how successfully retail chain store owners and managers in Bangladesh are satisfying customer expectations (Rana et al., 2014).

24.2.7 PHYSICAL EVIDENCE

The term "physical evidence" refers to the appearance, security, décor, and atmosphere of the facilities. Physical evidence has an effect on how customers perceive an organization's service delivery. Additionally, Thi and Thuung (2016) were unable to demonstrate that service and physical appearance have a favorable effect on customer satisfaction in their study. Miriti (2016) conducted research to determine the effect of marketing strategy on consumer preferences for categorized retail brands in Nairobi. The study discovered that location has a beneficial effect on consumers' preference for private retail brands in Nairobi. Kapferer (2011) stated that customer happiness may be quantified in a variety of ways, including price, service efficiency, employee attitude, overall firm performance, and proximity. Physical evidence has an effect on the perception customers make about an organization's service (Bitner, 2012). Architectural evidence, such as atmosphere, background music, seating comfort, and the physical architecture of a service facility, all have a significant impact on consumer satisfaction with a service experience (Bitner, 2012). The employees' outer appearance, particularly their clothing and grooming, may provide hints (Lovelock, 2006).

24.2.8 CONCEPTUAL FRAMEWORK

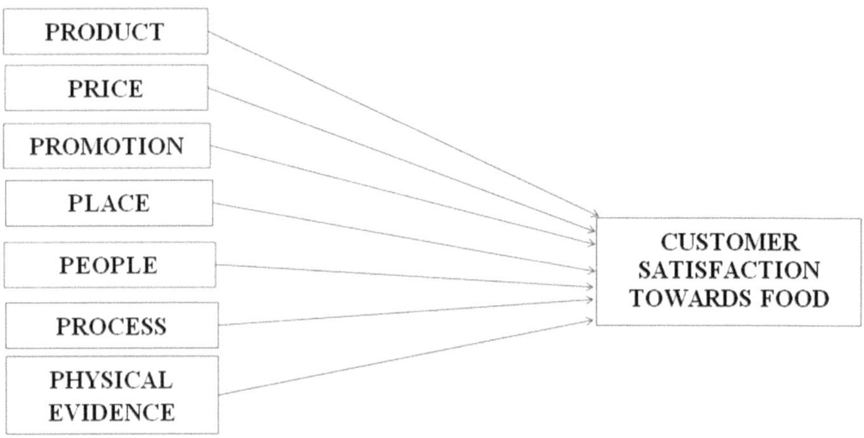

FIGURE 24.1 Framework of 7P's marketing mix and customer satisfaction.

24.3 METHODOLOGY

The quantitative study was executed for three weeks using online questionnaires that were distributed and collected in online forms targeting MSU students. A total of 300 online questionnaires were answered and recorded. However, 26 sets of questionnaires were removed upon data entry due to incomplete response. The structured questionnaire involves four sections. The questionnaires were divided into three sections, A, B, and C. Section A was about background information, Section 24.3 focused on the quality dimensions involving 7P's and Section 24.4 focused on measuring customers' satisfaction level after using the food delivery apps. The responses were thoroughly checked for statistical analysis while incomplete response was not included. Researcher made questionnaire checklist to gather the needed data from the respondents. The questionnaire is based on the previous study about Customers Satisfaction & Marketing Factors According To 7P's In Mamak Restaurant. Frequency, standard deviation, descriptive statistics, multiple regression analysis method were used in analyzing the data received by using SPSS19.

24.4 FINDINGS

24.4.1 RELIABILITY ANALYSIS

The Cronbach coefficient alpha is used to determine the internal reliability of the questionnaire instrument's constructs. All of the numbers can be deemed adequate and reliable because they exceed 0.60, as Nunnally and Berstein (1994) attest. Given the rationale, it is believed that the online questionnaire utilized in this study has a high degree of internal consistency (reliability).

24.4.2 RESPONDENT'S BACKGROUND

The demographic characteristics of the study are divided into gender, age, marital status, education level, and occupation to provide general information about the respondents.

TABLE 24.1 Results of Cronbach's Alpha Analysis for Each Variables of the Study.

Items	Number of items	Cronbach's Alpha
Overall Items	42	.964
Product	5	.705
Physical evidence	6	.797
People	4	.733
Promotion	5	.783
Price	5	.829
Place	4	.775
Process	5	.814
Customer satisfaction	8	.829

Source: Study data.

TABLE 24.2 Summary of Demographic Profiles.

Gender		
• Male	59	22.3
• Female	205	77.7
Total	264	100
Age		
• 18–21	44	16.7
• 22–25	94	35.6
• 26–29	80	30.3
• 30 and above	46	17.4
Total	264	100
Marital status		
• Single	198	75.0
• Married	63	23.9
• Divorced	3	1.1
Total	264	100
Education level		
• SPM	12	4.5
• Diploma	65	24.6
• Degree	164	62.1
• Master	20	7.6
• Phd	3	1.1
Total	264	100

TABLE 24.2 *(Continued)*

Salary (RM)		
• Less than RM1000	118	44.7
• RM1001–RM2000	41	15.5
• RM2001–RM3000	39	14.8
• RM3001–RM4000	30	11.4
• RM4001–RM5000	11	4.2
• RM5001 and above	25	9.5
Total	264	100
Frequency of using mobile food delivery apps per month		
• 1–2 times per month	170	64.4
• 3–4 times per month	74	28.0
• 5–6 times per month	6	2.3
More than 6 times per month	14	5.3
Total	264	100
Mobile food delivery apps preference		
• Grabfood	64	24.2
• Food Panda	140	53.0
• Dah Makan	7	2.7
• Bungkus it	4	1.5
• Deliver eat	2	0.8
• Others	47	17.8
Total	264	100

Source: Study data.

Based on Table 24.1, most of the respondents who took part in the survey were females with the frequency of 205 respondents. This comparison is for more than half of the total numbers of respondents (77.7%) compared to male participants that consist of 59 respondents (22.3%). Besides that, most of the respondents who took part in the online questionnaire aged from 22 to 25 with the frequency of 94 respondents (35.6%) and the least ones are respondents aged 18–21 with the frequency of 44 respondents that comprise of 16.7% of the whole respondents. Apart from that, the data collected from the respondents' response showed that most of the participants are single with a frequency of 198 people (75.0%) while 23.9% of the respondents comprising of 63 people are married. As we can see from Table 24.1, degree student dominated the respondent composition which account for frequency of 164 people followed by diploma students with a frequency of 65 (24.6%). In the analysis of number of participants according to the salary, the largest group of participants have salary below 1000 which account for 44.7% of total respondents which is 118 people followed by respondents with salary 1001–2000 with a frequency 41 (15.5%). With regard to the frequency of using mobile food delivery apps per month, about 2.3% or 6 of total respondents use mobile food delivery apps 5–6 times a month which is the lowest number from the total of the respondents. The highest number of participants uses the mobile food delivery apps 1–2 times per month dominating about 64.4% or 170 and trailing behind are the participants who use mobile food delivery apps 3–4 times a month which accounts about 28% or 74 students. Participants who used more than 6 times per month represent only 5.3% or 14 of total participants.

Meanwhile, students who prefer to use Food Panda represent more than half of total respondents which is 53.0% or 140. Grabfood comes as the second most preferred mobile food delivery apps with a frequency of 64 or 24.2% of total respondents and only 0.8% respondent which comprise of two people prefer to use DeliverEat.

24.4.3 DESCRIPTIVE STATISTIC

TABLE 24.3 Summary of Demographic Profiles.

Construct	Mean	Standard Deviation
Product (Food)		
1) PRO1: Food delivery apps offered various food selections.	4.09	0.88
2) PRO2: The foods sold in food delivery apps are tasty.	3.90	0.74
3) PRO3: The food in food delivery apps looks appetizing.	4.23	0.69
4) PRO4: The foods in food delivery apps are safe to consume.	3.93	0.79
5) PRO5: The food selection in food delivery apps is from well-known brand.	3.91	0.85
Physical Evidence		
6) PE1: The restaurants and menus can easily be browsed from the apps.	4.07	0.76
7) PE2: The food in online delivery apps looks appealing.	4.25	0.69
8) PE3: The information about the menu and restaurant are precise and easy to understand.	4.02	0.80
9) PE4: Food delivery apps have a familiar layout and friendly interface.	3.98	0.72

TABLE 24.3 *(Continued)*

Construct	Mean	Standard Deviation
10) PE5: Food delivery apps using a logo that is easy to recognize.	4.28	0.65
11) PE6: The menu selections in apps are categorized accordingly for customer's convenience.	4.14	0.71
People		
12) PEO1: It is easy to find a delivery man using food delivery apps.	4.03	0.78
13) PEO2: The delivery man is warm and helpful.	4.06	0.72
14) PEO3: The delivery man is competent in doing their job.	4.16	0.67
15) PEO4: The delivery man communicates effectively with the customer.	4.10	0.69
Promotion		
16) PROMO1: The online food promotion is always ongoing and up to date.	4.07	0.76
17) PROMO2: The food delivery apps always suggest the food to order.	3.89	0.88
18) PROMO3: Food delivery apps offer points to accumulate for future usage.	3.74	0.97
19) PROMO4: Food delivery apps actively offer promotions to customer.	4.03	0.80
20) PROMO5: Food delivery apps offer various vouchers for customer.	3.81	0.92
PRICE		
21) PRICE1: The price is stated clearly in their menu.	4.23	0.72
22) PRICE2: The bills are calculated accurately.	4.24	0.66
23) PRICE3: The food price is affordable.	3.56	0.88
24) PRICE4: The food prices on food delivery apps are reasonable.	3.58	0.89
25) PRICE5: The price of the food suits the quality of the food.	3.67	0.83
Place		
26) PLACE1: The outlet of the selected restaurant can be chosen according to customer's preference.	3.88	0.93
27) PLACE2: The distance between the restaurant and customer's place is always shown.	3.97	0.91
28) PLACE3: The online apps will always show the nearby restaurant.	4.10	0.76
29) PLACE4: This delivery covers the area that I live in	3.80	1.06
Process		
30) PROC1: The food delivery apps are responsive.	3.93	0.81
31) PROC2: Food delivery apps are easy to access anytime in a day.	4.05	0.78
32) PROC3: The payment system is easy.	4.16	0.75
33) PROC4: The ordering process is easy/not complicated.	4.19	0.72
34) PROC5: The progress of the order can easily be tracked.	4.13	0.71

TABLE 24.3 *(Continued)*

Construct	Mean	Standard Deviation
Customer Satisfaction		
35) CS1: The service that I have received from food delivery apps was met my expectation.	3.91	0.72
36) CS2: I am satisfied with the quality of the product that is offered by food delivery apps.	3.92	0.70
37) CS3: I am satisfied with food delivery apps buying process.	4.01	0.77
38) CS4: Overall I am satisfied with food service using food delivery apps.	4.04	0.75
39) CS5: I will do repurchase using food delivery apps.	4.07	0.75
40) CS6: I will recommend food delivery apps to my friends and family.	4.03	0.79
41) CS7: I rarely switch to other apps to order food online.	3.66	0.96
42) CS8: I will always prefer to order food online when needed.	3.86	0.96

Source: Study data.

24.4.4 CORRELATION

TABLE 24.4 Correlation Analysis.

Variables	1	2	3	4	5	6	7
1. Customer satisfaction	1.00						
2. Product	.680***	1.00	1.00	1.00	1.00	1.00	1.00
3. Physical evidence	.697***	.749***	.617***	.499***	.554***	.535***	.689***
4. People	.603***	.526***	.635***	.518***	.597***	.563***	
5. Promotion	.645***	.581***	.637***	.413***	.507***		
6. Price	.646***	.599***	.589***	.584***			
7. Place	.673***	.511***	.643***				
8. Process	.754***	.563***					

Note: ***$p<.001$

Source: Study data.

Based on the table above, the result shows that there is a positive relationship of product, physical evidence, people, promotion, price, place, process with customers' satisfaction. In addition, the correlation of each independent variable (product, physical evidence, people, promotion, price, place, process) is significant at 0.0001 level toward customer satisfaction.

24.4.5 MULTIPLE REGRESSION ANALYSIS

TABLE 24.5 Multiple Regression Analysis 7Ps and Customer Satisfaction.

Variables	B	SE B	B	Sig.	Tolerance	VIF
1. Product	.338	.091	.193***	.000	.401	2.50
2. Physical evidence	.041	.090	.027	.652	.305	3.28
3. People	.204	.096	.095*	.036	.529	1.89
4. Promotion	.228	.070	.152**	.001	.483	2.07
5. Price	.193	.075	.119*	.011	.500	2.00
6. Place	.232	.088	.132**	.009	.425	2.36
7. Process	.536	.084	.338***	.000	.386	2.56

Note: $R^2 = .726$, ***p<.0001, p**<.01, p* < .05

Source: Study data.

Based on the table above, the result shows that there is a positive relationship of product, physical evidence, people, promotion, price, place, process with customers' satisfaction. In addition, the correlation of each independent variable (product, physical evidence, people, promotion, price, place, process) is significant at 0.0001 level toward customer satisfaction.

24.5 DISCUSSION

Pearson's correlation study revealed a strong correlation between product and customer happiness, with a normalized beta of 0.680 for hypothesis H1. Additionally, the results of multiple regression analysis suggested that product contributed the second-largest share of variance in customer satisfaction. The findings of this study corroborate those of prior research conducted by Tsiotsou (2005), Jakpar (2012), Ghezelbash (2017), Nuridin (2013), and Jahanshahi (2011), all of which demonstrate that consumer satisfaction is determined by the quality of items given. Saleem's (2015) research demonstrates that items have an impact on consumer satisfaction.

Concerning the relationship between pricing and customer satisfaction, the evaluated relationship revealed a considerable strong correlation. Using Pearson's correlation analysis, price is found to have a strong link with customer satisfaction. When examined using multiple regression, pricing is found to contribute considerably to customer satisfaction. The findings of this study corroborate those of Jiang (2004), Gupta (2014), Kaiman (2013), and Razak

(2016), all of whom demonstrate that consumer happiness is determined by the price offered. Customers' contentment is influenced by price.

Following that, promotion was discovered to have a substantial link with customer satisfaction and to account for the third biggest portion of the variation in customer happiness seen in the regression model. The findings of this study corroborate Ping Pi's (2011), Alipour's (2018), and Wahab's (2016) research, indicating that promotion has an effect on consumer satisfaction. More precisely, Woo Park's (2013) study found that consumer satisfaction is influenced by promotions such as price discounts, direct gifts, and collected points.

In support of hypothesis H4, Pearson's correlation analysis revealed a strong correlation between location and customer happiness, with a normalized beta of 0.673. Following a multiple regression analysis, it was determined to be statistically significant in contributing to customer satisfaction. The findings of this study corroborate Alipour's (2018), Muala's (2012), and Kadhim's (2016) findings that location (location) has an effect on consumer satisfaction. Consumer satisfaction is affected by the location of a business.

With regard to hypothesis H5, a correlation between individuals and customer satisfaction was established. Customer satisfaction has a strong association with people. Multiple regression analysis suggested that humans should be included in the model to account for variance in customer satisfaction. The findings of this study corroborate those of Verma (2017), Alipour (2018), and Muala (2012), all of whom showed that employees had an effect on consumer happiness.

In support of hypothesis H6, Pearson's correlation analysis revealed a substantial correlation between process and customer satisfaction, with a standardized beta of 0.754. Additionally, multiple regression analysis revealed that processes accounted for the greatest portion of the variance in customer satisfaction. The findings of this study corroborate those of Andrian's (2015), Abouennaga's (2016), and Suri's (2016) research, all of which indicate that the procedure has an effect on consumer satisfaction.

24.6 CONCLUSION AND RECCOMMENDATION

The Covid-19 is our constraint because it prevents us from doing any of the movements; we planned in advance for our data gathering process due to Movement Control Orders (MCO). Due to the current economic situation, the previously intended physical questionnaire for our research project will need to be completely transformed into an online survey. Our solution for resolving this issue is for future researchers to be prepared for any eventuality and to always have backup plans for any study they intend to undertake, ensuring that the project can still be completed regardless of the conditions.

Additionally, our restriction is in obtaining responses to our questionnaire. Certain target demographics may be resistant to strictly online surveys, as not everyone has the time, willingness, or patience to participate in a survey at all times. As a result, the old method of conducting surveys via paper surveys is more suitable and convenient for respondents.

Our second constraint is that our sample is limited to university students residing in Shah Alam. This shows that the sample may not represent the general population of Shah Alam and hence may introduce bias into the sample. This constraint also lengthens the process of obtaining sufficient responses. This could be overcome by increasing the sample size for the targeted population in a subsequent study, resulting in a more complete representation of the people in the specified location.

Additionally, we discovered that the population of female students who participated in this survey was larger than the population of male students. This appears to be out of proportion and skewed. While a greater proportion of the sample from each gender than the overall population was able to present each group, the number of participants remains limited in comparison with the female group. Thus, a larger and more well-stratified sample size will be required in the future to eliminate bias and improve the study's accuracy.

Additionally, we saw that the bulk of answers were from the Faculty of Hospitality and Creative Arts. Given that we are members of this faculty, distribution of the online survey would almost certainly begin and expand within our faculty and would seldom reach respondents from other faculties. This demonstrates an unbalanced and disproportionate distribution of the survey, which results in a biased response. Thus, in the future study, information and surveys should be disseminated evenly.

KEYWORDS

- **food delivery apps**
- **marketing mix**
- **customer satisfaction**

REFERENCES

Abbasi, A. S. Impact of Employee Characteristics and Their Performance on Customer Satisfaction. *Sci. Int. (Lahore)* **2013,** *25* (2).

Abdullah, F.; Abdullah, T. F.; Abdullah, M. F. Effects of Marketing Mix on Customer Satisfaction: Empirical Study on Tourism Industry in Malaysia. *Int. J. Appl. Res.* **2016,** *2* (2), 357–360.

Alalwan, A. A.; Rana, N. P.; Dwivedi, Y. K.; Algharabat, R. Social Media in Marketing: A Review and Analysis of the Existing Literature, Telematics, and Informatics **2017,** *34* (7), 1177–1190.

Alipour, M. The Effects of the 7P Marketing Mix of Components on Sporting Goods Customer Satisfaction. *Int. J. Busi. Manage. Invent.* **2018,** *7* (1).

Ataman, M. B.; Mela, C. F.; van Heerde, H. J. Building Brands. *Market. Sci.* **2008,** *27* (6), 1036–1054.

Bolton, R. N.; Lemon, K. N. A Dynamic Model of Customers' Usage of Services: Usage as an Antecedent and Consequence. *J. Market. Res.* **1999,** *36*, 171–186.

Booms, Bernard H.; Bitner, Mary Jo Marketing Strategies and Organization Structures for Service Firms. *Market. Serv. Am Market. Assoc.* **1981,** 47–51.

Cengiz, E.; Yayla, H. E.; The Effect of Marketing Mix on Positive Word of Mouth Communication: Evidence from Accounting Offices in Turkey. *Innov. Market.* **2007,** *3* (4), 74–86.

Cohen, J. *Statistical Power Analysis for the Behavioral Sciences*, 2nd ed.; Routledge Academic: New York, 2013.

Fakhimi Azar, S.; Akbari Vanehabad, M.; Rasouli, R. Investigating the Impact of Marketing Mix (4P) on Increasing the Sale of Water Heaters of Ghaynarkhazar Company. *J. Beyond Manage.* **2011,** *4* (16), 79–106.

Fichman, R. G.; Santos, B. L. D.; Zheng, Z. Digital Innovation as A Fundamental and Powerful Concept in The Information Systems Curriculum, 2014.

Ghezelbash, S. Evaluating The Impact of Promotion Price, Product Quality, Service Quality, Customer Satisfaction and Repeating Purchase Incentives. *J. Internet Bank. Comm.* **2017,** *22* (S8).

Gupta, P. Impact of Price sensitivity on Customer Satisfaction: An Empirical Study in Retail Sector. *J. Human. Soc. Sci. (IOSR-JHSS)* **2014,** *19* (5).

Jahanshahi, A. A.; Gashti, M. A. H.; Mirdamadi, S. A.; Nawaser, K.; Khaksar, S. M. S Study the Effects of Customer Service and Product Quality on Customer Satisfaction and Loyalty, Vol. 1 (7). Centre for Promoting Ideas, USA, 2011.

Jain, M. K. An Analysis of Marketing Mix: 7Ps or more. *Asian J. Multidisc. Stud.* **2013,** *1* (4), 23–28.

Jiang, P. Online Customer Intention to Return Price Perception, Attribute-Level Performance, and Satisfaction Unfolding Over Time. *Eur. J. Market.* **2004,** *39* (1/2).

Kadhim, F. A. Effects of Marketing Mix on Customer Satisfaction: Empirical Study on Tourism Industry in Malaysia. *Int. J. Appl. Res.* **2016,** *2* (2).

Kaiman, H. Flat on Customer Satisfaction (Studies in Marunda flats of North Jakarta) Quality and Price of House Influence. *Int. J. Busi. Manage. Invent.* **2013,** *2* (7).

Kanoga, S.; Njugana, R.; Bett, S. The Effect of Place on Performance of Shopping Malls in Nairobi County Kenya. *J. Soc. Sci. Human.* **2015,** *1* (4), 381–390.

Kotler, P.; Armstrong, G. *Principles of Marketing*, 11th ed.; Upper Saddle River, NJ: Prentice-Hall, 2006.

Kurniawan, B. Factors Affecting Customer Satisfaction in Purchase Decision on Ticket Online: A Case Study in Air Asia.

Lewis, C. I. *An Analysis of Knowledge and Valuation*; Open Court: LaSalle, IL, 1946.

Lim, H. Understanding American Customer Perceptions on Japanese Food and Services in the U.S. UNLV Theses/Dissertations/Professional Papers/Capstones, Jan 12, 2010.

Mai, L-W.; Ness, M. Canonical Correlation Analysis of Customer Satisfaction and Future Purchase of Mail-Order Speciality Food. *Br. Food J.* **1999,** *101* (11).

Muala, A. Al. Assessing the Relationship Between Marketing Mix and Loyalty Through Tourists Satisfaction in Jordan Curative Tourism. *Am. Acad. Scholar. Res. J.* **2012,** *4* (2).

Murangiri, P. K. Role of Marketing Mix on Performance of Micro Finance Institutions in Kenya. *Strat. J. Busi. Change Manage*. **2014**.

Nunnally, J. C.; Bernstein, I. H. *Psychometric Theory*, 3rd ed.; McGraw-Hill: New York, 1994.

Nuridin. Effect of Service Quality and Quality of Products to Customer Loyalty with Customer Satisfaction as Intervening Variable in PT. Nano Coating Indonesia. *Int. J. Busi. Appl. Soc. Sci.* **2018,** *4* (1).

Ohei, K. Marketing Mix: Its Role in Customer Satisfaction in the South African Banking retailing. *Banks Bank Syst.* **2015,** *10* (1), 83–91.

Pallant, J. *SPSS Survival Manual: A Step by Step Guide to Data Analysis Using SPSS*; Sabon Publishers: Sydney, 2005.

Palmer, A. *Introduction to Marketing: Theory and Practice*; Oxford University Press, 2004.

Park, J-W. Investigating the Effects of Sales Promotions on Customer Behavioral Intentions at Dutyfree Shops: An Incheon International Airport Case Study. *JAIRM* **2013,** *3* (1).

Pi, W-P. Effects of Promotion on Relationship Quality and Customer Loyalty in the Airline Industry: The Relationship Marketing Approach. *Afr. J. Busi. Manage.* **2011,** *5* (11), 4403–4414.

Pimplapure, M. N. Leadership Development for Managing in Vuca World. *Paripex—Indian J. Res.* **2018,** *7* (6), 37–39.

Prabowo, G. T.; Nugroho. A. Factors That Influence the Attitude and Behavioral Intention of Indonesian Users Toward Online Food Delivery Service by the Go-Food Application. *Adv. Econ. Busi. Manage. Res.* **2018,** *72*, 204–210.

Saleem, A. Product Perceived Quality and Purchase Intention with Consumer Satisfaction. *Global J. Manage. Busi. Res. E Market.* **2015,** *15* (1).

Sekaran, U.; Bougie, R. *Research Methods for Business*, 5th ed.; Wiley: New York, 2010.

Statista. Platform-to-Consumer Delivery, 2021. https://www.statista.com/outlook/dmo/eservices/online-food-delivery/platform-to-consumer-delivery/malaysia

Tsiotsou, R. The Role of perceived product quality and Overall Satisfaction on Purchase Intentions. *Int. J. Consum. Stud.* **2006,** *30* (2).

Zardi, H.; Wardi Y.; Evanita, S. Effect of Quality Products, Prices, Locations and Customer Satisfaction to Customer Loyalty Simpang Raya Restaurant Bukit tinggi "Salero Nan Tau Raso". *Adv. Econ. Busi. Manage. Res.* **2018,** *64*.

Zhao, Y.; Bacao, F. What Factors Determining Customer Continuingly Using Food Delivery Apps During 2019 Novel Coronavirus Pandemic Period? *Int. J. Hospital. Manage.* **2020,** *91*, 102683. https://doi.org/10.1016/j.ijhm.2020.102683

CHAPTER 25

SUSTAINABLE DEVELOPMENT GOALS AND SAFE AND SUSTAINABLE TOURISM PRACTICES IN THE NEW NORMAL

ANDREIA MARIA ROQUE

Instituto de Desenvolvimento do Turismo Rural Brasil Rural, São Paulo, Brazil

ABSTRACT

In 2015, the United Nations (UN) at the General Assembly in New York approved this new document entitled, "Transforming Our World: The 2030 Agenda for Sustainable Development," a global action plan aimed at bringing governments, companies, academia, and civil society together to balance environmental, economic, social, and institutional themes in a harmonious way composed of 17 Sustainable Development Goals (SDGs) and 169 targets. In 2020, the "Conscious Rural Tourism Network" was created with the purpose of bringing together the Sustainable Development Goals and Safe and Sustainable Tourism Practices to address the pandemic that plagued the world. This article aims to relate from data collected with the network listing the relevance of the SDGs for tourism, believing that new paths must be taken to strengthen the concept of sustainable and regenerative tourism, a concept that has been evolving in parallel to that of sustainable development that promotes a respectful relationship between all actors involved. In the end, this article aims to reflect the role that tourism can assume toward meeting the challenges proposed by Agenda 2030.

Post-COVID Tourism and Hospitality Dynamics: Recovery, Revival, and Re-Start.
Umendra Narayan Shukla & Sharad Kumar Kulshreshtha (Eds.)

25.1 INTRODUCTION

In 2015, the representatives of the 193 UN member states meeting at the General Assembly in New York approved this new document now entitled, "Transforming Our World: The 2030 Agenda for Sustainable Development" composed of 17 Sustainable Development Goals (SDGs) and 169 goals that brought to the agenda topics such as Eradication of poverty; Sustainable agriculture and zero hunger; Health and wellness; Quality education; Gender equality; Sanitation and drinking water for all; Clean and affordable energy; Decent work and economic growth; Industry innovation and infrastructure; Reduction of inequalities; Sustainable city and communities; Responsible consumption and production; Action against global climate change: Life on the water; Earth life; Peace, justice and effective institutions, and finally the importance of institutional partnerships and means of implementation.

The set of SDGs established to guide actions in favor of maintaining the natural resources and ecosystems of Agenda 2030 are focused on achieving water security on the planet, adopting clean and accessible energy, sustainable conservation of the oceans, seas, and marine resources, as well as protecting terrestrial ecosystems and combating global climate change.

In 2022, the scenario imposed by the pandemic due to the emergence of the new coronavirus (Sars-Cov-2), which causes the Covid-19 disease, not only brought about major problems for the tourism production chain but also demonstrated the distancing of the activity with the Sustainable Development Goals.

25.2 METHODOLOGY

This study adopted as its research universe the participating members of the Conscious Rural Tourism network, a network in Brazil that brings together entrepreneurs, service providers, farmers, and related technicians, advocating to adopt corporate social responsibility policies for the safety of all of health and sustainability in favor of modern Brazilian rural tourism, responsible and professional, without losing rurality. Believing that measuring the reach of the SDGs strengthens the commitment and calls for confrontation, the network currently consists of 114 participants representing all regions of Brazil, annually from February to April, answering a semistructured questionnaire with 17 questions listing the relevance of the ODS for conscientious rural tourism seeking to understand the group's understanding to foster attitudes toward achieving the goals proposed by the 2030 Agenda. In the end, it was possible to recognize that SDG 03, SDG 05, SDG 08, SDG 12, and SDG 15 are those with greater involvement in the network.

25.3 RESULTS AND DISCUSSION

Believing that new paths must be taken by tourism to strengthen the concept of Sustainable and Regenerative Tourism, a concept that has been evolving in parallel to that of Sustainable Development that promotes a respectful relationship between all actors involved given the importance of people in this concept of sustainability; this article aims to reflect the role that tourism can assume toward meeting the challenges proposed by the Agenda 2030.

- **Goal 1.** End poverty in all its forms everywhere: Tourism can certainly generate income through job creation, but we can pursue sustainable tourism development by linking poverty reduction, promoting entrepreneurship and small businesses, and empowering disadvantaged groups, particularly young people and women.
- **Goal 2.** End hunger, achieve food security, improve nutrition, and promote sustainable agriculture: Rural Tourism can strengthen traditional agricultural activities, resulting in increased income for local communities that can lead to sustainable, regenerative, and conscious agriculture while adding value to the tourism experience. However, tourism, in general, can boost agricultural productivity by promoting the production, use, and sale of local products in tourist destinations and their full integration in the value chain.
- **Goal 3.** Ensure healthy lives and promote well-being for all at all ages: Tourism can and should have a direct effect on health and well-being, promoting experiences that bring good times and joy, but also the generation of income from the activity should aim to improve society's quality of life, reducing mortality and preventing diseases, among other items.
- **Goal 4.** Ensure inclusive and equitable quality education and promote lifelong learning opportunities for all: Tourism has the potential to promote inclusion, the values of a culture of tolerance, peace, and nonviolence, and all aspects of global exchange and citizenship, but also new tourism paths must invest in education and professional training promoting skills, standards, and certifications.
- **Goal 5.** Achieve gender equality and empower all women and girls: The new paths of tourism must make women hard not only through

jobs and income generation but also as a tool for empowerment and strengthening of potential with equal and differentiated chances.

- **Goal 6.** Ensure availability and sustainable management of water and sanitation for all: Tourism can play a key role in gaining access to and securing water, as well as hygiene and sanitation for all. Efficient use of water in the tourism industry, along with appropriate safety measures, wastewater management, pollution control, and technological efficiency, can be the key to safeguarding our most precious resource.
- **Goal 7.** Ensure access to affordable, reliable, sustainable, and modern energy for all: Tourism can be an agent of change toward the use of renewable energy sources, promoting sustainable investments, and promoting the idea of urgency for energy transition, thus contributing to the construction of innovative solutions in all locations that work as well as strengthening values in favor of the emergence of new behavior patterns after the Covid-19 Pandemic.
- **Goal 8.** Promote sustained and sustainable economic growth, full and productive employment, and decent work for all: Tourism is one of the driving forces of global economic growth. Giving access to decent work opportunities in the tourism sector, society especially young people and women, is certainly one of the most important actions. This theme represents serious challenges to be overcome to face the growing condition of social inequality on a global scale.
- **Goal 9.** Build resilient infrastructure, promote inclusive and sustainable industrialization, and foster innovation: The sector can encourage governments to improve their infrastructure by adopting more sustainable, resource-efficient, and cleaner attitudes as a way to attract tourists and other sources of investment. The tourism sector can play an important role in innovation processes and in changing mentality and behavior, contributing to the dissemination of new practices.
- **Goal 10.** Reduce inequality within and among countries: Tourism can be a powerful tool for community development and reducing inequalities when it involves the local population and all stakeholders in their development. Tourism, reinterpreted on a sustainable basis, can constitute an important way to change inequality and the concentration of income or opportunities for some, to the detriment of most of the global population.
- **Goal 11.** Make cities and human settlements inclusive, safe, resilient, and sustainable: Sustainable tourism has the potential to contribute to

advances in favor of strengthening sustainable spaces and communities, promoting the preservation of environmental, cultural, and social heritage with targeted and thought-out actions in favor of structures that are more consistent with a new way of thinking for tourists and the tourism service provider.

- **Goal 12.** Ensure sustainable consumption and production patterns: A tourism sector that adopts sustainable production and consumption practices can play a significant role in accelerating the global movement toward sustainability and seeks to develop sustainable production and consumption practices, including resource efficiency initiatives that result in economic, social, and environmental issues. Therefore, it is necessary to keep active a critical reflection on the modes of production and consumption and the search for new dynamics on sustainable bases in favor of more responsible choices, thus inducing new behaviors.
- **Goal 13.** Take urgent action to combat climate change and its impacts: Tourism causes mundane climates and is affected by them. The risks arising from global warming as one of the most visible consequences of unsustainable modes of production and consumption represent one of the main challenges to be faced in the horizon of 2030. In this sense, there are common actions that we must adopt as part of the reality of the universe of tourism as decreasing energy consumption and switching to renewable energy sources, believing that tourism can help address the most pressing challenges of our time.
- **Goal 14.** Conserve and sustainably use the oceans, seas, and marine resources for sustainable development: The development of tourism must be part of the integrated management of the coastal zone to help conserve and preserve marine ecosystems, as tourism when it takes place in these areas directly influences these dynamics and therefore must focus on postures centered on strategies for valuing local socio-biodiversity, from strategies to value the nature of society and development.
- **Goal 15.** Protect, restore, and promote sustainable use of terrestrial ecosystems, sustainably manage forests, combat desertification, and halt and reverse land degradation and halt biodiversity loss: The new path of tourism seeking to be an element to achieve the goals proposed by the SDGs goes beyond the conservation and preservation of biodiversity, seeking to innovate with actions aimed at respecting terrestrial ecosystems, with actions aimed at experiences and experiences that

enhance awareness and environmental education. Tourism can represent a way to reconnect everyone, contributing to behavioral changes aimed at protecting biodiversity.

- **Goal 16.** Promote peaceful and inclusive societies for sustainable development, provide access to justice for all, and build effective, accountable, and inclusive institutions at all levels: Tourism could contribute to inducing new behaviors, capable of transforming society itself, and reinventing itself as a possible way to build peaceful, more tolerant, fair, and sustainable societies. Believing in sustainable tourism that benefits and engages local communities can also provide a source of livelihood, strengthen cultural identities, and encourage business activities, thus helping to prevent violence and conflict to take root and consolidate peace in societies post-conflict.
- **Goal 17.** Strengthen the means of implementation and revitalize the global partnership for sustainable development: Due to the multiple sectors linked in the tourism production chain, it has the capacity to strengthen the partnerships that are the necessary basis for the development of a new tourism, promoting greater awareness of the role of tourism as an inducer of change and social transformation.

25.4 CONCLUSIONS

The big bet will be on tourism in line with the SDGs aimed at travel in nature and in regional destinations, which are also impacted by the pandemic crisis that we have faced in the last two years, seeks to rebuild itself with innovation and sustainability. More than ever, the SDGs should be recognized as a compass that brings us closer to new trends aimed at regional travel that bring contact with nature and value the local cultural daily life.

Highlights for the activities of Rural Tourism that have special care committed to adopting the protocols, concerned with environmental conservation, adoption of good practices, implementation of sustainability actions, and care for others. Tourism will overcome challenges, valuing human, economic, and environmental relationships. And in the very near future, maybe even today, it will not be enough just to look at the SDGs but to go further and bring a regenerative vision beyond sustainability. Nature travel consumer behavior has changed. Many travelers accustomed to traveling abroad, now unable to cross borders, will begin to consume regional tourism with a high degree of demand for quality. You will have to race against time to adapt.

KEYWORDS

- **sustainable development goals**
- **tourism**
- **conscious rural tourism**

REFERENCES

Aall, C. Sustainable Tourism in Practice: Promoting or Perverting the Quest for a Sustainable Journal of Sustainable Development? *Sustainability* **2014,** *6* (5), 2562–2583. http://dx.doi.org/10.3390/su6052562

Allen, C.; Metternicht, G.; Wiedmmann, T. National Pathways to the Sustainable Development Goals (SDGs): A Comparative Review of Scenario Modelling Tools. *Environ. Sci. Policy* **2016,** *66*, 199–207. https://doi.org/10.1016/j.envsci.2016.09.008

Hall, M. C. Constructing Sustainable Tourism Development: The 2030 Agenda and the Managerial Ecology of Sustainable Tourism. *J. Sustain. Tour.* **2019,** *27* (17), 1044–1060 https://doi.org/10.1080/09669582.2018.1560456

Sachs, J.; Schmidt-Traub, G.; Kroll, C.; Lafortune, G.; Fuller, G.; Woelm, F. *Sustainable Development Report 2020: The Sustainable Development Goals and Covid-19 Includes the SDG Index and Dashboards*; Cambridge University Press: Cambridge, 2021. https://doi.org/10.1017/9781108992411

Stockholm Resilience Centre. *Azote for Stockholm Resilience Centre*; Stockholm University, 2017. https://www.stockholmresilience.org/research/

United Nations. Transforming Our World: The 2030 Agenda for Sustainable Development, 2015. http://sustainabledevelopment.un.org/.../21252030.

Weitz, N.; Nilsson, M.; Davis, M. A Nexus Approach to the Post—2015 Agenda: Formulating Integrated Water, Energy, and Food SDGs. *SAIS Rev Int Aff* **2014,** *34*, 37–50. DOI: 10.1353/sais.2014.0022

CHAPTER 26

POST-COVID-19 DYNAMICS OF REGIONAL TOURISM MARKETS: FUTURE STRATEGIES FOR REVIVAL AND RE-START

BIVEK DATTA

Nepal College of Travel and Tourism Management, Lazimpat, Kathmandu, Nepal

ABSTRACT

COVID-19 has jolted the entire world and continues to challenge the world through its horrific effects. The containment of the virus looks a distant reality until the world has the vaccine to treat it. As of 18th October, 2021 there are 240,260,449 confirmed cases of the coronavirus with 4,890,424 confirmed deaths. The impact has been felt in 216 countries, areas, or territories (WHO, 2020). The travel and tourism industry is one of the worst hit sectors and could result in a loss of 50 million jobs worldwide (WTTC, 2020). It is also forecasted that it might take ten months for the industry to rebound and take shape again (WTTC, 2020). The travel restrictions imposed in all the global destinations are of the tune of 96% (UNWTO, 2020). The present study provides an insight into the impact and revival challenges of COVID-19. An earnest effort is put forward to understand the impact of COVID-19 on the tourism sector. The study also focuses upon the revival path suggested by experts during various webinars organized by the tourism department of Amity University, Noida. As the whole world is at a standstill because of the lockdown, the tourist and the tourism industry is in a sphere of escalated uncertainty and they are in dire need of the reliable and updated information

Post-COVID Tourism and Hospitality Dynamics: Recovery, Revival, and Re-Start.
Umendra Narayan Shukla & Sharad Kumar Kulshreshtha (Eds.)

pertaining to the tourism sector. The webinars organized through online platforms are being considered the safest medium to transmit vital information and revival strategies post COVID-19. Taking the present measures into consideration, some new measures have been suggested during the webinars that contribute to sustainability of the tourism sector. It is suggested that all the regional tourism markets should chalk out a viable recovery path post COVID-19 to reorient the tourism industry. This would need a large dose of economic injection to revive the tourism markets paralysed by the deadly coronavirus. A global approach is suggested by the various experts from the academia, industry, destination management companies, travel agents, and tour operators so that a complete transformation of the tourism industry may become a reality in the future. Post Covid-19 recovery road map is also discussed in respect of all the regional tourism markets.

26.1 INTRODUCTION

Post 1970 there was a paradigm shift in the tourism demand and these two traditional markets started receiving stiff competition from other regional tourism markets. Table 26.1 portrays a description of International Tourism Demand in UNWTO Regions in the last 69 years.

TABLE 26.1 International Tourism Demand by UNWTO Regions.

Regions	1950 (%)	1960 (%)	1970 (%)	1980 (%)	1990 (%)	2000 (%)	2010 (%)	2019 (%)
Europe	66.5	72.5	70.5	68.4	63.5	57.8	50.9	51
Americas	29.6	24.1	23	18.9	18.8	18.4	15.9	16
Asia&Pacific	1.0	1.3	3.6	7.8	12.1	16.6	21.7	24
Africa	2.1	1.1	1.5	2.5	3.4	3.9	5.3	5
Middle East	0.9	1.0	1.4	2.4	2.1	3.3	6.4	4

Source: Boniface and Cooper, 2005.

26.2 UNWTO REGIONAL TOURISM MARKETS

26.2.1 ASIA AND THE PACIFIC AND POST COVID RECOVERY ROADMAP

There are four subregions in Asia and Pacific, that is, North East Asia, South East Asia, South Asia, and Oceania. In the year 1950 the Asia and Pacific region had a meagre share of less than 1% of international tourist arrivals (UNWTO Tourism Highlights), but by 2019 this share has increased to 24% which is considered to be a commendable growth rate contributing immensely to the economic development of the region. This region is undisputedly the most dynamic of all the five regional tourism markets (UNWTO).

26.2.1.1 POST COVID RECOVERY ROADMAP

Asia and the Pacific region suffered the most as far as the regional tourism markets are concerned in 2020 with highest level of travel restrictions. The international tourist arrivals dipped down by 84% in 2020 in comparison to 2019 (UNWTO, 2020). The subregions of Asia-Pacific Region display different roads to recovery. For example, in South East Asia the recovery of Vietnam was higher displayed by a positive GDP growth in 2020 in comparison to Thailand and Singapore which witnessed sharp contractions in their economic activities. The slower recovery rate in the European Union and the United States of America could be a headwind for the recovery prospects in the Asia and Pacific Region. Sustainability and digital transformation could be the key drivers to recovery in the Asia and Pacific Region. Asia and the Pacific region with majorly young and digitally literate population could lead the way, thus increasing the transformation journey potential.

26.2.2 EUROPE AND POST COVID RECOVERY ROADMAP

Europe's natural instinct of attracting tourists from time immemorial has paid rich dividends, ably justified by its supremacy in international tourist arrivals and foreign exchange earnings. The flurry of successful expeditions suffered a jolt in the 1960s and the successive decades. However, the region has still managed to grow and maintain its number one position in terms of international tourist arrivals. This growth has a positive correlation to the economic integration process that began in 1958. The continent has come a long way since then and the journey can be traced to the establishment of European Economic Community and its transformation to the European Union in 1993.

As of date, 27 countries with varying economic structures form the European Union (https://europa.eu) and are the most preferred tourist destinations in Europe with a share of almost 40% in terms of international tourist arrivals. Thus, tourism has played a very important role in rejuvenating the European Continent.

26.2.2.1 POST COVID RECOVERY ROADMAP

European Union has released funds paving the way for recovery so as to reach 19 countries of the European Union. These recovery and resilient funds will aid in boosting the economies of many countries post COVID-19. On 5th October, 2021, Malta got the signal to utilize these funds to boost its economy. After signing the agreement with the European Commission, Malta will be eligible to receive 13% in prefinancing for boosting its economy.

The European Council has approved the national recovery and resilience plans of Czech Republic and Ireland. The Czech Republic can sign the agreement and become eligible to receive 13% in prefinancing though Ireland has not asked for the funding. Croatia, Cyprus, Lithuania, and Slovenia also got the permission to utilize the European Union recovery and resilience funds to stimulate their economies and recover from the COVID-19 consequences. Austria, Belgium, Denmark, France, Germany, Greece, Italy, Latvia, Luxembourg, Portugal, Slovakia, and Spain have got the permission for the utilization of European Union recovery and resilience funds to stimulate their economies and recover from the COVID-19 consequences.

Next Generation European Union is the initiative taken by the European Union to enhance Europe's post Covid-19 recovery. The recovery plan is a futuristic project to assist the countries in managing the economic impact of COVID-19. The plan is to develop a green, digital, and more resilient Europe. The funding will help the Schengen countries and other member states and recover from crisis caused by the pandemic. The European Union has finalized €750 billion that will assist the member nations to recover from the post pandemic crisis. It was decided collectively by the member states of the European Union that bonds will be sold and the money will be provided to the countries that are the hardest hit by COVID-19. Italy was provided with €250 billion from this newly institutionalized recovery funds. This new futuristic project is aimed at providing additional funds to projects aimed at environment and rural development. It is now up to the member states of EU where these funds can be allocated. In May 2020, European Commission reiterated to include a package that gives a roadmap and passage for recovery of tourism through reopening of internal borders, improving the connectivity, and enhancing the tourism and hospitality services. As EU has limited powers pertaining to tourism and therefore can only support, coordinate, and supplement the member states in respect of tourism, European

Union also came up with the idea of EU digital COVID certificate so that it can be used by all member states to ensure that the travelers arriving at the destinations are vaccinated against the pandemic. Another plan initiated by EU is to have public health corridors to ensure and help to inform the tourists pertaining to safety protocols between countries.

26.2.3 AMERICAS AND POST COVID RECOVERY ROADMAP

After 1970 the share of this region started receding further as it received stiff competition from other regional tourism markets especially Asia and Pacific. In Americas, the United States of America has been a dominant tourism powerhouse in terms of both Inbound and Outbound forms of Tourism. Canada is also a known inbound and outbound tourist destination, whereas Mexico is an emerging inbound destination in the Americas (UNWTO, 2019). There are varying patterns of tourist movement in USA and Canada and less economically vibrant regions in the Central portions of North America, the Caribbean, and the South American subregion.

26.2.3.1 POST COVID RECOVERY ROADMAP

Americas are heavily dependent on tourism as a major economic activity. Americas have also understood the importance of new strategies and digitalization that will enhance and stimulate tourist experiences further. The region has to focus on safety, authenticity, and hygienic protocols to retain the trust of the prospective travelers. The time has come for the Americas is to go digital and travelers will also go online. Hospitality services will provide digital menus and mobile capabilities. Therefore, investments in enhancing technologies will be a key factor to develop the tourism sector. Destinations need to highlight their health and hygiene capabilities and need to appraise the prospective travelers about the health and hygiene protocols. For example, Mexico has been on the forefront in ensuring that the destination is ready for the tourists. Destinations also need to regroup and improve their competitiveness and make themselves distinct from others. For example, Peru has come up with new travel routes focusing on its culture, adventure treks, archaeological sites, and museums to open up in the new normal post COVID-19. Americas focus to target millennial travelers as they focus on personalization and technology. They also seek unique travel experiences such as engaging in hiking, sightseeing, and other adventure tourism activities that Americas are eyeing for.

26.2.4 AFRICA AND POST COVID RECOVERY ROADMAP

The major tourist-receiving countries in North Africa include Morocco and Tunisia. In the Sub-Saharan African subregion, South Africa is the most visited nation because of its cultural and natural attractions and the only country in this subarea which receives more than 10 million foreign tourists. The share of Africa in terms of International Tourist Arrivals has increased by a meagre 3% in the last 69 years (UNWTO, 2019).

26.2.4.1 POST COVID RECOVERY ROADMAP

There are many challenges post COVID that Africa needs to address, like:

- Policies should be practical in its approach to consider for implementation
- Focus on Intra-African travel
- Stress on regional agenda rather than on national, as there are different priorities of every region.
- Making domestic strategies that have domestic implementation for the local inhabitants
- Aligning the strategies of CFTA (African Continental Free Trade Area, depository of African Union Commission)
- Start creating supply chain through cultural tourism, value-added product for product chain by getting the regional government involved in the whole process.
- Looking for training programmes that focus on training of officials of CFTA
- Embed all the strategies, with solutions

In the global context, sustainability is one of the most significant aspects for tourism development and conservation. And in global economic downtime, pandemics have significantly affected the African continent along with other adversities like poverty, health issues. Still, they are fighting back with zeal and vigor. A competitive tourism sector is required for Africa. Ensuring safety and security protocols regarding entry requirements is the need of the hour to ensure travelers' confidence. Tourism is a major economic driver for many African countries and they are heavily dependent on tourism. Some countries have started reopening their borders, but the situation remains awful as waves of the pandemic continue to wrath through the African continent.

26.2.5 *MIDDLE EAST AND POST COVID RECOVERY ROADMAP*

Religious Tourism is visible in Saudi Arabia because of the holy cities of Mecca and Medina. United Arab Emirates has also positioned itself as a prominent tourist destination by building iconic landmarks and artificial islands within the desert. The share of Middle East in terms of International Tourist Arrivals has shown a minute increase from 1% in 1950 to 4% in 2019 (UNWTO, 2019). Tourism has been halted by many political disturbances and economic stagnation and has had a major impact on their travel and tourism industry. The tourism industry of Egypt is volatile and is affected by various political and economic factors.

26.2.5.1 POST COVID RECOVERY ROADMAP

COVID-19 has jolted the tourism industry globally and Middle East is not left untouched. A stronger travel industry needs to emerge post pandemic. The economic impact has been severe for Middle East as many economies are heavily reliant on tourism. Many countries in the Middle East have taken initiatives to support the tourism industry in their countries. Exemption from tourism levies and other fee are implemented in Bahrain and United Arab Emirates. In Egypt, hotels have been reopened for domestic tourists at 50% capacity from June onward, provided they comply with a number of safety measures including testing workers, installing disinfection equipment, assigning a floor or building as a quarantine zone for suspected or positive cases, and implementing social distancing norms. Government in Jordan has given tax rebate to support the tourism industry. The government of Saudi Arabia has come with new tourism development fund in June to support the tourism industry.

26.3 IMPLICATIONS FOR THE FUTURE OF TOURISM AND REVIVAL STRATEGIES SUGGESTED FOR THE REGIONAL TOURISM MARKETS POST COVID-19 DURING WEBINARS

The following Webinars were conducted under the aegis of Amity Institute of Travel and Tourism, Amity University Noida, India:

TABLE 26.2 Experts Opinions and Outcomes Discussed

S.NO.	FOCUS OF THE WEBINAR	PANELLISTS	ISSUES AND OUTCOMES DISCUSSED
1	Revival of Tourism and Hospitality Industry: Post COVID-19 Challenges	1. Prof. Justin Matthew Pang (Vietnam) 2. Prof. Sanjay Nadkarni (UAE) 3. Dr. Mohit Vij (UAE) 4. Mr. Samir Thapa (Nepal) 5. Mr. Ankur Narang (India)	A competitive tourism sector is the need of the hour with prime focus on personalization and digitalization
2.	Revival of Tourism & Hospitality Industry Post COVID-19: Challenges –Global Perspective	1. Prof. Marina Ambrosecchio (Italy) 2. Prof. Urs Wagenseil (Switzerland) 3. Prof. Christophe Martin (France) 4. Prof. S. P. Bansal (India) 5. Prof. Ajit Sikand (Germany)	Promoting Sustainable and safe travel. Ensuring hygienic and COVID free environment to prospective tourists. working in unison, a collaborative approach must be followed; adapting what resources are available
3.	Revival Strategic Roadmap for Global Tourism post COVID-19: Issues and Challenges	1. Ms. Irina Hatcher (Vietnam) 2. Mr. Puneet Arora (U.K.)	Maintain social distancing and promote tailor-made tour packages to attract tourists. Promote sustainable and responsible tourism.
4.	Strategic Revival Challenges for Tourism and Hospitality Industry: Post COVID-19-African Perspective	1. Mr. Henry Oriokot (Uganda) 2. Ms. Amrita Rajcoomar (Mauritius) 3. Dr. Shem Wambugu Maingi (Kenya) 4. Ms. Tando Songwevu (South Africa) 5. Mr. Arvind Dhar (South Africa)	Promote destinations at national level, promote domestic tourism, focus on intra-African travel

Source: Author

A comprehensive overview of the experts' deliberations during the Webinars is mentioned below.

26.3.1 EUROPE

The COVID-19 has cast a dreaded spell across the European nations of Italy, Spain, France, United Kingdom, and Germany to a large extent. In all these countries, tourism is a major economic instrument.

"Safe Travel" is the need of the hour and, if not followed religiously, will prove to be disastrous in times to come. Italy has coined the term "Proximity Travel," which will eventually be the most important term for the industry. Proximity Travel relates to traveling locally and promoting local tourism, local community, traveling in smaller groups, respecting social distancing norms, experiencing boutique hotels, and sustainable tourism measures. The "4-T approach"; with Trade, Technology, Tourism, and Training being the four pillars of Italy's economy. Safe Hospitality/Tourism Manager is the new job that may come in due course of time. Italy is planning to promote its products as hygienic, safe, contaminant free, and most importantly Corona free. This was discussed and shared in the Webinar by Prof. Marina Ambrosecchio from the European University of Tourism. Milan, Italy.

Prof Martin, Dean, Tourism Institute, Saint-Marie de Chavagnes, Cannes, France stated in the same webinar that new plans must be implemented. Over the next few years, France will focus on

- Attracting domestic tourism,
- Promoting destinations at national level,
- Work on Q & A Certification in hotels; more flexibility will be given to these units,
- Good marketing of destination and defining the role of Destination Management Companies
- Promoting sustainable tourism,
- Defining and working on lines of crisis management in tourism sector.

Within this Pandemic, France must also work collaboratively on developing a new tourist attraction policy as well as setting new training for students.

Dr. Wagenseil, Tourism Program Director, Lucerne University of Applied Sciences and Arts, Switzerland pointed out that addressing Sustainable Development Goals is of prime importance to develop the tourist destinations. He was of the opinion that working in unison, a collaborative approach must be followed; adapting what resources are available with Switzerland

and promoting them to find the optimum utilisation for the economy of Switzerland.

26.3.2 ASIA AND PACIFIC

The country of Vietnam in South East Asia is presently out of the lockdown but is following the social distancing norms as shared by Ms. Irina in a Webinar. She, during her discourse, shared the apprehensions expressed by the tourists because of the border closure, security, and health challenges. The country of Vietnam is also promoting itself as a safe destination. They have also opened up new travel routes for the promotion and development of tourism and have also chalked out plans pertaining to flexibility of prices as well as dates of travel booking and cancellation. The impetus is also on promoting and allowing traveling to COVID-19-free destinations, but the major mantra justifiably is social distancing for sure. The government is focusing on tailor-made tours. The government of Vietnam's major priority is to provide masks and sanitizers for travelers. They have also ensured proper medical screening of the staff who are in contact with the tourists and are updating the tourists regularly on safety regulations pertaining to COVID-19.

For the revival of outbound travel, first, India needs to focus on the current situation of COVID in the destination and also on the country's norms on travel restrictions. Role of health certificate will take a lead, along with the travel documents required to travel internationally. Destination Management Companies (DMC) need to analyze the behavior and attitude change of the travelers to customize facilities and services as per the requirements. This very relevant point was shared by Mr. Puneet Saxena in a Webinar on 02.05.2020. There are certain changes that are likely to happen in the attitude of travelers post COVID; they would rather look into Sustainable Tourism and Responsible Tourism than Mass Tourism.

According to Dr. Pang, Singapore has shown the world how they can combat COVID-19 by thorough monitoring through testing kits to check the spread of COVID-19.

26.3.3 AFRICA

The current situation is quite alarming in Africa as 20 million jobs are estimated to be lost. Airlines are shut, borders are sealed, most of the international tourists arriving in Africa are from France, which is one of the worst hit nations due to the pandemic. Therefore, Africa is focusing on domestic travel instead of international travel, waiting for the African Government to come up with strategies to promote tourism in the domestic and international fronts. Though COVID-19 has so far been a pain point and bane for all of us, it is actually a boon, a blessing in disguise for Nature which got a chance to revive itself, wildlife is enjoying uninterrupted movement across the Wild Parks and Sanctuaries. And, on the top of everything, it gave humans a chance to appreciate the technological advancements like Video Conferencing, Artificial Intelligence, and Virtual Reality. Africa is an education hub, financial center, and a core area for Medical Tourism. Amid all the negativity of the pandemic, Mauritius is still hoping for positivity and regaining the same old charm among tourists internationally and domestically. The greatest challenge that Mauritius is facing right now is that there is no air connectivity, either international or national. Thus, it is expected of the African Government to come up with some strategic solutions so that it can host international tourists while following all the precautions and safety protocols. There has been little relief from COVID-19 in Mauritius as few people are allowed to go to offices in order to strictly follow the precautions and Government advisories. Government is supportive and providing financial and monetary support to the citizens.

Ms. Tando deliberated that Africa basically needs workable economic plans and strategies to come up from this crisis and minimize the impact of this pandemic. Cultural tourism is vast and it covers a wide arena like art, culture, heritage etc., domestic tourists could be made aware of the vibrant culture, and diversities within Africa and Customised Tour Packages are to be arranged that cater to the needs of domestic tourists. Focus has to be shifted on domestic tourists because there is no way Africa can expect international tourists for at least 6 months.

In an informative discourse by Dr. Shem, African Union studies have estimated that Africa is likely to lose billions of dollars and million jobs due to this crisis. Some of the interventions have been more cosmetic in nature. There is a need to come up with a sound strategy and plan. Tourism as a leisure activity is not economically viable for local communities of Africa. Here, tourism is predominantly nature/wildlife based and this unique feature needs to be showcased to the world. Promoting these aspects of tourism would be economically viable for host community and guests would enjoy the peace and serenity of the environment.

In the current scenario, 80% of the air-planes are grounded, entertainment parks, restaurants, hotels, and many other places of touristic importance are closed indefinitely. To come up with the strategy that would fast revive the adversities created by pandemic, is quite vital.

Mr. Dhar reiterated that some such solutions for the revival of tourism can be e-commerce and use of technology. Past trends have shown that business travel creates actual business for tourism industry, be it accommodation, travel or entertainment, which is likely to be bland in near future as there would be reduced travelling for business purposes. Measures like cost-cutting on many in-flight services and flight tickets at hiked prices could be implemented to overcome losses. There is need to build confidence among tourists through information dissemination. Customer acquisition and retention strategies have to be in place.

26.3.4 MIDDLE EAST

According to Gulf Cooperation Council, it is expected that the tourism revenue could decline between $14–$17 billion (excluding airlines), assuming that the pandemic lasts for two quarters. Upto 400,000 tourism-related jobs in the region could be lost leading to unemployment levels at an all-time high. Government needs to restrategize and prepare marketing and promotional campaigns. They should set clear policies post COVID-19 pertaining to tourism. They need to offer financial support to rebuild the confidence of the tourism sector. They should make strategic investments especially for the aviation industry which is badly affected.

Middle East is facing a lot of problems in the tourism sector and nations are working on four stages, that is, crises, prerecovery, recovery, and transformation. This was shared by Dr. Vij. According to him, partnership between the private and government sectors needs to be strengthened. Stay Home Safe and Travel Tomorrow should be promoted. The industry should educate people pertaining to maintaining hygienic standards as we need to live with the pandemic for a long period of time.

26.3.5 AMERICAS

United States of America, which earns the maximum from Tourism, that is, 214 billion USD (UNWTO, 2019), is hard hit by the pandemic as most cases affected by COVID-19 are reported in the United States of America globally. Maximum cases are reported across New York which houses prime tourist attractions of the world. Tourism is also largely affected in the tourist-receiving countries of Canada and Mexico. The revival path looks tough especially for the United States of America which is battling one of the most hard-hit pandemics in ages. In Americas, 92% of the destinations have adopted COVID-19-related restrictions since January, 2020 (UNWTO, 2020). To combat the deadly Coronavirus, the United States of America has taken some strict measures by sealing its northern and southern borders to nonessential travel, advised the American citizens not to visit overseas locations, and urged the Americans to return home, as discussed by Mr. Puneet Arora during the webinar. Similar suits are followed by Canada which is resorting to allowing only its citizens and permanent residents to return home. Similar cases of international travel restrictions are also reported from Central American nations of Panama, Honduras, Guatemala, and El Salvador.

26.4 CONCLUSIONS

It is believed the spatio-temporal pattern of international tourism will be uneven post COVID-19 in all the five regional tourism markets of UNWTO and there will be uneven demand in all these regional markets. Many destinations will revisit themselves and focus more on domestic tourism rather than International Tourism. They will resort to more sustainable forms of tourism. Considering and evaluating the resilience of the tourism industry in the last seven decades, there is a glimmer of hope in all the regional tourism markets to bounce back as before and devise strategic revival techniques to overcome the barriers of travel restrictions. Many destinations have started promoting domestic tourism on a larger scale and resorting to Staycations which is the need of the hour post COVID-19 period.

COVID-19 has devastated the lifestyle of the people. The tourism and the hospitality industry is the most hard hit in all the five regional tourism markets. No continent is spared by the tremors of the COVID-19 and overcoming this economic hardship will take a lot of time. UNWTO Regions have shown great resilience in the last 70 years which is well reflected by various data sets published by various international organizations such as UNWTO, WTTC, WEF, PATA, OECD, IATA, and ICAO. In the time of crises caused by COVID-19, all the UNWTO Regions have been totally shaken by the pandemic, but at the same time all the UNWTO regional tourism markets are ever optimistic. They are all aware of the fact that the path of recovery is challenging but the tourism industry will rebound as it has done in the past.

KEYWORDS

- **COVID-19**
- **impact**
- **strategic revival**
- **strategies**
- **regional tourism markets**
- **webinars**

REFERENCES

Al-Tawfiqef, J. A.; Zumlaad, A.; Memis, Z. A. Travel Implications of Emerging Coronaviruses: SARS and MERS-CoV. *Travel Med. Infect. Dis.* **2014,** *12* (5), 422–428.

Amity Alumni Association. Industry Revival Post COVID 19 Discussed in Webinars at Amity University, 2020. http://alumni.amity.edu/happening-details.aspx?Event=10100&Events=Events

Amity Alumni Association. Revival Strategic Roadmap for Global Tourism Post COVID 19: Issues and Challenges, 2020. http://alumni.amity.edu/happening-details.aspx?Event=10160&Events=Events

Boniface, B. and Cooper, C. (2005) Worldwide Destinations: The Geography of Travel and Tourism, 4th edn. Elsevier Butterworth-Heinemann.

City Air News. Industry Revival Post COVID 19 Discussed in Webinars at Amity University. 2020. https://www.cityairnews.com/content/industry-revival-post-covid-19-discussed-in-webinars-at-amity-university

COVID-19: Putting People First 2020. https://www.unwto.org/tourism-covid-19

European Commission Joint Research Centre, Behavioural Changes in Tourism in Times of Covid-19: Employment Scenarios and Policy Options, JRC Science for Policy Report, 2020.

Girard, M. P.; Tam, J. S.; Assossou, O. M.; Kieny, M. P. The 2009 A (H1N1) Influenza Virus Pandemic: A Review. *Vaccine* **2010,** *28* (31), 4895–4902.

Hall, M. Crisis Events in Tourism: Subjects of Crisis in Tourism. *Curr. Iss. Tour.* **2010,** *13* (5), 401–417.

http://nawaat.org/portail/2020/05/06/covid-19-in-tunisia-tensions-arise-between-municipalities-and-central-government/

http://www.euro.who.int/en/publications/html/report-on-the-health-of-refugees-and-migrants-in-the-who-european-region-no-public-health-without-refugee-and-migrant-health-2018/en/index.html

http://www.mti.gov.eg/English/MediaCenter/News/Pages/MTI-imposes-3-month-ban-on-export-of-anti-infection-supplies-1.aspx

http://www.oecd.org/coronavirus/policy-responses/building-resilience-to-the-covid-19-pandemic-the-role-of-centres-of-government-883d2961/

http://www.oecd.org/coronavirus/policy-responses/investment-in-the-mena-region-in-the-time-of-covid-19-da23e4c9/

https://openknowledge.worldbank.org/bitstream/handle/10986/33634/COVID-19-Crisis-Through-a-Migration-Lens.pdf?sequence=1&isAllowed=y.

https://read.oecd-ilibrary.org/view/?ref=131_131938-b9ys3suiav&title=COVID-19-Crises-and-Fragility

https://www.byblosbank.com/Library/Assets/Gallery/Publications/CountryRiskWeeklyBulletin/Country%20Risk%20Weekly%20Bulletin%20628.pdf

https://www.gfmag.com/magazine/february-2020/wave-m-megadeals-mena

https://www.iemed.org/observatori/arees-danalisi/arxius-adjunts/anuari/med.2018/FDI_MENA_region_Carril_Milgram_Paniagua_Medyearbook2018.pdf

https://www.oecd.org/coronavirus/policy-responses/theterritorial-impact-of-covid-19-managing-the-crisis-across-levelsofgovernment-d3e314e1/ (accessed on 15 May 2021).

https://www.unescwa.org/sites/www.unescwa.org/files/escwa-covid-19-economic-cost-arab-region-en.pdf

https://www.unwto.org/

https://www.unwto.org/taxonomy/term/5

https://www.wttc.org/about/media-centre/press-releases/press-releases/2020/latest-research-from-wttc-shows-an-increase-in-jobs-at-risk-in-travel-and-tourism

Huber, C.; Finelli, L.; Stevens, W. The Economic and Social Burden of the 2014 Ebola Outbreak in West Africa. *J. Infect. Dis.* **2018,** *218* (5), 698–704.

IMF. The Great Lockdown: Worst Economic Downturn Since the Great Depression. Monetary Fund 2020. https://www.imf.org/en/News/Articles/2020/03/23/pr2098-imf-managing-director-statement-following-a-g20-ministerial-call-on-the-coronavirus-emergency

International Monetary Fund World Economic Outlook, April 2020-The Great Lockdown, 2020. https://www.imf.org/en/Publications/WEO/Issues/2020/04/14/weo-april-2020

Joo, H.; Maskery, B. A.; Berro, A. D.; Rotz, L. D.; Lee, Y. K.; Brown, C. M. Economic Impact of the 2015 MERS Outbreak on the Republic of Korea's Tourism-Related Industries. *Health Security* **2019,** *17* (2), 100–108.

NEPAD. Harnessing Africa's Engine for Growth: The role of NEPAD the Tourism Action Plan. NEPAD 2018. https://www.nepad.org/news/harnessing-africas-engine-growth-role-nepad-tourism-action-plan

OECD. OECD Economic Outlook **2020,** *2020* (2): Preliminary Version, OECD Publishing, Paris. https://doi.org/10.1787/39a88ab1-en. 1 December 2020

OECD. The Territorial Impact of COVID-19: Managing the Crisis across Levels of Government.

Petersen, E.; Wilson, M. E.; Touch, S.; McCloskey, B.; Mwaba, P.; Bates, M.; Dar, O.; Mattes, F.; Kidd, M.; Ippolito, G.; Azhar, E. I.; Zumla, A Rapid Spread of Zika Virus in the Americas—Implications for Public Health Preparedness for Mass Gatherings at the 2016 Brazil Olympic Games. *Int. J. Infect. Dis.* **2016,** *44*, 11–15.

Shepard, D. S.; Undurraga, E. A.; Halasa, Y. A.; Stanaway, S. J. The Global Economic Burden of Dengue: A Systematic Analysis. *Lancet Infect. Dis.* **2016,** *16* (8), 935–941.

Siu, A.; Wong, Y. R. Economic Impact of SARS: The Case of Hong Kong. *Asian Econ. Papers* **2004,** *3* (1), 62–83.

The Economic Times. COVID-19 Pandemic Could Cost Millions of Jobs in Global Tourism Industry: UN Body, 2020. https://economictimes.indiatimes.com/news/international/business/covid-19-pandemic-could-cost-millions-of-jobs-in-global-tourism-industry-un-body/articleshow/75194635.cms

UNWTO. World Tourism Barometer **2020,** *18* (6). https://doi.org/10.18111/wtobarometereng

WHO. Climate Change and Human Health: Global Environmental Change. World Health Organization, 2020. https://www.who.int/globalchange/environment/en/

World Bank. Understanding Poverty. World Bank, 2020. https://www.worldbank.org/en/topic/poverty/overview

World Economic Forum. This Is How Coronavirus Could Affect the Travel and Tourism Industry 2020. https://www.weforum.org/agenda/2020/03/world-travel-coronavirus-covid19-jobs-pandemic-tourism-aviation

WTTC. Impact of the Ebola Epidemic on Travel & Tourism; WTTC, 2018.

WTTC. Latest Research from WTTC Shows a 50% Increase in Jobs at Risk in Travel & Tourism, 2020.

CHAPTER 27

EFFECTIVE TOURISM EMPLOYEE MANAGEMENT PRACTICES AFTER COVID: EXTENDING THE VALUE OF CORPORATE SOCIAL RESPONSIBILITY FROM EXTERNAL TO INTERNAL

CINDIA CHING CHI LAM[1], UT LON (BILLY) IM[1], and ERDAN MA[2]

[1]*School of Tourism Management and School of Hospitality Management, Macao Institute for Tourism Studies, Macao SAR, China*

[2]*School of Tourism and Hospitality Management, Yunnan University of Finance and Economics, China*

ABSTRACT

The outbreak of COVID-19 has revolutionized global business structures, in particular, the tourism and hospitality industries. Employees of these industries are found to suffer from forced furlough, no-pay leave, and changes in roles and responsibilities, which have resulted in substantial stress. The stress from work often affects the wellbeing and work fulfillment of the employees, which has escalated their escapist and anti-crisis behavior. Confronted with the abrupt challenges from the pandemic, the management has to identify new practices that can lead both the employees and the enterprise to surf through the crisis. In this study, corporate social responsibility (CSR) that is normally perceived to generate nonoperational value at strengthening corporate image is found to serve also as an effective tool that mediates the effect of work stress on the undesirable work behaviors of employees. This finding broadens the knowledge on the value of CSR, when management

Post-COVID Tourism and Hospitality Dynamics: Recovery, Revival, and Re-Start.
Umendra Narayan Shukla & Sharad Kumar Kulshreshtha (Eds.)

can further utilize CSR in their strategic plans and when it can cushion the struggles with the employees, especially during a crisis.

27.1 INTRODUCTION

27.1.1 THE PANDEMIC

The outbreak of COVID-19 (pandemic) was first reported by the end of 2019 and has since then affected the globe for around 2 years. Its effects are undeniably far-reaching from business practices to consumer expectations, resulting in the new normal. As of the fourth quarter of 2021, a total of 244 million coronavirus cases were reported, with deaths of around 5 million people in the world (Worldometer, 2021). Comparing the forecast of the International Monetary Fund to the actual figures, the pandemic has brought more than a 7% decrease in gross domestic product (GDP) and a change in inflation of −0.74% around the world (Outlook, 2019, 2020). The tourism industry was one of the most hard-hit industries, mainly due to the travel and custom barriers. There was massive winding up of business and employees suffered from dismissal, forced furlough, changing roles, and alike, which have resulted in considerable work and financial stress.

Macao is the special administrative region (SAR) of China and the only city of China with legalized gambling. Though a small city with less than 33 square kilometers and only 670 thousand population, it has been welcoming around 40 million visitors per year before the pandemic (Macao Statistics and Census Service, 2020b). At the extreme period, this number has dropped by more than 99%. The gambling and tourism industries that used to contribute more than 90% of the public revenue has experienced a plunge of more than 96% in business (Macao Gaming Inspection and Coordination Bureau, 2020; O'Connor, 2019). From a normal unemployment rate of less than 2%, the figure has surged to around 3% in 2021 (Macao Statistics and Census Service, 2020a). It is important to apprehend that this figure has not included employees on forced furlough, cutting of salary, or who suffered from a decrease in fringe benefits.

To subsist from the pandemic, governments and large corporations in Macao have collaborated to implement remedy actions at assisting different segments of the economy in getting through the collapsed business environment and financial situation. The remedy and recovery actions can be deliberated into high-end policies and strategic operational levels. From the time of early outbreak, the Macao SAR government has set up the Novel Coronavirus Emergency Coordination Center that maintained a closed political connection with nearby areas on the exchange of information and implemented collaborated entry policies to control the spread of the pandemic (Imprensa Official, 2020). The Center has set up many accessible communication channels through social media, traditional television broadcasting, newspaper, and government websites to disseminate news and information to the public. The transparency of information has settled the nerves of the public and avoided the generation of horrors that may disturb the already rocking society. The government has also set up consumer goods pricing stability policies to avoid perverse increases in price, to avoid further financial burdens on the public (Macao Government Consumer Council, 2020).

In the financial aspect, the government has launched financial aids policies for both residents and local industries (Macao Financial Services Bureau, 2020). The supporting aid covered various life aspects that included subsidizing electricity and water fees, consumption vouchers, and tax reduction. They were set with the objectives to alleviate the household burden, augment consumption, and encourage local spending. For the commercial segment, the government has collaborated with banks and financial institutions to implement interest subsidization, exemption of profit tax, reduction in rents, and start capital projects to create job positions. For the gambling concessionaires and large enterprises, they have utilized their strengths to support the SMEs (Ma et al., 2021). Besides the donation of medical necessities, they have implemented creative policies like SME prepayment programs to prepay SME on future purchase; set up platforms for SME to market their goods and services and replace international purchases by local suppliers; send their professional cleansing teams to sterilize the shops, offices, storehouse, and delivery vans of local SMEs; and offer internship and job placement for graduates and work with universities to arrange management sharing to students. The financial institutions have also contributed through the prompt approval of financial grants, exemption of service charges, offering low-interest loans and extending the repayment period, as well as setting up e-commerce platforms for local SEMs and offering online cross-border financial systems for businesses to expand the channels of sales. These large enterprises have imposed the aforementioned actions to respond to the call of corporate social responsibility (CSR) by the Macao SAR government. However, even upon these remedial actions, the sizeable drop in business and visitors still have brought destructive effects to the industry and its employees.

27.1.2 THE NEED ON NEW MANAGEMENT PRACTICES

To survive the new normal, new management practices need to come into play, which is guided by management cognition and their mentality. The pandemic that has caused death on a staggering scale is a mega-crisis that has impacted the psychology and changed the behavior of employees. Crisis is termed as circumstances where people are confronted with genuine dangers (Caplan, 1961; Collins and Collins, 2005; Rosenthal and Pijnenburg, 1991) which exceed their capacity to handle (James and Gilliland, 2012). These circumstances are well solid for individuals to preserve inward adjustments with the environment, such as plagues and seismic tremors that would actuate mental stress and clutter, which will result in uneasiness, misery, and other negative feelings (Topper and Lagadec, 2013; Wu et al., 2021). In resonance with the self-defense mechanism, stress resulting from external inducement can lead to cognitive and behavioral reactions, counting numerous unusual and substitute activities that are inadvertent. When confronted by too many unpleasant circumstances, individuals tend to mobilize their resources to manage with the stretch to realize rebalance (Power, 2015).

Employees who were confronted by unforeseen external stressors have to adjust regarding their wellbeing, security, and work. Within the working environment, numerous work schedules have been drastically changed involuntarily during the pandemic. Copious work and life issues also transpired concurrently, especially salary alterations that directly impacted the quality of life and even the continuation of living of the employees, which have caused stress among the employees. The insecurity of the employees often influences their fondness toward their organization (Ornell et al., 2020; Yu et al., 2021). At such instances, employees are found to be more cognizant toward the extent of corporate social responsibilities practised by the corporations (Barakat et al., 2016). Socially responsible corporations will help to uproar the assurance of employees when facing questionable and emergency circumstances and may initiate positive employee behavior. The employees are more likely to contribute to society, tend to be less elude, and are more positive to exercise citizenship roles (Ahmad et al., 2020).

Amid the pandemic, employees are confronted by tremendous work stress that affects health, job security, and organization commitment (Filimonau et al., 2020; Yu et al., 2021). The study was thus carried out to investigate if the perception of corporate social responsibility can mediate the negative effects of work stress on the employees. The target samples were employees of the tourism industry in Macao, who have suffered from all the downside of the pandemic. A total of 872 samples working in five major tourism and hospitality segments were included in the study, namely travel agencies, hotels, casinos, restaurants, and retail. The level and type of employee stress were examined; and the perceived benefits of corporate social responsibility (CSR) and the type of CSR information expected by the employees were identified. The study found that CSR can effectively mediate the negative effect of work stress on employee work habits during a crisis. This has extended the value and contribution of CSR. Generally, the investment in CSR is associated with qualitative return on the improvement of external reputation and stronger association with employees. Through this study, the value of CSR has extended to stabilize the internal work habits of employees during times of crisis through the mediation of work stress. This will allow the management to better understand the potential return on CSR investment and identify another tool that can maintain the momentum and alleviate the nerves of employees during a crisis.

27.2 EMPLOYEE STRESS AND ITS NEGATIVE EFFECTS

Stress is caused by the experience of dangers and their effects; it is a response to a circumstance that is perceived as undermining and with no suitable coping responses (Seiffge-Krenke, 2013). Intense stress can alter the curing practices and stereotypes of employees (Yau and Potenza, 2013), with a focus on those who will have direct interaction with the customers. The pandemic is an external stimulus that acts as crisis stress forcing employees to respond. When employees are burdened by work stress, they will reassess their work behaviors, frame adapting strategies, and identify adapting practices that may alter their original work behaviors (Hanton et al., 2012). The process is multifarious and convolute, yet employees possess the remarkable internal capability to respond and adapt to stress (Lazarus, 1991).

The pandemic crisis has imposed noteworthy passionate impacts on the employees, who may feel on edge, perplexed, and miserable. These befuddling states influenced the efficacious adapting practices of the employees that had the potential to result in incapable decisions on harmful practices to counter the uneasiness and torment. The changed practices may instigate both short-term and long-term impacts that may modify the employees' statement of faith and conviction, resulting in an irreversible change in practices and demeanor. As shown in Table 27.1, 94.2% of employees were affected by the pandemic, 74.1% of them indicated the effects were significant, and among whom 37.5% of them even reported the effects were severe. The pandemic has rocked the globe and has affected nearly all business segments. These significant percentages reflected the effect of the pandemic did not only rest within the business level but had also influenced the employees, and the effects are substantial.

TABLE 27.1 The Extent of Pandemic Effect on Work (in Percentage).

Not affected by the pandemic at all	5.8
Slightly affected by the pandemic	20.1
The pandemic has caused obvious effects	36.6
Work was seriously affected by the pandemic	26.5
The pandemic has imposed very serious effects	11.0

Among the effects on employees (Table 27.2), it is found that the majority suffered from forced no-pay leave (45.9%); the same situation exists for both gambling and non-gambling employees, as well as tourism and hospitality employees working in SME, but not for those working in large corporations. Employees working in large corporations reported being most seriously suffered from additional workload (47.8%). This may relate to the varying employee structure of the different industry segments.

In Macao, there are local-employee protection legislative policies that restrict job positions in casinos to be offered only to residents; while large corporations in comparison hired a high proportion of expatriates than SME, due to the quota system also implemented by the government that allows enterprises to employ expatriates based on the number of residents they have employed (Macao Daily, 2010). Therefore, large corporations will employ a much higher number of expatriates than other business segments. During the pandemic, the expatriates are the foremost group of employees terminated by the enterprises because they used to be employed on an annual basis due to employment policies (Allin Media, 2020). Hence, the contracts of many expatriates were not renewed during the pandemic and their workload will be assigned to employees who stay with the enterprises. Although not the most reported negative effect by employees of large corporations, forced no-pay leave is still just the next in line (23.8%), illustrating the widespread of this effect among employees of all business segments.

Following forced no-pay leave, salary reduction is found to be another major effect. Both forced no-pay leave and salary reduction result in a drop in the amount of salary received and this will add to the financial burden of employees, especially those with families, children, and with mortgages. The financial burden will create stress, and this will be exemplified if the workload of employees is increased. Additional workload is also one of the most common adverse effects of the pandemic on employees. The significant drop in visitors undeniably affects the tourism and hospitality industries and the loss of business has prolonged for more than 2 years during COVID-19. That was a long period that was out of expectation by nearly all businesses and the supportive economic measures of the government could not provide adequate assistance the more prolonged the entry and travel measures were imposed. Under such situations, to prevent the closing down of the business, employees are forced to be on furlough, if not no-pay leave, and a reduction on fringe benefits. Their part of the workload will then be shared with colleagues at work, thus imposing stress on both parties.

The statistics also reflected that some employees suffered from a forced change in their roles, which is probably related to the multi-tasking requirement resulting from a decrease in manpower. This change in new roles will add on the responsibilities of the employees, who may be required to complete duties that fall beyond their expertise of knowledge. The employees thus will need to learn new skills and knowledge, while to the extreme, may not be able to complete the assigned tasks effectively. Under such instances, the employees will need to tolerate the stress of incompetence and the stress of being discharged from the job when comparatively capable employees will typically be maintained when the enterprise is forced to reduce the number of human resources. It is worthy to note that the situation on the change of roles is less conspicuous for employees of the gambling industry (1.7%), which is probably attributable to the specific skills and nature required for the work. Employees in the gambling industry need to have strong knowledge in the different types of gambling games and be trained on the vast number of gambling operations, which may not be easily replaced.

TABLE 27.2 Negative Effects of COVID-19 on Employees of Different Segments.

	Overall	Gambling	SME	Large
Forced no-pay leave	**45.9**	**44.8**	**42.9**	23.8
Salary reduction	27.5	25.9	33.4	18.7
Additional workload	20.9	16.4	21.9	**47.8**
Fringe benefits cut	19.6	15.1	21.0	10.3
Forced furlough	13.4	7.3	18.3	20.2
Forced change in roles	4.6	1.7	5.9	3.7
Worsened collegial relations	4.0	3.4	5.6	3.0

All numbers in percentage.

Large, large corporations; SME, small-and-medium enterprises. (Group A companies having more than 100 employees.)

Stress is found to feature a noteworthy negative relationship with work fulfilment ($r=-0.215$, $p<.01$) and applies a negative impact ($\beta= -0.195$, $p<.01$). Employees will have lower work fulfilment when there is stress. Stress will fortify the elude propensities, therefore, stressful employees are more likely to be late from work, have higher absence rates and allows more excuses to escape from responsibilities ($\beta= +0.117$, $p<.01$). In addition, they have a higher intention of having anti-crisis conduct when interacting with customers ($\beta= +0.091$, $p<.01$). They are also found to be more interested in the health status of their colleagues during the pandemic and whether they strictly comply with the health measures. Stressful employees will try to lessen their interaction with the people. The decrease in fulfilment may influence their emotional steadiness, resulting in lowering their passion for work.

The increase in the level of stress is also found to increase the intention to exercise more escape behaviour ($r=+0.151$, $p<.01$) regarding responsibilities, creativity, and mutual support between colleagues. This is probably related to the negative effect of stress on work fulfilment, when a decrease in work fulfilment will enhance escape behavior ($\beta= -0.127$, $p<.01$). This diminution in work fulfilment may cause employees to become touchier and demanding. These may be communicated through their mediation on the practices of their colleagues, like limiting the way of communication, being inquisitive on their well-being circumstances, and imposing more rules directions during interactions. Such changes in propensities may occupy the focus of the employees from their work and devastate the concordant working climate or even debilitate the relations with their customers.

27.3 EFFECTIVE ROLE OF CORPORATE SOCIAL RESPONSIBILITY

The pandemic has changed the business environment and has created a situation that will corrode work fulfilment and add on employee stress. There is a minimal chance that employers can maintain the enthusiasm of the employees using the existing management practices. Therefore, the current research was carried out to identify if CSR can help to relieve the negative effects of the pandemic crisis. CSR is the practice of social, natural, financial, lawful and morale obligations by corporations (Carroll and Shabana, 2010). It is the ethical conduct of a corporation to the society and customarily utilized as a corporate governance strategy that considers the needs of different stakeholders. Being one of the key stakeholders of corporations, employees are found to be mostly supportive of socially responsible practices (Aguinis and Glavas, 2019). CSR can drive employee engagement and in turn resulted in incremental efficiency and inner notoriety, whereas diminishing loss. It can also improve the connection and bonding between the employer and employees that will benefit the healthy operation of the corporation.

The management of socially responsible corporations is aware of the strength of employee endorsement to the sustainable accomplishment of the corporation. They will involve in activities that enhance the work environment, guarantee work security, and improve employee welfares. They believe that employee satisfaction will increase their devotion and commitment, thus magnifying work performance and customer satisfaction. From past studies, employees' conviction that the corporation is socially responsible is associated with their work fulfilment (Lee and Chen, 2018). Corporations involved in socially responsible activities can ensue valuable success in terms of work satisfaction and decreasing employee turnover. Employees working in socially responsible corporates are more likely to bear a sense of ethical equality and generate belief in the corporation, which will reduce the potential of taking negative actions at adverse circumstances. They behold CSR are corporate social actions that have embraced the fundamental expectations of the stakeholders when fulfilling the legal and moral requirements (Carroll and Shabana, 2010). These corporations are believed to have a balance between financial goals and the needs of society and have achieved such goals without compromising society's resources or that of future generations.

The findings of this study reinforced that employee of the tourism and hospitality industries perceived highly on the value of CSR (Table 27.3). The majority of employees perceived that CSR could improve the reputation and sustainability of the enterprise. These are the values of CSR on the external environment when it can carve an affirmative image for the enterprise in the society. The investment in CSR has been widely questioned in the past, especially when the quantitative return on investment is involved. Over time, the importance of a balance between achievement on monetary profits and contribution to society becomes a universal concern. It is perceived as an ethical act of an enterprise that can gain customer support. This transformation in customer perceived value, together with the establishment of international standards on CSR, like the Global Reporting Initiative (GRI) on sustainability reporting (Global Reporting Initiative, 2021) and the Environmental Social Governance (ESG) on sustainable and ethical investment (Hong Kong Stock Exchange, 2021), have demonstrated the genuine value of CSR in respect to the non-operational (external) level of the enterprises.

Aside from external values, it is also found that CSR is perceived to have operational (internal) value at increasing the ethical standard and loyalty of employees (Table 27.3). This is critically important in the contemporary era when ethical operations and management are considered key attributes for the sustainability of an enterprise. Through international efforts, international associations have widely set ethical standards and rules, and they are then requested to be taught by universities as fundamental liberal arts knowledge for students (Sobczak et al., 2006). This ensures that management will possess an ethical mentality and balance between financial and societal needs. The current findings reinforced the support of employees on the exercising of CSR by the enterprises. The perception on the value of CSR by employees can reflect how CSR has been affecting the internal business environment of enterprises. The impact of CSR on employees can be exercised through two major means—through the reporting on the involvement of socially responsible activities by the enterprise in the sustainability reports and through the direct involvement of employees in socially responsible activities and projects of the enterprise. The sustainability reports can provide an overall picture of the actions taken and activities involved in the different socially responsible areas, so employees can better understand the enterprise's involvement in those areas and their respective impact. The direct involvement of employees in the activities and programs can provide a more significant impact when the employees can have an in-depth understanding of the level of investment in terms of different resources. Employees will also be better convinced of the needs and objectives of the socially responsible activities, which can help to modify their ethical standards when these will normally be reflected in their decision-making and work processes. A better understanding of the employees on the CSR positions of an enterprise can improve employee loyalty when they perceive themselves to be more attached to society in an ethical relation.

TABLE 27.3 Employee Perception on the Value of CSR (in Percentage).

	Total	Slightly agree	Agree	Strongly agree
Increase enterprise reputation	91.7	9.9	40.8	41.1
Sustainability of the enterprise	90.3	9.5	42.1	38.6
Increase employee ethical standard	88.0	16.6	39.0	32.3
Increase employee loyalty	87.0	15.4	39.9	31.8

Furthermore, CSR is found to essentially intervene the impact of stress on work fulfilment (from $\beta = 0.195$ to $\beta = -0.043$) and escape behaviors (from $\beta = +0.117$ to $\beta = +0.015$). It is imperative to note that escape behaviors will contrarily influence self-discipline and emphatically influence anti-crisis behaviour (from β=+0.091 to β= −0.015). Hence, the effectiveness of mediating the adverse effects can lower the potential on the obstructions of employee conduct, which may diminish the degree of self-discipline. Likewise, CSR essentially intervenes in the impact of stress on anti-crisis behavior. The relationship between the diverse behavior changes, work fulfilment, and the perceived value of CSR are in reverberation, when the equality of the relationships is associated. Highly fulfilled employees who are perceived highly on CSR will be grounded with the deliberation to be more self-disciplined and may be less inclined to come late for work, absent or show avoidance from work responsibilities.

From the statistics, employees of the tourism and hospitality industries expected to be well informed on various types of CSR information (Table 27.1), among which fringe benefits provided by the enterprise, training, and development that have been carried out and work health and safety measures were the top three types of information valued by employees. Furthermore, employees also desired to be informed about the incentive reward schemes and the promotion system. It is appealing to find that although employees are interested in the changes in the rate of the overall salary, but they are being valued lower than the fringe benefits and incentive reward schemes. This may be due to the operational arrangement when individual salary increase is usually attached to either to performance or based on a flat rate applied to the same bracket of employees, so the value of this information may not be comparable to fringe benefits or incentive reward schemes that are detached from the basic salaries. Results also showed that employees valued information on employee turnover rate and the profiles of the employee group. During the pandemic, there has been massive dismal of employees in some enterprises, therefore employees may want to understand better the extent of employees that were retained and whether those retained are residents or expatriates in order to determine if the enterprises are exercising corporate social responsibilities. The extent of CSR information valued by the employees further supported the influence of CSR on the employees, illustrating the internal value it has on the enterprise.

27.4 CONCLUSION

The COVID-19 pandemic had brought sizeable challenges to enterprises in terms of cascaded drop in business and the erosion of employee loyalty when massive dismissal of human resources was reported in different business segments. Management has to identify new practices that can maintain the stability and enthusiasm of the employees, so they can focus the resources to combat the pandemic crisis. The effectiveness of CSR at mediating the destructive effect of work stress on undesirable work behavior has provided management with an additional tool to plan ahead of the crisis and be prepared with more robust contingency remedies. The understanding of the usefulness of CSR will further enable the justification on its investment when the value of CSR can now broaden from external coverage to cover the extent of internal employees.

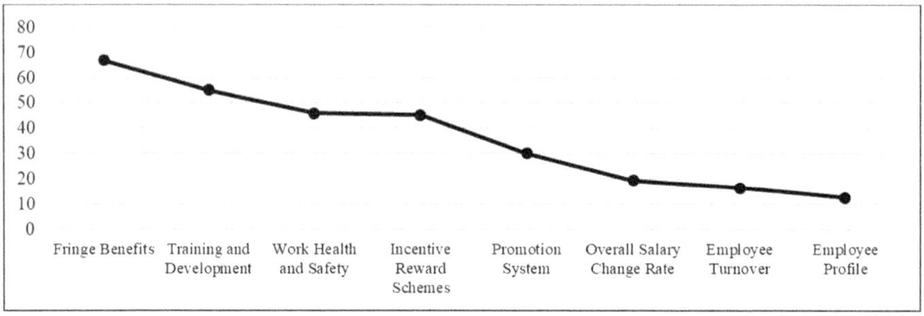

FIGURE 27.1 Types of CSR information valued by employees (in percentage of employees).

KEYWORDS

- **COVID-19 pandemic**
- **corporate social responsibility**
- **employee management practices**
- **hospitality**
- **tourism**

REFERENCES

Aguinis, H.; Glavas, A. On Corporate Social Responsibility, Sensemaking, and the Search for Meaningfulness Through Work. *J. Manage.* **2019,** *45* (3), 1057–1086.

Ahmad, R.; Ahmad, S.; Islam, T.; Kaleem, A. The Nexus of Corporate Social Responsibility (CSR), Affective Commitment and Organisational Citizenship Behaviour in Academia: A Model of Trust. *Empl. Relat. Int. J.* **2020,** *42* (1), 232–247.

Allin Media. Leave Clearance and No Pay Leave for Employees in Gaming Sector; Possible Layoff of Overseas Labor and Contractors [Online] 2020. https://www.allinmedia.com.hk/2020/02/26/%E7%96%AB%E6%83%85%E6%8B%96%E7%B4%AF%E5%8D%9A%E4%BC%81%E5%93%A1%E5%B7%A5%E9%9D%A2%E8%87%A8%E6%B8%85%E5%B9%B4%E5%81%87%E6%94%BE%E7%84%A1%E8%96%AA%E5%81%87-%E6%81%90%E7%8F%BE%E5%A4%96%E5%83%B1%E5%8F%8A/

Barakat, S. R.; Isabella, G.; Boaventura, J. M. G.; Mazzon, J. A. The Influence of Corporate Social Responsibility on Employee Satisfaction. *Manage. Decis.* **2016.**

Caplan, G. *An Approach to Community Mental Health*; Grune and Stratton: NY, 1961.

Carroll, A. B.; Shabana, K. M. The Business Case for Corporate Social Responsibility: A Review of Concepts, Research and Practice. *Int. J. Manage. Rev.* **2010,** *12* (1), 85–105.

Collins, B. G.; Collins, T. M. *Crisis and Trauma: Developmental-Ecological Intervention*; Lahaska Press, 2005.

Filimonau, V.; Derqui, B.; Matute, J. The COVID-19 Pandemic and Organisational Commitment of Senior Hotel Managers. *Int. J. Hosp. Manage.* **2020,** *91*, 102659. DOI: https://doi.org/10.1016/j.ijhm.2020.102659

Global Reporting Initiative. The Global Standards for Sustainability Reporting [Online] 2021. https://www.globalreporting.org/standards

Hanton, S.; Wagstaff, C. R.; Fletcher, D. Cognitive Appraisals of Stressors Encountered in Sport Organizations. *Int. J. Sport Exerc. Psychol.* **2012,** *10* (4), 276–289.

Hong Kong Stock Exchange. HKEX Listing Regulations-Environmenta, Social and Governance [Online] 2021. https://www.hkex.com.hk/Listing/Rules-and-Guidance/Environmental-Social-and-Governance?sc_lang=en

Despacho do Chefe do Executivo n.º 23/2020, 2020.

James, R. K.; Gilliland, B. E. *Crisis Intervention Strategies;* Brooks/Cole 2012.

Lazarus, R. S. *Emotion and Adaptation*; Oxford University Press: New York, 1991.

Lee, L.; Chen, L. F. Boosting Employee Retention Through CSR: A Configurational Analysis. *Corp. Soc. Responsib. Environ. Manage.* **2018,** *25* (5), 948–960.

Ma, E.; Lam, C. C. C.; Im, U. L. B. The Effects of Pandemic on a Tourism City and the Impacts of Government Policies in Macao. In *Virus Outbreaks and Tourism Mobility*; Emerald Publishing Limited, 2021.

Macao Daily. Protection on Priority of Employment Opportunity to Local Residents, 2010.

Macao Financial Services Bureau. Anti-Pandemic Fund Scheme [Online] 2020. https://info.dsf.gov.mo/infography.html

Macao Gaming Inspection and Coordination Bureau. Monthly Gross Revenue From Games of Fortune [Online] 2020. http://www.dicj.gov.mo/web/en/information/DadosEstat_mensal/2020/index.html

Macao Government Consumer Council. Macao Price Information Platform [Online] 2020. https://www.consumer.gov.mo/Advert/mobile-app.aspx?lang=zh

Macao Statistics and Census Service. Labour Force [Online] 2020a. https://www.dsec.gov.mo/en-US/Statistic?id=301

Macao Statistics and Census Service. Visitors Arrivals [Online] 2020b. https://www.dsec.gov.mo/en-US/Statistic?id=402

O'Connor D. Casino Taxes Account for Nearly 90 Percent of Macau Government Budget [Online] 2019. https://www.casino.org/news/casino-taxes-account-for-nearly-90-percent-of-macau-government-budget/#:~:text=Expertise%3A%20Commercial%20Gaming%2C%20Entertainment%2C%20Politics.%20Casino%20taxes%20in,the%20Macau%20government%20comes%20from%20a%20casino%20floor.

Ornell, F.; Schuch, J. B.; Sordi, A. O.; Kessler, F. H. P. "Pandemic Fear" and COVID-19: Mental Health Burden and Strategies. *Brazilian J. Psychiatry* **2020,** *42* (3), 232–235.

Outlook, W. E. Global Manufacturing Downturn, Rising Trade Barriers {Online] 2019. https://www.imf.org/en/Publications/WEO/Issues/2019/10/01/world-economic-outlook-october-2019

Outlook, W. E. A Long and Difficult Ascent [Online] 2020. https://www.imf.org/en/Publications/WEO/Issues/2020/09/30/world-economic-outlook-october-2020

Power, M.; Dalgleish, T. *Cognition and Emotion: From Order to Disorder*, 3rd ed.; Psychology Press, 2015.

Rosenthal, U.; Pijnenburg, B. *Crisis Management and Decision Making: Simulation Oriented Scenarios*; Springer Science & Business Media, 1991.

Seiffge-Krenke, I. *Stress, Coping, and Relationships in Adolescence*; Taylor & Francis, 2013.

Sobczak, A.; Debucquet, G.; Havard, C. The Impact of Higher Education on Students' and Young Managers' Perception of Companies and CSR: An Exploratory Analysis. *Corporate Governance* **2006.**

Topper, B.; Lagadec, P. Fractal Crises–A New Path for Crisis Theory and Management. *J. Contingencies Crisis Manag.* **2013,** *21* (1), 4–16.

Worldometer. COVID-19 Coronovarius Pandemic [Online] 2021. https://www.worldometers.info/coronavirus/

Wu, J.; Wu, Y.; Tian, Y. Temporal Associations Among Loneliness, Anxiety, and Depression During the COVID-19 Pandemic Period. *Stress Health* **2021,** 38, 90–101.

Yau, Y. H.; Potenza, M. N. Stress and Eating Behaviors. *Minerva Endocrinol.* **2013,** *38* (3), 255.

Yu, J.; Park, J.; Hyun, S. S. Impacts of the COVID-19 Pandemic on Employees' Work Stress, Well-Being, Mental Health, Organizational Citizenship Behavior, and Employee-Customer Identification. *J. Hosp. Mark. Manage.* **2021,** 1–20.

INDEX

D

E

X